Vicary's
Anatomie of the Bodie of Man

Early English Text Society
Extra Series, No. LIII

The Anatomie
of
The Bodie of Man

by

Thomas Vicary

edited by

Frederick J. Furnivall and Percy Furnivall

Part I

(all published)

EARLY ENGLISH TEXT SOCIETY

Extra Series, 53

KRAUS REPRINT CO.
Millwood, New York

UNIVERSITY PRESS

Great Clarendon Street, Oxford OX2 6DP
United Kingdom

Oxford University Press is a department of the University of Oxford.
It furthers the University's objective of excellence in research, scholarship,
and education by publishing worldwide. Oxford is a registered trade mark of
Oxford University Press in the UK and in certain other countries

© The Early English Text Society 1888

The moral rights of the authors have been asserted

Database right Oxford University Press (maker)

First Edition published in 1888

All rights reserved. No part of this publication may be reproduced,
stored in a retrieval system, or transmitted, in any form or by any means,
without the prior permission in writing of Oxford University Press,
or as expressly permitted by law, or under terms agreed with the appropriate
reprographics rights organization. Enquiries concerning reproduction
outside the scope of the above should be sent to the Rights Department,
Oxford University Press, at the address above

You must not circulate this book in any other form
and you must impose this same condition on any acquirer

Published in the United States of America by Oxford University Press
198 Madison Avenue, New York, NY 10016, United States of America

British Library Cataloguing in Publication Data
Data available

Library of Congress Cataloging in Publication Data
Data available

Extra Series, 53

ISBN 978-0-85-991755-1

The Anatomie
of
The Bodie of Man

BY

Thomas Vicary

SERJEANT OF THE SURGEONS TO HENRY VIII, QUEEN MARY, EDWARD VI, AND
QUEEN ELIZABETH; MASTER OF THE BARBER-SURGEONS' COMPANY; AND CHIEF
SURGEON TO ST. BARTHOLOMEW'S HOSPITAL, LONDON, 1548-62

THE EDITION OF 1548
AS RE-ISSUED BY THE SURGEONS OF ST. BARTHOLOMEW'S IN 1577

WITH A LIFE OF VICARY, NOTES ON SURGEONS IN ENGLAND,
BARTHOLOMEW'S HOSPITAL, AND LONDON, IN TUDOR TIMES,
AN APPENDIX OF DOCUMENTS, AND ILLUSTRATIONS

EDITED BY

FREDK. J. FURNIVALL, M.A., Hon. Dr. Phil.

AND

PERCY FURNIVALL,
A STUDENT OF ST. BARTHOLOMEW'S

PART I

LONDON:
PUBLISHED FOR THE EARLY ENGLISH TEXT SOCIETY
BY HUMPHREY MILFORD, OXFORD UNIVERSITY PRESS,
AMEN HOUSE, E.C. 4.

First Impression . . . *1888*
Second Impression . . *1930*

𝕰𝖝𝖙𝖗𝖆 𝕾𝖊𝖗𝖎𝖊𝖘, LIII.

FORETALK.

TILL Mr. W. H. Cross, the Clerk, and Dr. Norman Moore, the Warden, of St. Bartholomew's, publish Part I of their Records of the Hospital, we cannot complete the Life of THOMAS VICARY, for our Forewords to his *Anatomie*. But as the Text and Appendix in this Part of our book need a short temporary Foretalk, with a sketch of Vicary's Life, we give it here.

The first tidings of Vicary (who was probably born between 1490 and 1500) are, that he was 'a meane practiser (had a moderate practise) at Maidstone,' and was not a traind Surgeon. In 1525 he is Junior of the three Wardens of the Barbers' or Barber-Surgeons' Company in London. In 1528 he is Upper or first Warden of the Company, and one of the Surgeons to Henry VIII, at £20 a year. In 1530 he is Master of the Barber-Surgeons' Company, and is appointed—in reversion after the death of Marcellus de la More—Serjeant of the Surgeons, and Chief Surgeon to the King. This Headship of his Profession, Vicary takes in 1535 or 1536, together with its pay of £26 13s. 4d., and holds it (under Edw. VI, Q. Mary, and Q. Elizabeth) till his death in 1561 or 1562. He is the Paget of his great Tudor time.

In 1535, a fresh Grant is made to Vicary of either his old twenty pounds a year, or a fresh one: p. 114, below. In 1539, Vicary gets from Henry VIII a beneficial lease for 21 years of the Rectory-house, tithes, &c. of the dissolvd Boxley Abbey in Kent, close to Maidstone; and as he is a person of influence with the King, a rich Northamptonshire squire, Anthony Wodehull, who has an infant daughter, and is probably a patient of the chief Court Surgeon, appoints Vicary as one of the Trustees of his Will (proved Oct. 11, 1542), with a view (no doubt) to the protection of his girl's property and person during her nonage. In 1541, as the acknowledgd Head of his profession, Vicary is appointed the First Master of the newly amalgamated Companies of Barbers and Surgeons, and is painted—with other Surgeons, Barbers and Physicians—by Holbein. In this year 1541, he also gets a beneficial lease for 60 years, from Sir Thos. Wyat, the poet, of lands in Boxley, Kent. In 1542, he and his son William (also probably a Surgeon) are appointed by Henry, Bailiffs of Boxley Manor, &c. in Kent, with yearly salaries of £10 each. In Sept. 1546-7, Vicary is again Master of the united Company of Barbers and Surgeons. In Dec. 1547, he marries his second wife, Alice Bucke.

In 1546-7, Henry VIII handed over Bartholomew's (with other Hospitals, &c.) to the City of London. He gave it a small endowment (nominally £333 odd) out of tumble-down houses, which he charged with pensions to parsons. The balance of the endowment was but enough to keep, as patients, 'thre or foure harlottes, then being in chyldbedde.' So the City set to work, raisd £1000 for repairs, fittings, &c., practically re-opend the Hospital, for 100 patients, and, on 29 Sept. 1548, appointed Chief-Surgeon VICARY as one of the 6 new Governors of the Hospital to act with the 6 old ones. Vicary must soon after have become Resident Surgical Governor of the Hospital. He was re-appointed annually; he is given the old Convent Garden in June 1551; and in June 1552 is made 'one of the assistants of this house for the terme of his lyffe' (extract by Dr. N. Moore). He has 3 Surgeons under him, at £18 (1549), and then £20 (1552) a year each. The Hospital finds him a Livery gown, and repairs his house. He holds his appointment till his death, late in 1561, or early in 1562. That to him is due part of the Hospital organization, and some of the beautiful unselfish spirit shown in the City 'Ordre' for Barts in 1552, we do not doubt. This 'Ordre' no one can read without admiring.

In Sept. 1548, Vicary was, for the 4th time, elected Master of the Barber-Surgeons. In 1548 too, he publisht his *Anatomie*—the first in English on the subject,—but whether this was after or before he was made a Governor of Barts, we cannot say. The book was reprinted by the Surgeons of Barts in 1577, with a few Forewords; and from the unique copy of that issue, the earliest now known, our reprint is made, with added head-lines and side-notes. Frequently supplemented, Vicary's little *Anatomie* held the field for 150 years. (Unluckily the biographical details of an Italian doctor in one of the added Treatises, have been lately set down to Vicary.)

Foretalk. vii

In 1553, Queen Mary made a special grant to Vicary of the Arrears of his Chief Court-Surgeons' Annuity of £26 13s. 4d. which he came into in 1536, on De la More's death or resignation. In 1554 he was appointed Surgeon to Mary's husband, K. Philip; and in 1555, Philip and Mary re-granted to Vicary—his son William being doubtless then dead—the Bailiffship of the Manor of Boxley, &c., and the 2 Annuities of £10, which Henry VIII had granted to Vicary and his son in 1542. Year by year Vicary quietly workt on, doing his duty to the sick poor at Barts, and in the Barber-Surgeons' Company. He had saved money enough by March 1557-8, to lend his brother-in-law, Thos. Dunkyn, yeoman of St. Leonard's, Shoreditch, £100, which he secures in favour of his nephew Thomas Vicary, of Tenterden in Kent, clothier; and possibly about this time he buys of Jn. Joyce a house and some land next to Boxley Church, in Kent, which he devises to his nephew Stephen Vicary, son of his brother William, late of Boxley. In Sept. 1557-8, he is, for the 5th and last time, Master of the Barber-Surgeons' Company.

On Jan. 27, 1560-1, Vicary makes his Will; and he probably dies late in 1561, or early in 1562, as the last payment to him of his Annuity of £20 is in Sept. 1561, and his Will is proved by his widow on April 7, 1562. Where he is buried, we have not yet been able to find. Shortly before his death he was (says Mr. S. Young) named in a Commission of Queen Elizabeth's to the Barber-Surgeons' Company to press Surgeons for her military service.

We hope in our Forewords to give further details about Vicary and his life and times. Some are in the Appendix in this Part I, which also contains particulars about Barts not printed before. These we commend to our readers' attention. The illustrations will help to realize the London, Bartholomew's, and Kent of the good old Surgeon's day. For any corrections, information, suggestions, and notes for our Part II (which will contain a full Index), we shall be grateful. We desire only to do justice to the old Worthy of Kent, and the noble Hospital for which he and his fellow-citizens of London workt in so generous a spirit. May our successors 350 years hence be able to say of us Victorians as we can of Vicary and the Londoners of his Tudor time: like Englishmen, they tried to do their duty!

3 St. George's Square, Primrose Hill, London, N. W.
14 June, 1888.

CONTENTS.

Portrait of **Vicary**, after Holbein, by Mr. Austin Young.
Early Plans of Bartholomew's, from Aggas (ab. 1560), & Ryther (1604).
Norden's Map of London, 1593, by Van den Keere.
Map of **Vicary's** Road from London to Maidstone and Boxley, ab. 1575.
Map of the Neighbourhood of London, ab. 1575.
Ships, &c. from Saxton's Maps, ab. 1575.

	PAGES
Foretalk (Temporary)	v—vii
Vicary's Anatomie	1—86
Appendix: (Contents p. 87-8)	
I. Grants to **Vicary** by Kings and Queens	89—98
II. Payments to **Vicary**, &c. by Kings and Queens ...	99—122
III. Extracts from City Records as to Barts, **Vicary**, the Plague, London Vagabonds, &c.	123—178
IV. **Vicary's** Bailiff's Accounts of Boxley Manor ...	179—180
V. Dunkyn's £100 Mortgage to **Vicary**, 7 March 1557-8	181—186
VI. **Vicary's** Will, 27 Jan. 1560-1 (proved, 7 April 1562)	187—194
VII. Henry VIII's Statutes relating to Surgeons ...	195—209
VIII. Supplement to the Statutes: the Surgeons' Compromise with the City as to serving on Quests, &c.	210—219
IX. Ten Recipes by Hen. VIII and his Physicians. Poem, 'What Veins to bleed in,' &c.	220—230
X. Payments to Surgeons in 1529-43, by Hen. VIII, &c.	231—235
XI. Pay of Army and Navy Surgeons to Hen. VIII ...	236—237
XII. Some of Hen. VIII's Payments to Holbein and to Players	238—242
XIII. The 185 Freemen of the Barber-Surgeons' Company in 1537; the nos. of the other City Companies	243—246
XIV. Ordinances of the Barber-Surgeons of London, A.D. 1529, &c.	247—268
XV. Ordinances of the Barber-Surgeons of York, A.D. 1592, &c.	269—288
XVI. Order for the Government of **Barts**, A.D. 1552 ...	289—336

PLANS OF
BARTHOLOMEW'S
IN
1560 & 1604

(For 1593, see Norden or Van den Keere.)

WEST SMITHFIELD AND BARTHOLOMEW'S HOSPITAL.
Reduced from Ralph Aggas's Map of London, ab. 1560 (issued 1603).

T. Smith Field.
W. Charter House.
V. V. City Walls.

West Smithfield and St. Bartholomew's Hospital, slightly enlarged from Augustine Ryther's Map of London, 1604.

VICARY'S ROAD FROM LONDON TO MAIDSTONE AND BOXLEY.

(From Christopher Saxton's Map of Kent, Surrey, &c. (1578-9), with the Roads inserted, and other Additions, by Philip Lea, after 1690. The names Dulwich and Bulwell are inked in by a modern hand.)

LONDON AND ITS NEIGHBOURHOOD.
From Christopher Saxton's Map, ab. 1573. (*Pall Mall Gazette* block.)

Elizabethan Ships, Whale, and Dolphin from Christopher Saxton's Maps, 1573-9.
(From the *Pall Mall Gazette* blocks.)

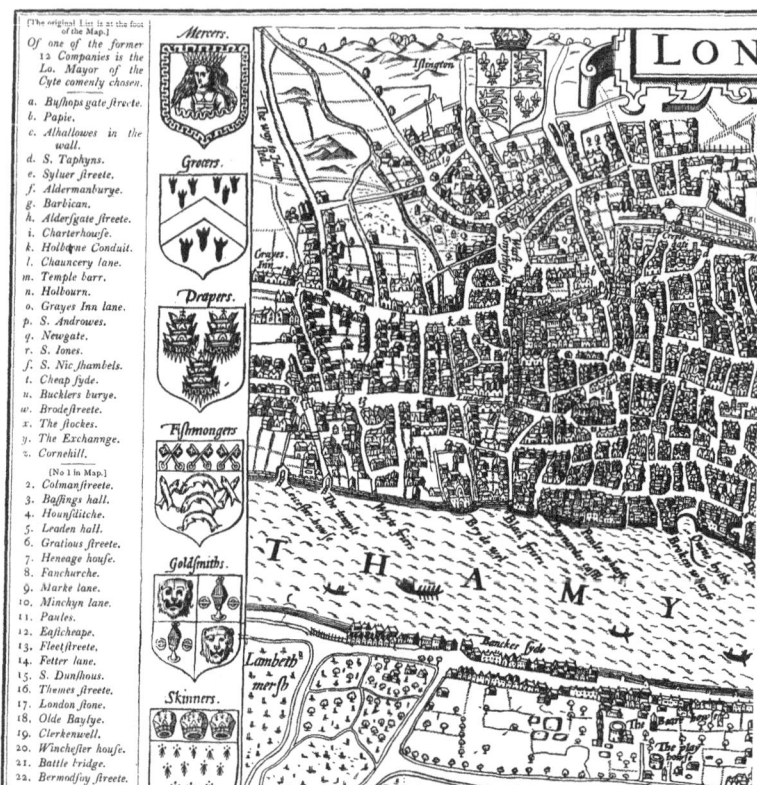

[The original List is at the foot of the Map.]

Of one of the former 12 Companies is the Lo. Mayor of the Cyte comenly chosen.

a. Bushopsgate streete.
b. Papie.
c. Alhallowes in the wall.
d. S. Taphyns.
e. Sylver streete.
f. Aldermanburye.
g. Barbican.
h. Aldersgate streete.
i. Charterhouse.
k. Holborne Conduit.
l. Chauncery lane.
m. Temple barr.
n. Holbourn.
o. Grayes Inn lane.
p. S. Androwes.
q. Newgate.
r. S. Iones.
s. S. Nic shambels.
t. Cheap syde.
u. Bucklers burye.
w. Brodestreete.
x. The stockes.
y. The Exchannge.
z. Cornehill.

[No 1 in Map.]
2. Colmanstreete.
3. Bassings hall.
4. Houndsditche.
5. Leaden hall.
6. Gratious streete.
7. Heneage house.
8. Fanchurche.
9. Marke lane.
10. Minchyn lane.
11. Paules.
12. Eastcheape.
13. Fleetstreete.
14. Fetter lane.
15. S. Dunsthous.
16. Themes streete.
17. London stone.
18. Olde Baylye.
19. Clerkenwell.
20. Winchester house.
21. Battle bridge.
22. Bermodsioy streete.

Ioannes Nurden Anglus descripsit anno 1593.

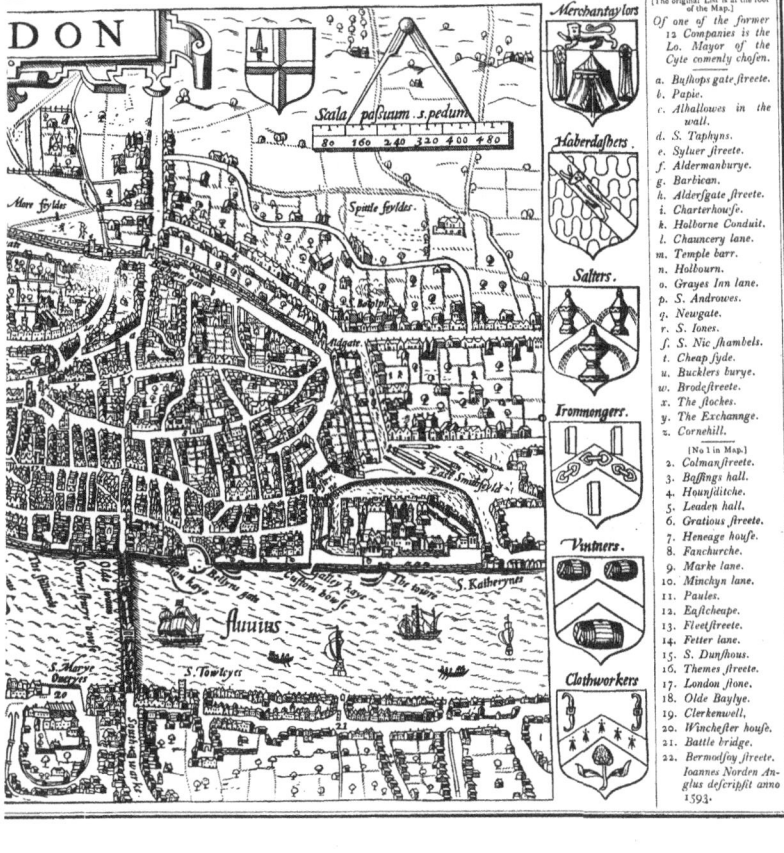

A profitable Treatise of the Anatomie of mans body:

Compyled by that excellent Chirurgion, M. Thomas Vicary, Esquire, Seriaunt Chirurgion to king Henry the eyght, to king Edward the .vj. to Queene Mary, and to our most gracious Soueraigne Lady Queene Elizabeth, and also cheefe Chirurgion of S. Bartholometwes Hospital.

Which work is newly renewed, corrected, and published by the Chirurgions of the same Hospital now beeing.
An. 1577.

¶ Imprinted at London, by Henry Bamforde.

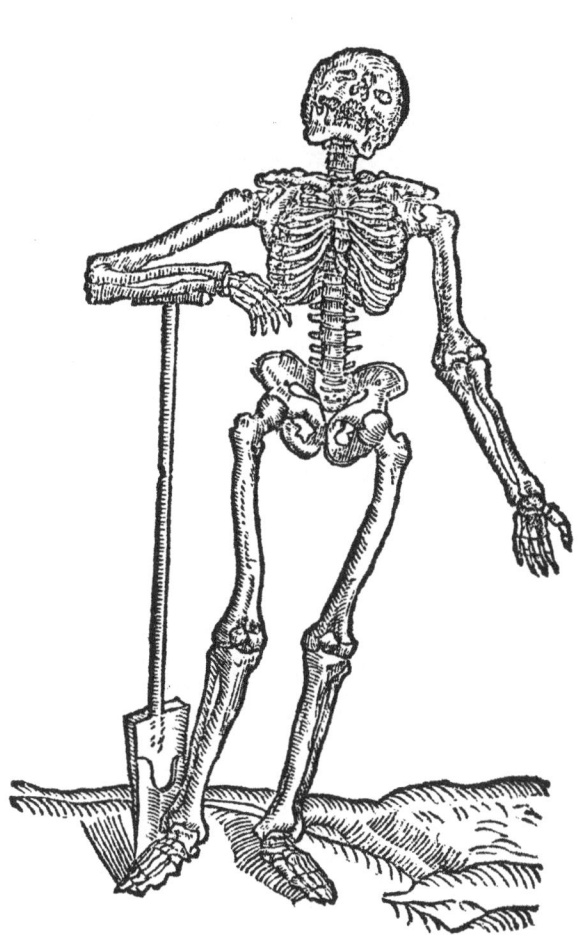

Nowe he that is the perfect guyde,
    doth knowe our helpes were here alone,
By homely style it may be spyde,
    for rules in Rhetorike haue we none:
Our heads doo lacke that fyled phrase,
    whereon fine wittes delight to gase.
If any say we deserue heere blame,
    we pray you then amende the same.

TO THE RIGHT

Worshipful, Sr. Rouland Haiwarde, Knight, President of little Saint Bartholomewes in 𝔚est Smithfeelde, Sr. Ambrose Nicholas, Knight, Maister Alderman Ramsay, vvith the rest of the worshipful Masters and Gouernours of the same, William Clowes, Wil. Beton, Richard Story, and Edward Bayly, *Chirurgions of the same Hospital, wishe health and prosperitie.*

[1577.]

He People in times past did prayse and extoll by Pictures and Epigrames the famous dedes of all sutche persons vvho so euer in any vertuous qualitie or Liberal Science excelled. *Sulpitius Gallus* among the *Romanes* was highly renovv*med for his singuler cunning in Astronomie, by vvhose meanes *Lucius Paulus* obteyned the victorie in his vvarres against *Percius*. *Pericles* also among the Athenians vvas had in great admiration and honour for his profounde knovvledge in Philosophie, by vvhom the vvhole Citie of *Athens* vvas from care and vvoe deliuered, vvhen they supposed their dest[r]uction to be neare at hand, by a blacke darknes of some admiration hanging ouer their Citie. Howe honorably vvas *Apelles* the Paynter esteemed of mightie king *Alexander*, by whom onely he desired to be

Folk of old praised those who excelled in virtue or science.

* sign. ¶ iij back.

Pericles was honoured by the Athenians;

Apelles by King Alexander.

6　The Epistle Dedicatorie. 1577. Envy of Physic.

[margin: The Wise Greeks honoured Surgerie.]

paynted. But amongst all other Artes and Sciences, vvhose prayse in tymes past flourished and shined most brightly, Chirurgerie among the vvise *Grecians* lacked not his prayse, honour, and estimation. For dyd not that worthy and famous captayne of the Greekes, *Agamemnon*, loue dearely and revvarde bountifully both *Podalerius* and *Machaon*, through vvhose cunnings skill

*[margin: * sign. ¶ iiij.]*

in Surgerie, thousands of vvorthy *Greekes vvere saued aliue and healed, vvho els had dyed and perished. And further heere to speake of *Philoneter*, of *Attalus*, of *Hiero*, of *Archelaus*, and of *Iula*, kinges of famous memorie, vvho purchased eternal prayse by their study

[margin: But now, envy condemns Physic and despises Surgery, till pain comes.]

and cunning in Phisicke and Surgery. But novve in these our dayes enuie so ruleth the rost, that Phisicke should be condemned, and Surgerie despised for euer, but that sometime payne biddeth battayle, and care keepeth skirmishe, in suche bytter sorte, that at the last this Alarum is sounded out: Novve come Phisicke, and then helpe Surgerie! Then is remembred the

[margin: Jesus, the Son of Sirach, says, 'Honour the Physician and Surgeon.']

saying of *Iesus the sonne of Sirache*, which is notable:[1] 'Honour the Phisition and Chirurgeon for necessitie, vvhom the almightie God hath created, because from the hyest commeth Medicine, and they shall receyue

---

[1] 'The Wisdome of Iesus, the sonne of Sirach, called Ecclesiasticus,' ch. xxxviii. (1) Honour the Physition with that honour that is due vnto him, because of necessitie: for the Lord hath created him. (2) For of the most High commeth healing, and he shall receiue giftes of the King. (3) The knowledge of the Physition lifteth vp his heade, and in the sight of great men he shall be in admiration. (4) The Lorde hath created medicines of the earth, and he that is wise will not abhorre them. (5) Was not the water made sweete with wood, that man might knowe the vertue thereof? (6) So hee hath given men knowledge, that hee might bee glorified in his wonderous workes. (7) With such doeth he heale men, and take away their paines. (8) Of such doeth the apothecarie make a confection, and yet he can not finish his owne workes: for of the Lorde commeth prosperitie and wealth ouer all the earth. (9) My sonne, fayle not in thy sickenesse, but pray vnto the Lorde, and he will make thee whole. (10) Leaue off from sinne, & order thine hands aright, and clense thine heart from all wickednes. (11) Offer sweete incense ... (12) Then give place to the Physition: for the Lorde hath created him: let him not go from thee, for thou hast neede of him. (13) The houre may come, that their enterprises may haue good successe. (14) For they also shall pray vnto the Lorde, that he would prosper that which is giuen for ease, and their phisike for the prolonging of life.'—*Apocrypha*, 1583.

The Epistle Dedicatorie. 1577. Thomas Vicary.

gyftes of the King.' VVherefore vve exhort the vvyse man, that he in no tyme of prosperitie and health neglect these noble *Artes and mysteries of Phisicke and Chirurgerie, because no age, no person, no countrey can long time lacke their helpes and remedies. VVhat is it to haue landes and houses, to abounde in siluer and golde, to be deckt with pearles and Diamondes, yea, and to possesse the vvhole vvorlde, to rule ouer Nations and countreys, and to lacke health, the cheefest Iuel[1] and greatest treasure of mans lyfe and delight.

Consider then, vvee beseeche your vvorships, vvhat prayses are due to suche noble Sciences, whiche onely vvorke the causes of this aforesayde health, and hovve muche the vveale publique are bounde to al them whose cares and studies daylye tendeth too this ende. Amongst vvhom heere is to be remembred *Master Vicary*, Esquire, Seriaunt Chirurgion to Kinges and Queenes of famous memorie: VVhose learned vvorke of the Anatomie is by vs, the forenamed Surgions of Sainct *Bartholomewes in Smithfeelde, nevvly reuiued, corrected, & published abroad to the commoditie of others, who be Studentes in Chirurgerie: not vvithout our great studies, paynes, and charges. And although we do lack the profound knowledge and sugred eloquence of the Latin and Greeke tongues, to decke and beautifie this vvorke, yet we hope the studious Reader shal thereby reape singuler commoditie and fruite, by reading this little Treatise of the Anatomie of mans body, the vvhich is onely grounded vpon reason and experience, which are two principal rootes of Phisicke and Surgerie, As it is graunted by *Galen* in his thirde Booke, *De Methode medendi:* and vve vvho dayly worke and practise in Surgerie, according to the deepnes of the Arte,—aswel in greeuous vvounds, Vlcers, and Fistules, as other hyd and secrete diseases, vpon the

*sign. ¶ iiij, bk.

What shall it profit a man,

if he has the whole world, and yet lacks Health, man's greatest treasure?

Among those to whom praise is due,

is Thomas Vicary, whose *Anatomie* we Surgeons of Barts have newly published.
* sign. ¶ v.

It is grounded on Reason and Experience.

And we who daily treat grievous wounds

[1] jewel

body of man, dayly vsed by vs in S. Bartlemews Hospital and other places, &c., Those poore *and greeued creatures, aswell men and women as children,—do knowe the profite of this Art to be manyfolde, and the lacke of the same to be lamented. Therfore Galen truely vvriteth, saying, That no man can vvorke so perfectly as aforesayde, vvithout the knovvledge of the Anatomie: For (sayth he) it is as possible for a blinde man to carue and make an Image perfect, as a Chirurgion to vvorke[1] without errour in mans body, not knovving the Anatomie. And further, for as muche as your Worships are very careful for those poore and greeued creatures within the Hospital of S. Bartlemewes, &c., vvhereof Master Vycarie vvas a member, VVe are therefore novve encouraged to Dedicate this little vvorke of the Anatomie, beeing his and our trauayles, to you as Patrons of this Booke, to defende agaynst the rauening Iavves of enuious Backbyters, vvhiche neuer cease by all vnlawful meanes to blemishe and de†face the vvorkes of the learned, expert, and vvel disposed persons. Finally, vve do humbly craue of your Goodnesses to accept in good part this Treatise concerning the Anatomie, as the fruites of our studies and labours, vvhereby wee shal be muche better encouraged to set foorth hereafter other profitable vvorkes for the common vvealth. Heerein yf your VVisedomes doo vouchsafe to heare our requestes, and to alovve these our dooinges, as dyd noble Amasus, king of Egypt, accept the labours of his payneful Artificers, we haue not onely to thanke your VVorships for so dooing, but also to pray alway for you vnto the almightie God to requite your goodnesses, receyuing you into his protection and keeping. Amen.

¶ To the Reader.

DEARE Brethren,[1] and freendly Readers, we haue here, according to the trouth and meaning of the Author, set forth this needeful and necessarie worke concerning the *Anatomy of mans body*, beeing collected and gathered by master *Thomas Vicary*, and nowe by vs the Chirurgions of *Sainct Bartholomewes* Hospital, reuiued, corrected, and published. And albeit this Treatise be small in Volume, yet in commoditie it is great and profitable. Notwithstanding, if the thinges therein conteyned be not discretly and wisely studied and applyed, according to the true meaning of the Author, Wee haue to tell you hereof, that therein is great peryl, because, through ignoraunt Practicio*ners, not knowing the Anatomie, commonly doth ensue death, and seperation of soule and body. Furthermore, whereas many good and learned men in these our dayes, doo cease to publishe abroade in the Englishe toung their workes and trauayles, it is, for that if any one fault or blemishe, by fortune be committed, eyther by them or the Printer escaped, they are blamed, yea, and condemned for ignoraunt men, and errour-holders. But nowe we here cease from these poyntes to trouble the gentle Reader with longer discoursing, for whose sakes and commodities wee haue

Readers, we issue this Anatomie of Thos. Vicary, revised by us.

Though small, it is valuable, but needs discreet use.

* sign. ¶ vij.

Many men will not print in English now,

to avoid blame for chance mistakes.

[1] All that follows (save 'Vicarie to his Brethren') is in black letter.

10 The Bart.'s Surgeons of 1577 to the Reader.

<small>Do you correct our faults gently, and speak kindly of the Author.
* sign. ¶ vij, bk.</small>

taken these paynes: Wishing that men more skylful and better learned, woulde haue borne this burthen for vs. Crauing onely this muche at your handes, for to correct our faultes fauourably, and to reporte of the Author courteously, who *sought (no doubte) your commodities onely, and the profite of the common-Wealth, without prayse and vayneglory of him selfe.   4

<small>We commit you to God's keeping.</small>

Thus wee, the Chirurgions aforesayde,   8
commit you to the blessed keeping
of Almightie God, who always
defende and increase your
studies and ours.   12
Amen.

<small>Lord, make our Rulers protect godly Arts!</small>

O Lorde which made the loftie Skyes,
worke in our Rulers hartes,
Alwayes to haue before their eyes   16
safegarde to godly Artes.

¶ Thomas Vycarie to his Brethren practising Chirurgerie. [1548]   [sign. ¶ viij.]

EEREAFTER foloweth a little treatise of the Anatomie of mans body, Made by Thomas Vycarie, Citizen and Chirurgion of London, for all suche young Brethren of his Felowship practising Chirurgerie. Not for them that be expertly seene in the Anatomie: for to them Galen, the Lanterne of all Chirurgions, hath set it foorth in his Canons, to the high glory of God, and too the erudition and knowledge of al those that be expertly seene and learned in the noble Science of Chirurgerie. And because al the noble Philosophers wryting vppon Chirurgerie *doo condemne al suche persons as practise in Chirurgerie, not knowing the Anatomie, Therefore I haue drawen into certayne Lessons and smal Chapters a parte of the Anatomie, but touching a part of euery member particulerly: Requiring euery man that shal reade this little Treatise, to correct and amende it where it shal be neede, and holde me excused for my bolde enterprise, and accept my good wyl towards the same.

This book is for young Surgeons, not expert ones.

They must know Anatomy; and so I haue described the parts of the body.

* sign. ¶ viij, bk.

¶ *A breefe Treatife of* the[1] Anatomie of mans body: Compyled by me *Thomas Vycarie* Esquire, and Sargeant Chirurgion to king Henry the eyght, for the use and commoditie of al Vnlearned Practicioners in Chirurgerie.

[1548]

[CHAPTER I.]

IN the name of God, Amen! Heere I shal declare vnto you shortly and breefly the sayinges and the determinations of diuers aunciont Authors, in three poyntes, very expedient for al men to knowe, that entende to vse or exercise the mysterie or arte of Chirurgerie. The first is, to knowe what thing Chirurgerie is: The Second is, how that a Chirurgion should be chosen: And the thirde is, with what properties a Surgion should be indued.

THE fyrst is, to know what thing Chirurgerie is. Heerein I doo note the saying of *Lamfranke*, whereas he sayth, Al thinges that man would knowe, may be knowen by one of these three thinges: That is to say, by his name, or by his working, or els by his very being and shewing of his owne properties. So then it followeth, that in the same manner we may know what Chirurgerie is by three thinges. First, by his name, as thus, The Interpreters write, that Surgerie is deriued oute of these wordes, *Apo tes chiros, cai tou ergou*, that is too bee vnderstanded, A hand working,

[1] *orig.* of the

Ch. I.] *Surgery is Hand-working in Man's Body.* 13

and so it may be taken for al handy artes. But noble
Ipocras sayth, that Surgerie is hande working in mans *It is hand-working in man's body,*
body; for the very ende and profite of Chirurgerie is
4 hande working. Nowe the seconde manner of knowing
what thing Chirurgerie is, it is the saying of *Anicen* to
be knowen by his beeing, for it is verely a medecinal *and also a medicinal science.*
science: and as Galen sayth, he that wyl knowe the
8 certentie of a thing, let him not busy him selfe to
knowe only the name of that thing, but also the working
and the effect of the same thing. Nowe the thirde
way to knowe what thing Chirurgerie is, It is also to
12 be knowen by his beeing *or declaring of his owne * sign. A. ij.
properties, the which teacheth vs to worke in mannes
body with handes: as thus, In cutting or opening *It is cutting, healing, and removing excrescences.*
those partes that be whole, and in healing those partes
16 that be broken or[1] cut, and in taking away that is
superfluous, as warts, wennes, skurfulas, and other
lyke. But further to declare what Galen sayth Surgery is, It is the laste instrument of medecine: That is *It is the last resource, after Diet and Medicine,*
20 to say, Dyet, Pocion, and Chirurgerie: of the whiche
three, sayth he, Dyet is the noblest and the most vertuous. And thus he sayth, whereas a man may be
cured with Dyet onely, let there be geuen no maner of
24 medicine. The seconde instrument is Pocion: for and
if a man may be cured with Dyet and Pocion, let there
not be ministred any Chirurgerie. The thirde and
laste Instrument is Chirurgerie, through whose vertue
28 and goodness is remoued and put away many greeuous *and removes grievous diseases which they cannot touch.*
infirmities and diseases, which might not have bene
remoued, nor yet put away, neither with Diet nor with
Pocion. And by these three meanes it is knowen what
32 thing Chirurgerie is. And this suffiseth †for vs for † sign. A. ij, back.
that poynt. Nowe it is knowen what thing Chirurgerie
is, there must also be chosen a man apt and mete to
minister Surgerie, or to be a Chirurgion. And in this

[1] *orig.* broke nor

14 The qualities of a Surgeon: good looks, &c. [Ch. I.

A Surgeon must be a temperate and well-made man.

poynt al Authors doo agree, that a Chirurgion should be chosen by his complexion,[1] and that his complexion be very temperate, and al his members wel propor-

One with an ugly face can't have good manners.

tioned. For Rasis[2] sayth, Whose face is not seemely, it is vnpossible for him for to haue good manners. And Aristotle, the great Philosopher, writeth in his Epistles to the noble king Alexander (as in those Epistles more playnely doth appeare) howe hee should choose al suche persons as should serue him, by the forme and shape of the face, and al other members of the body. And furthermore they say, he that is of an euill complexion, there must needes folowe like conditions. Wherefore it agreeth that a Chirurgion must be both of a good and temperate complexion, as is afore rehearsed. And

He must also keep God's commands, and have a steady hand.

principally, that he be a good lyuer, and a keeper of the holy commaundements of God, of whom commeth al cunning and grace, and that his body be not quaking,

* sign. A. iij.

and his hands *stedfast, his fingers long and smal, and not trembling; and that his left hand be as ready as his right hande, with al his lymmes able to fulfil the good workes of the soule. Nowe here is a man meete to be made a Chirurgion. And thoughe he haue al these good qualities before rehersed, yet is he no good Chirurgion, but a man very fitte and meete therfore. Now then, to knowe what properties and conditions this man must haue before he be a perfect Chirurgion. And I doo note foure thinges moste specially that euery

He must be learned, expert, clever, and well-mannered.

Chirurgion ought for to haue: The first, that he be learned; the seconde, that he be expert; the thirde, that he be ingenious; the fourth, that he be wel manered. The first (I sayde), he ought to be learned, and that he knowe his principles, not onely in Chirurgerie, but also in Phisicke, that he may the better defende his Surgery. Also he ought to be seene in

---

[1] Disposition, habit of body. See p. 18 below.
[2] See Forewords: *Anatomie* section.

Ch. I.] *Surgeons must know Anatomy, & not drink.* 15

natural Philosophie, and in Gra*m*mer, that he speake congruitie in Logike, that teacheth him to proue his proportions with good reason. In Rethorike, that teacheth him to speake seemely and eloquently: also in Theorike, that teacheth *him to knowe thinges natural and not natural, & thinges agaynst nature. Also he must knowe the Anatomie; for al authors write against those Surgions that worke in mans body, not knowing the Anatomie; for they be likened to a blind man that cutteth in a vine tree, for he taketh more or lesse than he ought to doo. And here note wel the saying of Galen, the prince of Philosophers, in his Estories, that it is as possible for a Surgion not knowing the Anatomie, to worke in mans body without error, as it is for a blind man to carue an Image & make it perfyt. The .ij. I said, he must be expert; for Rasus sayth, he ought to knowe and to see other men work, and after to have vse and exercise. The thirde, that he be ingenious or wittie, for al thinges belonging to Chirurgerie may not be written, nor with letters set foorth. The fourth, I sayde, that he muste be wel manered, and that he haue al these good conditions here folowing: First, that he be no spousbreaker, nor no drunkarde. For the Philosophers say, amongst all other thinges beware of those persons that followe dronkennes, for they be accompted for †no men, because they liue a life bestiall: wherfore amongst al other sortes of people, they ought to be sequestred from the ministring of medicine. Likewise a Chirurgion must take heede that he deceiue no man with his vague promises, for to make of a smal matter a great, because he woulde be counted the more famous. And amongst other thinges, they maye neither be flatterers, nor mockers, nor priue backbyters of other men. Likewise they muste not be proude, nor presumptuous, nor detracters of other me*n*. Likewise they ought not to

He must know Natural Philosophy, Grammar, Logic, and Rhetoric,

* sign. A. lij, ox.

and specially Anatomy,

as Galen says,

He must not be an adulterer or drunkard,

† sign. A. liij. for drunkards are no men, but live a beastial life.

He must not deceive,

flatter,

be proud.

or covetous,	be too couetous, nor no nigarde, & namely[1] amongst their freendes, or men of worship; but let them be
but free in word and deed.	honest, curteous, and free, both in worde and deede. Likewise they shal geue no counsayle except they be asked, and then say their aduise by good deliberation; and that they be wel aduised afore they speake, chefly
He must keep his Patients' secrets.	in the presence of wise men. Likewise they muste be as priuie and as secrete as anye Confessour, of al thinges that they shal eyther heare or see in the house of their Pacient. They shal not take into their cure any maner
* leaf A. iiij, bk.	of person, except he wyl be obedient vnto *their preceptes; for he can not be called a pacient, vnlesse he be
He must tend the poor as well as the rich,	a sufferer. Also that they doo their diligence aswel to the poore as to the riche. They shal neuer discomfort their pacient, and shall commaunde all that be about him that they doo the same; but to his freendes, speake truthe, as the case standeth. They must also be bolde in those thinges whereof they be certayne, and as dreadfull in al perilles. They may not chide with the
must be pleasant, and not tempt women.	sicke, but be alwayes pleasaunt and mery. They must not couet any woman by waye of vylanie, & specially in the house of their Pacient. They shal not, for couetousnes of money, take in hande those cures that be
He mustn't promise cures by a certain day,	vncurable, nor neuer set any certaine day of the sickemans health, for it lyeth not in their power: folowing the distinct counsayle of Galen, in the amphorisme of Ipocras, saying, *Oportet seipsum non solum.* By this, Galen meaneth, that to the cure of euery sore there
for that depends first on God.	belongeth foure thinges: of which, the first and principal belongeth to God, the second to the Surgion, the thirde to the Medicine, and the fourth to the Pacient. Of the whiche foure, and if any one doo fayle, the
† sign. B. j.	†Pacient can not be healed: then they, to whom belongeth but the fourth parte, shal not promise the whole, but bee first wel aduised. They must also be

[1] specially

Ch. II.] *The Anatomy of the eleven simple Members.* 17

gracious and good to the poore; and of the rich take liberally for bothe. And see they neuer prayse themselues, for that redoundeth more to their shame and discredite, then to their fame and worship: For a cunning and skilful Chirurgion neede neuer vaunt of his dooings, for his works wyll euer get credite ynough. Likewise, that they despise no other Chirurgion without a great cause; for it is meete that one Chirurgion should loue another, as Christe loueth vs all. And in thus dooing, they shall increase both in vertue & cunning, to the honour of God and worldly fame. To whome he bring vs al. Amen!

He must be kind to the poor, and make the rich pay for them.

He must not despise other Surgeons.

[CHAPTER II.]

¶ Heereafter foloweth the Anathomie of the simple members.

AND if it be asked you how many simple members[1] there be, it is to be answered, eleuen, and two that be but superfluities of members; *and these be they, Bones, Cartylages, Nerues, Pannykles,[2] Lygaments, Cordes,

The 11 simple Members, and 2 superfluities.
* sign. B. j, bk.

[1] Hear Bartholomeus de Glanvilla (*de Proprietatibus Rerum*, in John de Trevisa's English) on Members simple: 'Avicen sayth / that membres bene bodyes made of the fyrst medlyng of humours. Other, as it is sayde secundum Johannem / a membre is a stedfaste and sadde [firm, solid] partie of a beaste, composed of thynges that bene lyke other [or] vnlyke, and is ordeyned to som specyall offyce / And by that that it is called a stedfast partie, it is distingued from the partie that is not stedfaste / as a spirite. In that that hit is sayde to be made of thynges that ben lyke and vnlyke, hit is vnderstonde double dyuersite of membres, simple or vnlyke, and compouned or of office. For the membres ben called / membres lyke and simple / the whose partes be of the same kynde with the holle: as euery partie of bloudde is bloudde, and so of other. And suche symple membres and lyke, ben rather [earlier] in kynde, than the membres or limmes of office: for the simple ben partis of the limmes that ben composed... The membres and the limmes ben composed to se, to fele, and to meue, and ben instrumentis of the soule, as hond, fote, and eien, and other such that ben nedeful in diuers qualitees to the werkynge of the soule.' Bk. V. ed. Berthelet, 1535 ('the chef d'œuvre of Berthelet's press'), sign. F. ij.

[2] 'Pannicle (*panniculus*, dim. of *pannus*), fine cloth, a little piece of cloth.

VICARY. C

3 *

Arteirs, Weines, Fatnes, Fleshe and Skinne; and the superfluities be the heares[1] & the nayles. I shal beginne at the Bone, because it is the fundation, and the hardest member of al the body. The Bone is a consimile[2] member, simple and spermatike,[3] and colde and dry of complexion,[4] insencible, and inflexible; and hath diuers formes in mans body, for the diuersitie of helpings. The cause why there be many bones in mans body, is this: Sometime it is needeful that one member or one lymme should moue without another. Another cause is, that some defende the principal members, as dothe the bone of the brest, and of the head: and some to be the fundation of diuers partes of the body, as the bones of the Ridge,[5] and of the legges: and some to fulfyll the hollowe places, as in the handes and feete, &c.

The Grystle is a member simple and spermatike, next in hardnes to the bone, and is of complexion colde and drye, and insencible. The grystle was ordeyned for sixe causes or profites that I fynd in it: The first is, that the continual mouing of the hard bone might not be done in a iuncture, but that the grystle should be a meane betweene the Lygament and him: The seconde is, that in the time of concussion or oppression, the softe members or limmes

1st Member: The Bone.

Its functions.

2nd Member: The Gristle:

the 6 causes for which it was designed.

The fleshy pannicle (*panniculus carnosus*), the fleshy membrane or skin, which lies next under the fat of the outward parts, and is the fourth covering that enwraps all the body, from the head to the sole of the foot.'—1681. Blount, *Glossographia.*

[1] Excrement (outgrowth), as Shakspere calls Armado's moustache in *Love's Lab. Lost*, V. i. 112.

[2] Uniform in structure. L. *consimilis*, alike in all parts. Ital. *consimile*, all alike.—Florio. See p. 23, below.

[3] *Spermatick Parts and Vessels*, (in *Anat.*) are those Arteries and Veins which convey the Blood to the Testicles; also those Vessels thro' which the Seed passes: Also all whitish Parts of the Body, which by reason of their Colour, were anciently thought to be made of the Seed; as Bones, Sinews, Gristles &c.—Kersey's *Phillips's New World of Words*, 1706.

[4] *Complexion*, the Colour of the Face, the Natural Constitution, or Temperature of the Body.—Kersey's *Phillips.*

[5] Spine. A.Sax. *hrycg*, the back of a man or beast, a ridge, roof.

Ch. II.] *The Gristle, Ligaments, Sinews & Nerves.* 19

should not be hurt of the harde: The thirde is, that
the extremitie of bones and Ioyntes that be gristly,
might the easelyer be folded and moued together without hurt: The fourth is, for that it is necessarie in
some meane places to put a grystle, as in the throte
bowel[1] for the sounde: The fyfth is, for that it is
needeful that some members be holden vp with a
grystle, as the liddes of the eyes: The sixth is, that
some limmes haue a sustayning and a drawing abrod,
as in the nose and the eares, &c. *{The uses of Gristle.}*

The Lygament is a member consimple, simple, and
spermatike, next in hardnes to the gristle, and of complexion cold and dry, and is flexible and insensible,
and byndeth the bones together. The cause why he
is flexible and insencible is this: If it had bene
sensible, he mighte not haue suffred the labour and
mouing of the ioynts: and if it had not ben flexible of
his bowing, one lymme should not haue moued without
another. The se*conde profite is, that he be ioyned
with sinews, for to make Cordes & Brawnes: The
thirde helpe is, that he be a resting place to some
sinewes: The fourth profite is, that by him the
members that be within the body be sustayned, as the
matrix and kidneys, and diuers other, &c. *{3rd Member: The Ligament / binds the bones together. / * sign. B. ij, bk. / It joins with Sinews to make Tendons and Muscles.}*

The Sinew is a consimiler member, simple and
spermatike, meane betweene harde and softe, and in
complexion colde and drye, and he is both flexible and
sensible, strong and tough, hauing his beginning from
the braine, or from *Mynuca*, whiche is the marow of
the backe. And from the brayne commeth .vij. payre
of Nerues sensatiues; and from *Mynuca* commeth .xxx.
payre of Nerues motius, and one that is by him selfe,
that springeth of the last spondel. Al these senews
haue both feeling and mouing; in some more, and in
some lesse, &c. *{4th Member: Sinews. / Sinews start from the Brain or Spinal Cord. / 5th Member: Nerves. / Nerves of feeling spring from the Brain: those of motion from the Spinal Cord.}*

[1] Windpipe, 'wosen' or wesand. O.Fr. *boel*, L. *botellus*, a sausage, intestine.

6th Member:
Cords or Tendons

A Corde or Tendon[1] is a consimple or official member, compounde and spermatike, synowy, strong, and tough, meanly betweene hardenes and softnes, and meanely sensible and flexible, and in complexion colde and dry. And the Corde or Tendon is thus made:

are made from

* sign. B. 3.

The synewes *that come from the brayne & from Mynuca, and go to moue the members, is intermingled with the Lygamentes; and when the Synewes and Lygamentes are intermingled together, then is made a corde. And three causes I perceyue why the Cordes were made: The fyrst is, that the Synewe alone is so sensible that he may not suffer the great labour and trauel of mouing, without the felowship and strength of the Lygament that is insencible, and that letteth his great feeling, and bringeth him to a perfect temperaunce, and so the Cordes moue the limmes to the wil of the soule. And this Corde is associated with a simple flesh, and so therof is made a brawne or a muskel, on whom he might rest after his trauel. And this Brawne is called a Muskle. Then when this Corde is entred into this brawne, he is departed into many smal threeds, the whiche be called 'wylle.'[2] And this wyl hath three properties: The fyrst is in length, by whose vertue that draweth, it hath might: The seconde[3] in breadth, by whome the vertue that casteth out hath might: The third in thwartnes, in whom the vertue *that holdeth hath might: and at the ende of the Brawne those threedes be gathered to make another muskel, &c.

Sinews and Ligaments.

They move the limbs as the Will directs.

They combine with flesh and form Muscles,

which divide into fibres, and are called Will.

* sign. B. 3, bk.

7th Member:
Arteries.

Nowe I wyl begynne at the Artere.[4] This Artere

---

[1] *Tendon* (*Lat.* in *Anat.*) is a similar [homogeneous] nervous Part, joyn'd to Muscles and Bones, by which the voluntary Motion of the Members is chiefly performed.—1706. Kersey. *Official* must be 'having an office or function.'

[2] Seemingly identifying the fibres by which the Will acts, with the Will itself: 'so the Cordes moue the limmes to the wil of the soule.'

[3] *orig.* soconde.

[4] *Arteria* (Gr. in *Anat.*), an Artery: The Arteries are those hollow skinny

Ch. II.] *Arteries from Heart, & Veins from Liver.* 21

is a member consimyle, simple and spermatike, hollowe and synowy, hauing his springing from the hart, bringing from the harte to euery member, blood and
4 spirite of lyfe. It is of complexion cold and drye. And al these Arteres haue two cotes, except one that goth to the Lunges, and he hath but one cote that spreadeth abrode in the Lungs, and bringeth with him to the
8 Lunges blood, with the spirite of lyfe to nourishe the Lungs withal: and also that Artere bringeth with him from the lunges ayre to temper the fumous heate that is in the harte. And this Artere is he that is called
12 *Arteria venalis*, because he hath but one cote as a vayne, and is more obedient to be delated abrode through al the lunges, because that the blood might the sooner sweate through him: wheras al other Arters
16 haue two cotes, because one cote may not withstande the might and power of the spirit of life. Diuers other causes there be, which shal be declared in *the Anathomie of the brest, &c.

20 The Veyne[1] is a symple member, in complexion colde and drye, and spermatike, like to the Artere, hauing his beginning from the Lyuer, and bringing from the Lyuer nutritiue blood, to nourishe euery
24 member of the body with. And it is so to be vnderstanded, that there is no more difference betweene these two vessels of blood, but that the Artere is a vessel of blood spiritual or vytal. And the Veyne is a vessel
28 of blood nutrimental, of the which Veynes there is

Arteries spring from the heart, and carry life-blood to every limb.

All Arteries have 2 coats,

save Arteria Venalis, and that has but one.

* sign. B. 4.

8th Member · Veins.

Veins spring from the Liver,

and differ from Arteries only in carrying nutrimental blood, while Arteries carry vital blood.

Vessels like Veins, in which the most thin and hottest part of the Blood, together with the Vital Spirits, pass thro' the Body.—1706. Kersey.

- *Veins* (in *Anat.*) are long and round Canals or Pipes which consist of four Coats, *viz.* a Nervous, a Glandulous, a Muscular, and a Membranous one; their Office being to receive the Blood that remains after the Nourishment is taken, and to carry it back to the Heart to be revived and improved: These Veins have several Names according to the different Parts they pass thro'; as the *Axillary*, the *Basilick*, the *Cephalick*, the *Hepatick*, the *Pulmonary*, &c.—1706. Kersey.

22 The 2 chief Veins. Flesh & its Functions. [Ch. II.

The chief Veins are Vena Portæ and Venacelis or Vena Cava.

noted two most principal, of the which one is called *vena porta*;[1] the other is called *venacelis*, of whom it is too much to treate of now, vntyll we come to the anathomie of the wombe, &c.

9th Member: Flesh.

The flesh is a consimiler member, simple, not spermatike, and is ingendred of blood congeled by heate, and is in complexion hote and moyst. Of the which is noted three kindes of fleshes; that is to say, one is soft & pure fleshe: the seconde is muskulus, or hard & brawny[2] flesh: the thirde is glandulus, knotty, or kurnelly fleshe. Also the commodities of the fleshe be indifferent, for some be common to *euery kinde of fleshe, and some be proper to one maner of fleshe alone. The profytes of the fleshe be many; for some defende the bodye from colde, as dothe clothes: also it defendeth the body from harde thinges comming agaynst it: also through his moysture he rectifieth the body in sommer, in time of great heate. Wherefore it is to be considered what profitablenes is in euery kinde of fleshe by him selfe. And fyrst of simple and pure fleshe, whiche fulfylleth the concauities of voyde places, and causeth good forme and shape: and this fleshe is founde betweene the teeth, and on the ende of the yarde. The profite of the Brawny fleshe, or muskulus fleshe, shal be spoken of in the Anathomie of the armes. The profites of the Glandulus fleshe are these: First, that it turneth the blood into a cullour like to him selfe, as doth the fleshe of a womans paps turne the menstrual blood into mylke: secondly, the Glandulus fleshe of the Testikles turneth the blood into sparme: thirdly, the Glandulus flesh of the cheekes, that ingendreth the spittle, &c.

Of flesh are 3 kinds:
1. soft;
2. brawny;
3. knotty.

* sign. B. 4, bk.

Flesh is like clothes to the body.

Simple flesh fills up hollows.

Glandulous flesh gives blood its colour,

and makes spittle.

---

[1] *Vena Pòrtæ*, the Port-Vein, which takes Name from the two Eminences call'd by Hippocrates *pulai*, i. e. *Portæ* or Gates, between which it enters the Liver.—1706. Kersey. *Venacelis* is '*Vena Cava*, the largest Vein in the body.'—*ib.*

[2] *Brawny*, full of Brawn or Sinews.—Kersey.

Ch. II.] 3 kinds of Fat. Two kinds of Skin. Hair. 23

The next is of Fatnes, of the whiche *I finde three kindes: The firste is *Pinguedo*, and it is a consimilar[1] member, not spermatike, and it is made of a subtyl portion of blood congeled by colde: and it is of complexion colde and moyst, insencible, and is intermedled amongest the partes of the fleshe. The seconde is *Adeppes*,[2] and is of the same kinde as is *Pinguedo*, but it is departed from the fleshe besydes the skinne, and it is as an oyle, heating and moysting the skinne. The thirde is *Auxingia*,[3] and it is of kinde as the others be, but he is departed from the fleshe withinforth about the kidneys, and in the intrayles, and it helpeth both the kidneys and the intrayles from drying by his vnctiositie, &c.

Then come we to the Skinne. The Skinne is a consimile member or officiall, partely spermatike, strong and tough, flexible and sencible, thinne and temperate, Wherof there be two kindes: One is the Skinne that couereth the outwarde members: and the other the inner members, whiche is called a Pannicle, the profitablenesse of whome was spoken of in the laste Lesson: But the Skinne is properlye wouen *of Threedes, Nerues, Veynes, and Arteirs. And he is made temperate, because he should be a good deemer of heate from colde, and of moystnes from drynes, that there shoulde nothing noye nor hurt the body, but it geueth warning to the common wittes thereof, &c.

The Heyres of euery part of mans body are but

*sign. C. j.
10th *Member*: Fatness.
Fat is of 3 kinds:
1. blood congealed by cold;

2. an oil moistening the skin;

3. greasing the kidneys.

11th *Member*: Skin.
Skin is of 2 kinds:
1. external;

2. internal, membrane.

* sign. C. j, bk.

It warns the Wits, of hurtful things.

Superfluity I. Hair.

---

[1] *Similar Parts* or *Simple Parts*, (in *Anat.*) those Parts of the Body that are throughout of the same Nature and Frame; as the Flesh, Bones, Veins, Arteries, Nerves, &c.—1706. Kersey.

[2] *Adeps*, (*Lat.*) Fat, Tallow, Grease: Among Anatomists, it is consider'd as a similar Part of the Body, differing in this respect from *Pinguedo*, that it is a thicker, harder, and more earthy Substance, which flows from the Blood thro' peculiar Vessels into certain Baggs or Bladders that receive it.—1706. Kersey.

[3] *Axungia* (*axis, ungo*), that which besmears the axle, waggon-grease, fat.— Riddle. 'The Grease or Swarf in the Axle-tree of a Wheel; the Fat or Tallow of an Hog, Boars-Grease.'—Kersey.

Hair is made of fume from viscous matter.	a superfluitie of members, made of the grosse fume or smoke passing out of the viscoues matter, thickened to the forme of heyre. The profitablenesse of him is declared in the Anathomie of the head, &c.
Superfluity II. Nails come from fume too, and grow at the ends of fingers and toes.	The Nayles likewise are a superfluitie of members, engendred of great earthly smoke or fume resolued through the natural heate of humours, and is softer then the Bone, & harder then the Fleshe. In complexion they be colde and drye, and are alwayes waxing in the extremitie of the fyngers and toes. The vtilitie of them are, that by them a man shal take the better holde: also they helpe to clawe the body when it needeth: Lastly, they helpe to deuide thinges, for lacke of other tooles, &c.

[CHAPTER III.]

¶ *Heereafter foloweth the Anathomie of the compound members, and first of the head.*

* sign. C. ij.

The Head is the abode of the reasonable Soul.	BEcause the head of man is the habitation or dwelling place of the reasonable soule of man, therefore, with the grace of God, I shal fyrst speake of the Anatomie of the head. Galen saith in the seconde Chapter *De iuuamentis*,[1] and Auicen rehearseth the same in hys fyrst preposition and third Chapter, prouing that the Head of man was made neither for Wittes, nor yet for the Braynes, but onely for the eyes. For
Beasts with no heads have their wits in their breasts.	beastes that haue no heades, haue the orgayne or instrumentes of Wittes in their brests. Therefore God and nature haue reared vp the head of man onely for the eyes, for it is the hyest member of man: and as a beholder or watchman standeth in a highe Towre to geue warning of the Enemies, so doth the Eye of man

---

[1] Juvamentis, *orig.* iuuamentes.

Ch. III.] *5 things outside the Head: Hair; Skin.* 25

geue warning vnto the common Wittes, for the defence of all other members of the body. Nowe to our purpose. If the question be asked, how many things be there *conteyning on the head, and howe many thinges conteyned within the head? As it is rehearsed by Guydo, there bee but fyue conteining, and as many conteined: as thus, The Heyre, the Skinne, the Flesh, the Pannicles, and the Bone, neither rehearsing Veyne nor Artere. The which Anathomie can not be truly without them both, as thou shalt wel perceiue both in this Chapter, but specially in the next. And nowe in this lesson I shall speake but of Heire, Skinne, Fleshe, Veines, Pannicles, and Bones, what profite they doo to man, euery of them in his kinde.

* sign. C. ij, bk.

Guydo says there are 5 things outside the head, and 5 inside.

The 5 outside, are,

Of the Heire of the head (whose creation is knowen in the Anatomie of the simple members) I doo note foure vtilities why it was ordeyned: the fyrst is, that it defendeth the Brayne from too muche heate, and too muche colde, and many other outwarde noyances: The seconde is, it maketh the forme or shape of the head to seeme more seemelyer or beautyfuller. For if the head were not heyred, the face and the heade should seeme but one thing; and therefore the heyre formeth and shapeth the head from the face: The thirde is, that *by the cullour of the heyre is witnessed & knowen the complexion of the Brayne: The fourth is, that the fumosities of the brayne might assend and passe lyghtlyer out by them. For if there were a sad thing, as the skinne or other, of the same nature as the heyre is, the fumosities of the brayne might not haue passed throwe it so lightly, as it doth by the heyre.

1. Hair, which

protects the brain

and adorns the head,

* sign. C. 8.

and lets out the fumosities of the brain.

[¹ *lacestosus*, brawny.]

The Skinne of the head is more lazartus,¹ thicker, and more porrus than any other Skinne of any other member of the body. And two causes I note why: One is, that it keepeth or defendeth the brayne from too muche heate and colde, as doth the heyre: The

2. Skin, more muscular, thicker and more porous than on any other part,

other, it discusseth to the common wittes of al thinges that noyeth outwardly, for the heyre is insencible : The thirde cause why the skinne of the head is more thicker then any other skinne of the body, is this, that it keepeth the brayne the more warme, and is the better fence for the brayne, and it bindeth and keepeth the bones of the head the faster together.

to keep the brain warm.

Next followeth the Fleshe, the which is al Musculus or Lazartus fleshe, lying vpon *pericranium* without meane.[1] *And it is made of subtile Wylle, and of simple fleshe, Synewes, Veines, & Arteirs. And why the fleshe that is al musculus or lazartus in euery member of a mans body was made, is for three causes : the fyrst is, that by his thicknes he shoulde comforte the digestion of other members that lye by him : The seconde is, that through him euery member is made the more formelyer, and taketh the better shape : The thirde is, that by his meanes euery member of the body draweth to him norishing, the which others withholde to put foorth from them : as it shal be more playnlyer spoken of in the Anathomie of the wombe.

3. Muscular Flesh, made of thin fibres,

* sign. C. 3, bk.

to comfort the digestion of near members.

Next followeth *Pericranium*,[2] or the couering of the bones of the head. But heere it is to be noted of a Veyne and an Artere that commeth betweene the flesh and this Pericranium, that nourisheth the vtter part of the head, and so entreth priuily through the commissaries[3] of the skul, bearing to the Brayne and to his Pannikles nourishing : of whose substaunce is made bothe Duramater, and also Pericranium, as shall be

4. Pericranium,

with a vein and artery under it,

taking nourishment to the brain.

---

[1] Intervening medium. *Lazartus* is *lacertosus*, brawny, muscleful.

[2] *Pericranium*, (*Gr.* in *Anat.*) a Membrane or Skin that lies under the thick hairy Skin of the Head, and immediately covers the whole Scull, except just where the Temporal Muscles lie.—1706. Kersey.

[3] *Commissure*, a joyning close, or couching of Things together ; a Closure or Seam : In *Anatomy*, the Mould of the Head, where the Parts of the Scull are united.—1706. Kersey. For *Duramater*, see note 4, p. 28.

Ch. III.] *Duramater and the 7 Bones of the Skull.* 27

declared in the partes conteyned in the head. Here *it *sign. C. 4.
is to be noted of this Pannikle, Pericranium, that it
bindeth or compasseth al the bones of the head, vnto
4 whom is adioyned the Duramater, and is also a part of   Duramater is
his substaunce, howbeit they be separated, for Dura-   part of it, and is under the skull.
mater is nerer ye brayne, and is vnder the skull. This
Pericranium was made principally for two causes : one
8 is, that for his strong bynding together he should make
firme and stable the feeble commissaries or seames of
the bones of the head : The other cause is, that it
shoulde be a meane betweene the harde bone and the
12 softe fleshe.

Nexte is the Bone of the Pot of the head keeping   5. Skull, which
in the Braynes, of which it were too long to declare
their names after al Authors, as they number them
16 and their names ; for some name them after the Greeke
tongue, and some after the Arabian ; but in conclusion,
al is to one purpose. And they be numbred seuen   has 7 bones:
bones in the pan or skul of the head : the fyrst is
20 called the Coronal bone, in which is ye Orbyts or holes   i. Coronal,
of the Eyes, and it reacheth from the Browes vnto the
middest of the head, and there it meteth with the   *sign. C. 4, bk.
seconde bone called Occipissial,[1] a *bone of the hinder   ii. Occipital,
24 part of the head called the Noddel of the head, which
two bones, Coronal and Occipissiale, be deuided by the
Commissaries[3] in the middes of the head. The thirde   iii. iv. Parietal (side bones),
and fourth bones be called Parietales,[2] and they be the
28 bones of the sideling parts of the head, and they be
deuided by the Commissories[3] both from the foresayde
Coronal and Occipissial. The fyfth and the syxth

---

[1] *Occipitis Os*, the Occipital Bone, a Bone of the Scull, which lies in the hinder part of the Head ; being shap'd almost like a Lozenge, with its lower angle turned inwards.—1706. Kersey.

[2] *Parietals*, or *Parietal Bones*, (in *Anat.*) two Bones of the fore part of the Head, which are the thinnest in the Scull, and almost of a square figure.—1706. Kersey.

[3] Sutures. See note 3, p. 26.

28 Bones of Head. 5 things inside Head. [Ch. IV.

v. vi. Petrosa (temporal).

vii. Paxillary (Sphenoid), which wedges the others together.

These are the 7 Head-bones.

bones be called Petrosa[1] or Mendosa[2] : and these two bones lye ouer the bones called Parietales, on euery side of the head one, lyke skales, in whom be ye holes of the eares. The seuenth and last of the head is called Paxillarie, or Bazillarie ;[3] the whiche bone is, as it were, a wedge vnto all the other seuen bones of the head, and doth fasten them togeather. And thus be all numbred : the first is the Coronal bone, the seconde is the Occipissial, the thirde and the fourth is Parietales, the fyfth and the sixth is Petrosa or Mendosa, and the seuenth is Paxillari, or Bazillari. And this suffiseth for the fyue thinges conteyning. 4 8 12

[CHAPTER IV.]

* sign. D. J.

The 5 things inside the Head:

1. Duramater,

*⁕¶ In this Chapter is decla-
red the fiue thinges conteyned
within the head.* 16

Next vnder the bones of the head withinfoorth, the first thing that appeareth is Duramater; then is Piamater; then the substaunce of the Brayne; and then Vermy-formes and Retemirabile. But first to speake of Duramater,[4] whereof and howe it is sprong and made : First, it is to be noted of the Veine and 20

---

[1] *Petrosum Os,* (in *Anat.* i. e. the rocky Bone) the inner Process of the Bones of the Temples, so call'd by reason of its Hardness and Craggedness.—1706. Kersey.

[2] *Mendosa Sutura,* or *Squamen Sutura,* (in *Anat.*) a scaly joining together of Bones ; as in the Bones of the Temples, and those of the Fore-part of the Head.—Kersey.

[3] *Basilare Os,* (in *Anat.*) the same with *Sphenoides,* a Bone of the *Cranium* common both to the Scull and upper Jaw. It is seated in the middle of the Basis of the Scull, and is joyn'd to all the Bones of the *Cranium* by the *Sphenoidal Suture,* except in the middle of its Sides, where it is continued to the *Ossa Petrosa,* as if they were but one Bone.—Kersey. See note 2, p. 44.

[4] *Mater Dura,* or *Meninx Crassa,* (in *Anat.*) a Membrane or Skin, that sticks close to the Scull on the inside, in some Places, and mediatly covers the Brain, and the *Cerebellum* or lesser Brain ; having four Cavities, or hollow Parts, which supply the place of Veins.—Kersey.

Ch. IV.] *Tough Dura-Mater, & tender Pia-Mater.* 29

Arteire that was spoken of in the laste Chapter before, howe priuyly they entred through the commissoris or seames of the head, and there, by their vnion together, *made of the Vein and Artery comming through the seams of the head.*
4 they doo not onely bring and geue the spirite of lyfe and nutriment, but also doo weaue them selues so togeather, that they make this pannicle Duramater. It is holden vp by certayne threedes of him selfe comming *Its fibres run into the Pericranium.*
8 through the sayd commissories, running into Pericranium or pannicle that couereth the bones of the head. And with the foresayde Veyne and Arteire, and these threedes comming from Duramater, is *wouen and made ** sign. D. j, bk.*
12 this Pericranium. Also, why this panicle Duramater is set from the skul, I note two causes: the first *It is kept apart from the skull.* is, that if the Duramater shoulde haue touched the skul, it shoulde lightly haue bene hurt with the
16 hardnes of the bone: The seconde cause is, that the matter that commeth of woundes made in the head pearsing the skul, shoulde by it the better be defended and kepte from Piamater, and hurting of the brayne.
20 And next vnto this panicle there is another pannicle called *Pia mater*,[1] or meeke mother, because it is so *2. Pia mater,* softe and tender vnto the brayne. Of whose creation it is to be noted as of Duramater, for the original of
24 their fyrst creation is of one kind, both from the Hart and the Lyuer, and is mother of the very substaunce of the brayne. Why it is called Piamater, is, for because it is so softe and tender ouer the brayne, that it nour- *which nourishes and*
28 isheth the brayne and feedeth it, as doth a louing *feeds the Brain, as a mother does* mother vnto her tender childe or babe; for it is not so *its child.* tough and harde as is Duramater. In this panicle Pia mater, is much to be noted of the great number of

---

[1] *Mater Pia*, or *Meninx Tenuis*, a Skin which immediately clothes the Brain and *Cerebellum*. It is extremely full of Blood-Vessels; and design'd, as some think, to keep in the Spirits there bred, and to prevent their flying away. These Skins are call'd *Matres*, i. e. Mothers, by the *Arabians*, as if all the other Membranes of the Body took their rise from, or were propagated by them.— 1706. Kersey's *Phillips's New World of Words*.

4

30 Of Pia-Mater, and Cells of the Brain. [Ch. IV.

*sign. D. ij.
It has many Veins and Arteries, and enwraps the brain.

Some of these Veins, &c. go into the brain,

and turn the vital spirit into animal.

The 2 Membranes over the Brain,

*sign. D. ij, bk. are, 1 hard, the other soft to protect it.

3. The Brain.

It is divided into 10 cells, and 3 Ventricles,

Veynes and Arteirs that are planted, rame*fying throughout al his substaunce, geuing to the brayne both spirite and lyfe. And this Pannicle doth circumuolue or lappe al the substaunce of the brayne: and in some places of the brayne the Veynes and the Arteirs goo foorth of him, and enter into the diuisions of the brayne, and there drinketh of the brayne substaunce into them, asking of the hart to them the spirite of lyfe or breath, and of the Lyuer, nutriment. And the aforesayde spirite or breath taketh a further digestion, and there it is made animal; by the elaboration[1] of the spirite vital, is turned and made animall. Furthermore, why there bee moo pannicles ouer the brayne then one, is this: If there had beene but one pannicle onely, eyther it must haue beene harde, or soft, or meane betweene both. If it had beene harde, it should haue hurt the braine by his hardnes: if it had beene soft, it shoulde haue beene hurt of the harde bone: and yf it had beene but meanely neyther hard nor soft, it should haue hurt the braine by his roughnes, and also haue beene hurte of the harde bone. Therefore God and nature haue ordeyned two Pannicles, the *one harde, and the other softe: the harder to be a meane betweene the softe and the bone; and the softer to be a meane betweene the harder and the braine it selfe. Also these Pannicles be colde and dry of complexion, and spermatike.

Next is the Brayne, of which it is marueylous to be considered and noted, how this Piamater deuideth the substaunce of the Brayne, and lappeth it into certen selles or diuisions, as thus: The substaunce of the braine is diuided into three partes or ventrikles, of which the foremost part is the moste:[2] the seconde or middlemost is lesse: the third or hindermost is the least. And from eche one to other be issues or pas-

[1] *Orig.* eleboration.   [2] B*ig*gest.

Ch. IV.] *Ventricles of the Brain, & their Powers.* 31

sages that are called *Meates*,[1] through whom passeth the spirit of life too and fro. But here ye shal note that euery Ventrikle is diuided into two partes; and in euery parte God hath ordeyned and set singular and seueral vertues, as thus: First, in the foremost Ventrikle God hath founded and set the common Wittes, otherwise called the fyue Wittes, as Hearing, Seeing, Feeling, Smelling, and tasting. And also there is in one part of this Ventrikle, the vertue *that is called Fantasie, and he taketh al the formes or ordinaunces that be disposed of the fiue wittes, after the meaning of sensible thinges: In the other parte of the same Ventrikle is ordeyned and founded the Imaginatiue vertue, the whiche receyueth of the common Wittes the fourme or shape of sensitiue thinges, as they were receyued of the common wittes withoutfoorth, representing their owne shape and ordinaunces vnto the memoratiue vertue. In the middest sel or ventrikle there is founded and ordeyned the Cogitatiue or estimatiue vertue: for he rehearseth, sheweth, declareth, and deemeth those things that be offered vnto him by the other that were spoken of before. In the thirde Ventrikle, and last, there is founded and ordeyned the vertue Memoratiue: in this place is registred and kept those things that are done and spoken with the senses, and keepeth them in his treasurie vnto the putting foorth of the fyue or common wittes, or orgaynes, or instrumentes of animal workes, out of whose extremities or lower partes springeth Mynuca, or marowe of the spondels: of whom it shall *be spoken of in the Anatomie of the necke and backe. Furthermore, it is to be noted that from the foremost Ventrikle of the brayne springeth seuen payre of sensatiue or feeling senews, the which be produced to the Eyes, the Eares, the

Side notes:
- each divided into 2 parts.
- In the foremost Ventricle are the Five Wits;
- * sign. D. iij.
- also the Fancy,
- and the Imagination.
- In the 2nd or middle Ventricle is Thought.
- In the 3rd Ventricle is the Memory.
- * leaf D. iij, bk.
- From the foremost Ventricle spring 7 pair of sensitive Sinews.

[1] L. *Meatus*, a Passage or Way; also the Pores of the Body.—Kersey.

Nose, the Toung, and to the Stomack, and to diuers other partes of the body: as it shal be declared in their anatomies.

Also it is to bee noted, that aboute the middest ventrikle is the place of Vermiformis,[1] with curnelly fleshe that filleth; and Retemirabile,[2] or wonderful caule vnder the Pannicles, is sette or bounded with Arteirs onely, whiche come from the harte, in the whyche the vitayle spirite, by his great labour is turned and made animal. And ye shal vnderstande, that these two be the best kept partes of al the body; for a man shal rather dye, then any of these should suffer any manner of greefes from withoutfoorth; and therefore God hath set them farre from the hart. Heere I note the saying of Haly Abbas,[3] of the comming of smal Arteirs from the hart, of whom (sayth he) is made a marueylous net or caule, in the which caule is inclosed the *Brayne, and in that place is layde the spirite of feeling; from that place hath the spirite of feeling his first creation, and from thence passeth to other members, &c. Furthermore, ye shal vnderstand that the brayne is a member colde and moyst of complexion, thinne, and meanely[4] viscous, and a principal member, and an official member, and spermatike. And fyrst, why he is a principal member, is, because he is the gouernour or the treasurie of the fyue wittes: And why he is an official member, is, because he hath the effect of feeling and stering: And why he is colde and moyst, is, that he shoulde, by his coldnes and moystnes, abate and temper the exceeding heate and drought that com-

---

[1] *Vermiformis Processus*, (in *Anat.*) a prominence or bunching Knob of the *Cerebellum*, or lesser Brain, so nam'd from its Shape.—1706. Kersey.

[2] *Rete*, (*Lat.*) a Net. . . In *Anatomy*, the same as *Omentum*, or the Caul. . .
Rete Mirabile, a fine *Plexus*, or Weaving together of many small Arteries in the Brain, especially of brute Beasts; so call'd by reason of its admirable Structure.—Kersey.

[3] See the account of him in the Forewords.   [4] moderately.

Ch. IV.] *Galen's advice. The Brain and the Moon.* 33

meth from the harte: Also, why he is moyst, is, that it should be the more indifferenter and abler to euery thing that shoulde be reserued or gotten into him: 4 Also, why it is soft, is, that it should geue place and fauour to the vertue of stering: And why it is meanely viscous, is, that his senewes should be strong and meanely toughe, and that they shoulde not be letted in 8 their working throughe his ouermuche hardnes. Heere Galen *demaundeth a question, which is this, Whether that feeling and mouing bee brought to Nerues by one or by diuers? or whether the aforesayde thing be 12 brought substancially or radically. The matter (sayth he) is so harde to searche and be vnderstoode, that it were much better to let it alone and passe ouer it. Aristotle, intreating of the Brayne, sayth: The Brayne 16 is a member continually mouing and ruling al other members of the body, geuing vnto them both feeling and mouing; for if the Brayne be let,[1] al other members be let: and if the Brayne be wel, then al other 20 members [of] the body be the better disposed. Also the brayne hath this propertie, that it moueth and followeth the mouing of the Moone: for in the waxing of the Moone, the Brayne followeth vpwardes; and in the 24 wane of the Moone, the brayne discendeth downwardes, and vanisheth in substaunce of vertue: for then the Brayne shrinketh togeather in it selfe, and is not so fully obedient to the spirit of feeling. And this is 28 proued in menne that be lunatike or madde, and also in men that be epulentike,[2] or hauing the falling sicknesse, that *be moste greeued in the beginning of the newe Moone, and in the latter quarter of the Moone. 32 Wherefore (sayth Aristotle) when it happeneth that the Brayne is eyther too drye or too moyst, then can it not worke his kinde: for then is the body made colde: then are the spirites of lyfe melted and resolued away:

Causes of the qualities of the Brain.

* leaf D. 4, back.
Galen's wise advice about a puzzling question:

Let it alone!

The Brain rules all the other members of the body.

It follows the moving of the Moon,
rising and falling with it,

as lunatics and epileptics prove.
* sign. E. j.

---

[1] hindered, stopt. [2] epileptic.

and then foloweth feebleness of the wittes, and of al other members of the body, and at the laste death.

[CHAPTER V.]

¶ *Heereafter foloweth the*
Anatomie of the Face.

The Forehead

THE Front or the Forhead conteyneth nothing but the Skinne and Musculus fleshe, for the panicle vnderneth it is of Pericranium, and the bone is of the Coronal bone. Howebeit there it is made broade, as yf ther were a double bone, whiche maketh the forme of the Browes. It is called the Forhead or Front, from one Eare to the other, and from the rootes of the Eares of the head before, vnto ye browes. But the cause why the Browes were set *and reared vp, was, that they shoulde defende the Eyes from noyaunce withoutfoorth : And they be ordeyned with heare, to put by the humour or sweat that cometh from the head. Also the Browes do helpe the Eyeliddes,[1] and do beautifie and make fayre the face; for he that hath not his Browes heyred, is not seemely. And Aristotle sayth, that ouer measurable Browes betokeneth an enuious man : Also high browes and thicke betokeneth hardnes : and browes with little heare betokeneth cowardnes : and meanly, signifieth gentlenes of hart. Incisions about these partes ought to be done according to the length of the body, for there the Muscle goeth from one Eare to the other. And there, if any incision should be made with the length of the Muscle, it might happen the Browe to hang ouer the Eye without remedie, as it is many times seene, the

stretcnes from ear to ear.

* sign. E. j, back.

It protects the eyes, and helps the Eyelids to adorn the face.

The Brows mark men's characters.

Incisions in these parts must be made lengthwise.

---

[1] See Shakspere on eyelids : *Lucrece,* 366-9 ; *The Tempest,* 'fringed curtains of thine eyes,' &c., but specially *Cymbeline,* II. ii. 19-23 :
'. . . her lids,
. . . these windows, white and azure, lac'd
With blue of heaven's own tinct.'

Ch. V.] *The Eyelids, Ears; Eyes and their Coats.*

more pitie! The browes be called *Supercilium* in Latin; and vnder, is the Eye liddes, which is called *Cilium*, and is garnished with heyres. Two causes I
4 finde why the eye-liddes were ordeyned: The fyrst is, that they shoulde keepe and defende the Eye from *duste and other outwarde noyances: the seconde is, when the eye is weery or heauy, then they should be
8 couered, and take rest vnderneath them. Why the heyres were ordeyned in them, is, that by them is addressed the formes or similitudes of visible thinges vnto the apple of the eye. *[The Eyelids keep the eye from dust, &c., * sign. E. ij. and take forms of things to the Apple of the Eye.]*

12 The Eare is a member semely and grystlye, able to be folden without, and is the orgayne or instrument of hearing: It is of complexion colde and drye. But why the eare was set vp out of the head, is this, that the
16 soundes that be very fugitiue should lurke and abyde vnder his shadowe, tyl it were taken of the instrumentes of hearing: Another cause is, that it should keepe the hole that it standeth ouer, from thinges falling in, that
20 might hinder the hearing. The senewes that are the Orgayns or Instrumentes of hearing, spring each from the Brayne, from whence the seuen payre of senewes do spring; & when they come to the hole of the Eare,
24 there they writhe lyke a wyne presse; and at the endes of them there be like the head of a worme, or like a little teat, in whiche is receyued the sounde, and so caried to the common * wittes. *[The Ear is the organ of hearing, thro' sinews that come from the brain: these twist like a wine-press, and have a teat at their end, to receive Sound. * sign. E. ij, bk.]*

28 The Eyes be nexte of nature vnto the Soule: for in the Eye is seene and knowen the disturbances & greefes, gladnes and ioyes of the Soule, as loue, wrath, and other passions. The Eyes be the instrumentes
32 of sight. And they bee compounde and made of ten things: that is to say, of seuen Tunicles[1] or Cotes, and of three humours. Of the whiche (sayth Galen) *[The Eyes are made of 7 Tunicles and 3 Humours.]*

[1] *Tunick* or *Tunicle* (in *Anat.*), a little Coat, Membrane, or Skin, covering any part of the Body: of these there are four noted ones that belong to the Eye,

36 The 7 Coats and 3 Humours of the Eye. [Ch. V.

The Eye:

the Brayne and the head were made for the Eye, that they might be in the hyest place, as a beholder in a towre, as it was rehearsed in the Anatomie of the head. But diuers men holde diuers opinions of 4 the Anatomie of the Eyes: for some men accompt but three tunikles, and some sixe. But in conclusion, they meane all one thing: For the very truth is, that there be counted and reckoned seuen Tunikles, that is to say, 8

Names of its 7 Tunicles,

Sclirotica,[1] *Secondyna, Retyna,*[2] *Vnia, Cornua, Arania,*[3] and *Coniunctiua:*[4] and these three humours, that is to

and its 3 Humours.

say, *humor Vitrus, humor Albigynus,* and *humor Crystallinus.*[5] It is be knowen howe and after what maner 12 they spring. You shal vnderstande, that there spring-

* sign. E. iij.
From the front Ventricle of the Brain spring 2 hollow sinews,

eth of the brayny substaunce of his for*most Ventrikles, two senewes, The one from the right side, and the other from the left, and they be called the 16

viz. the *Corneous,* the *Uveous,* the *Vitreous,* and the *Crystalline,* to which there are as many Humours answerable.—1706. Kersey.

[1] *Sclerotes,* or *Sclerotica Tunica,* the horney Coat of the Eye. See *Cornea Tunica.*—Kersey.

[2] *Retiformis Tunica,* or *Retina,* one of the Tunicles or Coats of the Eye, which resembles the Figure of a Net, and is the principal Instrument of Sight. —Kersey.

[3] *Aranea Tunica* or *Crystallina* (*Lat.* in *Anat.*), a Coat of the Eye, that surrounds and encloses the Crystalline Humour; taking Name from its thin light Contexture, like that of a Cob-web.—Kersey.

[4] *Conjunctiva Tunica* (*Lat.* in *Anat.*), a Coat of the Eye, so call'd from its sticking close to it; the same with the '*Adnata Tunica,* the common Membrane or Coat of the Eye, otherwise call'd *Conjunctiva* and *Albuginea:* It arises from the Scull, grows to the outward part of the *Tunica Cornea;* and, that the *Visible Species* may pass there, leaves a round hollow space forward, to which is join'd another nameless Coat made up of the Tendons of those Muscles that move the Eye.'—1706. Kersey.

[5] *Vitreal* or *Vitreous,* belonging to Glass, Glassy; a Term in *Anatomy,* as the *vitreous Humour,* which is one of the three Humours of the Eyes, so nam'd from its resembling melted Glass. 'Tis thicker than the *aqueous Humour,* but not so solid as the Crystalline, and exceeds both in quantity.—1706. Kersey.

Aqueous Humour, or *the Watery Humour,* one of the Humours of the Eye which is the outmost, being transparent, and of no Colour: It fills up the space between the *Tunica Cornea* and the *Crystalline Humour.—ibid.*

Crystalline or *Icy Humour,* a white, shining Humour of the Eye, which is thicker than the rest, and the first Instrument of Sight.—1706. Kersey.

Ch. V.] The Eyes and their 2 Optic Nerves.

fyrst payre, for in the Anatomie they be the first paire of senewes that appeare of al seuen. And it is shewed by Galen, that these senews be hollowe as a reede, for
4 two causes. The fyrst is, that the visible spirit might passe freely to the Eyes: The second is, that the forme of visible thinges mighte freely be presented to the common wits. Nowe marke the gooing foorth
8 of these senewes: When these senewes goo out from the substaunce of the Brayne, he commeth through the Piamater, of whose substaunce he taketh a Pannicle or a Cote: and the cause why he taketh that
12 Pannicle, is to keepe him from noying;[1] and before they enter into the skul, they meete, and are vnited into one senewe the length of halfe an inche: and then they depart[2] agayne into two, and eche goeth
16 into one eye, entring through the brayne panne; and these senews be called *Nerui optici*. And three causes I finde why these Nerues are ioyned in one before they passe into the Eye: First, if it happen any diseases in
20 one eye, the other *should receyue all the visible spirite that before came to bothe: The seconde is, that all thinges that we see shoulde not seeme two: for if they had not beene ioyned together, euery thing shoulde
24 haue seemed two, as it doth to a worme, and to other beastes: The thirde is, that the Senewe might stay and helpe the other. But herevpon Lamfranke[3] accordeth muche, saying, that these two Senewes come together
28 to the Eyes, and take a Panikle both of Piamater and of Duramater; and when they enter into the Orbyt of the Eye, there the extremities are spread abroade, the which are made of three substances: that is to say, of
32 Duramater, of Piamater, and of Nerui optici. There

Side notes: taking the power of sight to the Eyes, and bring ing back what they see, to the Wits. / One sinew goes into each Eye. / The 2 are called *Nervi optici*. / * sign. E. lij, bk. / This is to prevent every one seeing double. / The extremities spread abroad,

---

[1] annoying, getting hurt.   [2] part, separate.
[3] An eminent surgeon: see the account of him in Hamilton's *Hist. of Medicine*, i. 364. He was a native of Milan, and died in France about 1300. His *Chirurgia Magna et Parva* was first printed in 1490.—Cooper.

38 *The Coats and Three Humours of the Eye.* [Ch. V.

and take 3 Coats, be ingendred three Tunikles or Cotes, as thus: Of the substance that is taken from Duramater, is ingendred the fyrst cote that is called Secondina: and of Nerui optici is ingendred the third cote that is called Retina: and eche of them is more subtiller then other, & goeth about the humours without meane. And it is to be *which are divided, and form G.* vnderstoode, that eche of these three Tunicles be diuided, and so they make sixe: That is to *say, iij. *• sign. E. iv.* 8 of the partes of the brayne, and three of the parts outwardes, and one of Pericranium that couereth the bones of the head, whiche is called Coniunctiua. And thus you maye perceyue the springing of them, as thus: of Duramater springeth Sclirotica and Cornua: of Piamater springeth Secondina and Vnia: and of Nerui optici springeth Coniunctiua.

Three Humours are in the eye, Nowe to speake of the humors, which be three; and their places are the middle of the Eyes: Of the whiche *Vitreous,* the fyrst is Humor Vitrus, because he is lyke glasse, in colour very cleare, redde, liquit, or thinne; and he is in the inward side next vnto the brayne: and it is thin, because the nutritiue blood of the Crystalline might passe, as water through a sponge should be clensed and made pure, and also that the visible spirite mighte the lightlyer passe through him from the Brayne. And he goeth about the Crystaline humour, vntil he meete with *Aqueous,* Albuginus humour, which is set in the vttermost parte of the Eye. And in the myddest of these humours *Crystalline,* Vitrus and Albuginus, is set the Crystalline humour, in *• sign. E. iv, bk.* whiche is set principally the syght of *the Eye. And *and each is wrapt in a Membrane.* these humours be separated and inuolued with the Pannicles aforesayde; betweene euery Humour a Panicle. And thus is the Eye compound and made. But to speake of euery Humour and euery Pannicle in his due order and course, it would aske a long processe, and a long Chapter: and this is sufficient for a Chirurgion.

Nowe to begin at the Nose: You shall vnderstande

Ch. V.] *The Nose, Nostrils, and Profits of them.* 39

that from the Brayne there commeth .ij. Senews to the holes of the brayne pan, where beginneth the concauitie of the Nose; and these two be not properly
4 senewes, but organes or instrumentes of smelling, and haue heades lyke teates or pappes, in whiche is receyued the vertue of smelling, and representing it to the common wittes. Ouer these two is set Colatorium,[1]
8 that we cal the Nose-thrils: and it is set betweene the Eyes, vnder the vpper part of the Nose. And it is to be noted, that this concauitie or ditche was made for two causes: The fyrst is, that the ayre that bringeth
12 foorth the spirite of smelling, might reste in it tyll it were taken of the organ or instrument of smelling: The seconde cause is, that *the superfluities of the Brayne might be hydden vnder it vntill it were clensed. And
16 from this concauitie there goeth two holes down into yͤ mouth, of which there is to be noted three profites: The fyrste is, that when a mans mouth is close, or when he eateth or sleepeth, that then the ayre might
20 come through them to the Lunges, or els a mans mouth should alwayes be open: The seconde cause is, that they helpe to the relation of the forme of the Nose: for it is sayd, 'a man speaketh in his Nose,' when any
24 of these holes be stopped: The thirde cause is, that the concauitie might be clensed by them when a man snuffeth the Nose, or draweth into his mouth inwardly. The Nose is a member consimple or official, appearing
28 without the face, somewhat plicable, because it shoulde the better be clensed. And it is to be perceyued that it is compounde and made of skinne and Lazartus fleshe, and of two bones standing in maner triangle-
32 wise, whose extremities be ioyned in one part of the Nose with the Coronal bone, and the nether extremities

marginal notes: The Nose has 2 Sinews or organs of smell, with heads like teats. It has Nostrils, * sign. F. j. and 2 holes into the Mouth. When a man speaks in his Nose. The Nose is bendable, and made of musculous flesh, 2 bones, and 2 gristles.

---

[1] *Colatorium*, a Strainer, or *Cribrum Benedictum*, the blessed Sieve, put by the ancients 'in the Reins, and thro' which they would have the Humour call'd *Serum* strain'd into the Ureters.'—1706. Kersey, *Cribrum Ben.* See p. 44, n. 1.

40 Muscles of the Nose. Temples and Cheeks. [Ch. V.

sign. F. j, back.

The Nose has 2 Muscles.

It should be of moderate size.

The Temples

sign. F. ii.

have inward dents to receive humour from the brain.

The Cheeks have 7 upper muscles, and 12 lower ones to the under jaw;

are ioyned with two grystles, and another that diuideth the Nose-*thrilles within, and holdeth vp the nose: Also there be two concauities or holes, that if one were stopped the other should serue: Also there is in the Nose two Muskles to helpe the working of hys office. And Galen sayth, that the Nose shapeth the Face moste; for where the Nose lacketh (sayth he), al the rest of the face is the more vnseemely. The Nose should be of a meane bignes, and not to exceede in length or breadth, nor in highnesse. For Aristotle sayth, yf the Nose-thrills be too thinne or to wyde, by great drawing in of ayre, it betokeneth great straightnes of hart and indignation of thought. And therefore it is to be noted, that the shape of the members of the body betokneth and iudgeth the affections and wyll of the Soule of man, as the Philosopher sayth.

The temples be called the members of the head, and they haue that name because of continuall mouing. And as the science of the Anatomie meaneth, the spirite vital is sente from the hart to the brayne by Arteirs; and by veynes and nutrimental blood, where the vessels pulsatiues be lightly hurt. Also the temples *haue dentes or holes inwardely, wherin he taketh the humour that commeth from the brayne, and bringeth the eyes asleepe; and if the sayde holes or dentes be pressed and wroung, then by trapping of the humour that continueth, he maketh the teares to fal from the Eye.

The Cheeks are the sideling partes of the face; and they conteyne in them Musculus fleshe, with Veynes and Arteirs; and aboute these partes be many Muscles. Guydo maketh mention of .vij. about the chekes & ouer lyp. And Haly Abbas sayth, there be .xij. Muscles that moue the nether Iawe, some of them in opening, and other some in closing or shutting, passing vnder the bones of yᵉ temples, And they be called

Ch. V.] *The Cheeks show Men's Dispositions, &c.* 41

Temporales:[1] And they be right noble and sensatiue; *also Temporal ones;*
of whose hurte is muche peril. Also there be other
Muskles for to grinde and to chewe. And to al these
4 Muscles commeth Nerues from the brayne, to geue them *and all have Nerves from the Brain.*
feeling and mouing: and also there commeth to them
many Arteirs and Veines, and cheefly about the temples, and the angles or corners of the Eyes, and the
8 Lippes. And as the Philosophers say, the cheefe
beau*tie in man is in the cheekes; and there the com- *sign. F. ii, bk.*
plexion of man is most knowen: as thus, if they be full, *The disposition of a man is known by his cheeks;*
ruddy, and meddled with temperate whitenes, and not
12 fat in substaunce, but meanely[2] fleshly, it betokeneth
hotte and moyst of compl[e]xion, that is, sanguin and *whether he's sanguine,*
temperate in culler. And if they be white coloured,
without medling of rednes, and in substaunce fat and
16 soft, quauering, it betokeneth excesse and superfluitie
of colde and moyst, that is flematike: And if they *phlegmatic,*
be browne in colour, or cytrin, yelowe, redde, and thin,
and leane in substaunce, betokeneth great drying and
20 heate, that is cholerike: And if they be as it were *choleric,*
blowen in colour, and of little fleshe in substaunce, it
betokeneth excesse and superfluitie of drynes and colde;
and that is melancolie. And as Auicen sayth, the Cheekes *or melancholy.*
24 doo not only shewe the diuersities of complexions, but
also the affection and wil of the hart: for by the affec- *They show his affections too.*
tion of the hart, by sodaine ioy or dreede, he waxeth
eyther pale or redde. The bones or bony partes, fyrste
28 of the Cheekes, be two: of the Nose outwardely, two: *The Cheeks have 10 bones,*
of the vpper Mandibile,[3] two: within the Nose, three:

---

[1] *Temporalis* (in *Anat.*), a Muscle of the upper Jaw, otherwise call'd *Crotaphites*, which, arising from part of the *Os Frontis, Sincipitis*, and *Sphenoides*, is inserted to the upper part of the *Processus Coronæ* of the lower Jaw. This Muscle with its Partner draws the lower Jaw upwards.—Kersey.

[2] moderately.

[3] *Mandibula* (in *Anat.*), the Mandible or Jaw, either Upper or Lower: The Upper consists of Twelve Bones, on each Side six; but the Lower at riper Years grows into one continued Bone, extremely hard and thick.—Kersey.

as thus : *one diuiding the Nosethrilles within ; and in ech Nosethril one ; and they seeme to be rowled like a wafer, and haue a holownesse in them, by whiche the ayre is respyred and drawen to the lunges, and the superfluitie of the brayne is purged into the mouthwardes, as is before rehearsed. But Guydo and Galen say, that there be in the face nyne bones ; yet I can not finde that the nether Mandible should be of ye number of those nyne, for the nether Mandible accompted there, proveth them to be ten in number ; Of which thing I wyl holde no argument, but remit it to the sighte of your Eyes.

The partes of the mouth are fyue, that is to say, the Lippes, the teeth, the toung, the Uuila, and the Pallet of the mouth. And first to speake of the lippes : they are members consimile or official, full of Musculus fleshe, as is aforesayde, and they were ordeyned for two causes ; one is, that they should be to the mouth as a doore to a house, and to keepe the mouth close tyl the meate were kindly chewed : The other cause is, that they should be helpers to the pronouncing of the speache. The teeth[1] are members *consimile or official, spermatike, and hardest of any other members, and are fastened in the cheke bones, and were ordeyned for three causes : First, that they should chewe a mans meate, or it should passe downe, that it might be the sooner digested : The seconde, that they should be a helpe to the speache ; for they that lacke their Teeth, doe not perfectly pronounce their wordes : the thirde is, that they should serue to beasts as weapons. The number of them is vncertayne ; for some men haue mo, and some lesse ; they that haue the whole number haue .xxxij., that is to say .xvj.

---

[1] *Dens* . . The Teeth are of three sorts, *viz.* the *Incisores,* or Cutters, which bite off the Morsel ; the *Canini* or Dog-teeth that break it ; and the *Molares* or Grinders that make it small.—Kersey.

Ch. V.] The Teeth, Tongue, and Uvula. 43

aboue, and as many beneath, as thus, two *Dvvallies*, (16 at top, 16 below;)
two *Quadripulles*,[1] two *Cannines*,[2] eyght *Morales*,[3] and
two *Cansales*.[4] The Toung is a carnous member, com- 3. the Tongue,
pounde and made of many Nerues, Lygamentes, Veines
and Arteirs, ordeyned principally for three causes:
The first is, that when a man eateth, the Toung mighte
helpe to turne the meate tyll it were wel chewed:
The seconde cause is, that by him is receiued the taste to receive taste, and speak;
of sweete and sowre, and presented by him to the com-
mon wittes: The thirde is, that by him *is pronounced * sign. F. iv.
euery speach. The fleshly parte of the toung is white,
and hath in him nine muskles; and about the roote with 9 muscles;
of him is Glandulus, in the whiche be two welles, and
they be euer ful of spittle, to temper and keepe moyst
the toung, or els it would waxe dry by reason of his
labour, &c. The Uuila[5] is a member made of a sponge- 4. the Uvula,
ous fleshe, hanging downe from the ende of the Pallet
ouer the goulet of the throte, and is a member in complexion colde and dry; and oftentimes when there
falleth rawnes or muche moystnes into it from the
head, then it hangeth downe in the throte, and letteth
a man to swallowe; and it is broade at the vpper ende,
and smal at the nether. It was ordeyned for diuers
causes: One is, that by him is holpen the sounde of to help speech,
speache; for where the Uuila is lacking, there lacketh
the perfect sound of speache: Another is, that it might
helpe the prolation[6] of vomites: another is, that by
him is tempered and abated the distemperaunce of the and temper the air that goes to
ayre that passeth to the Lunges: another is, that by the Lungs;

---

[1] ? the *Duals* the central Incisors; the *Quadruples* the side ones.
[2] *Caninus*. Doggish: currish: of a dogge.—Cooper. [3] *Molaris*. A cheeke tooth.—1578. Cooper. [4] ? *Casuales* (chance-teeth), or *Clausales*, shutting ones?
[5] *Uvula*, the little piece of red, spungy Flesh that hangs down from the Palate or Roof of the Mouth, between the two Glandules call'd *Amygdalæ*: Its use is to prevent the Air from ent'ring too cold into the Lungs, and to hinder any Liquor that is drunk from falling upon the Nostrils: It is otherwise termed *Ura, Uvigena*, and *Uvigera*; as also *Cion* and *Columella*.—Kersey.
[6] forth-casting, bringing up.

him is guyded the superfluities of the brayne that commeth from the coletures *of the Nose;¹ or els the superfluities should fal down sodenly into the mouth, the which were a displeasure. The Pallet of the mouth 4 conteyneth nothing els but a carnous Pannikle; and the bones that be vnderneath it haue two diuisions, One along the Pallet from the diuision of the Nose, and from the opening of the other Mandible vnto the 8 nether ende of the Pallet, lacking halfe an inch; and there it diuideth ouerthwart; and the first diuision is of the Mandible, and the seconde is of the bone called Paxillarie or Bazillarie,² that sustayneth and byndeth 12 al other bones of the head together. The skinne of the Pallet of the mouth is of the inner parte of the stomack and of Myre, and of Ysofagus,³ that is, the way of the meate into the stomacke. The way how to 16 know that such a pannicle is of that part of the stomacke, may be knowen when that a man is touched within the mouth, anone he beginneth to tickle in the stomacke; and the neerer that he shal touche vnto the 20 throte, the more it abhorreth the stomacke, and often times it causeth the stomacke to yeld from him that is within him, as when *a man doth vomite.

Also in the mouth is ended the vppermoste ex- 24 tremitie of the Wesande, which is called Myre or Isefagus. And with hym is conteyned *Trachia arteria*,⁴ that is, the way of the ayre, whose holes be couered with a lap like a tong, and is gristly, that the meate 28 and drinke mighte slyde ouer him into Isofagus: The

---

¹ *Colatoire du nez.* The spungie bone through which the sniuell passeth from the braine into the nosethrils.—1611. Cotgrave. See note, p. 39.

² *Os basilaire.* The Nape, or Nuke-bone; the bone wherby all the parts of the head are supported: some call it the Cuneall bone, because it is, wedge-like, thrust in betweene the bones of the head and th' upper Jaw.—1611. Cotgrave. (See note 3, p. 28, above.)   ³ See note 4, p. 47, below.

⁴ *Arteria trachea* or *Aspera* (*i. e.* the rough Artery), the Wind-pipe, a gristly Vessel, which consists of several Rings and Parts; its use being, to form and convey the Voice, to take in Breath, &c.—1706. Kersey's *Phillips.*

Ch. VI.] *The Neck; its 4 Parts and 7 Spondels.* 45

whiche grystle, when a man speaketh, it is reared vp, and couereth the way of the meate: and when a man swalloweth the meate, then it couereth the way of the 4 ayre, so that when the one is couered, the other is discouered. For if a man open the waye of the ayre when he swalloweth, if there fal a crum into it, he shal neuer cease coughing vntil it be vp agayne. And this 8 suffiseth for the necke.

which protects the air-passage.

[CHAPTER VI.]
¶ *Heereafter foloweth the*
Anatomie of the Necke.

12 THE Necke foloweth next to be spoken of. Galen proueth that the Necke was made for no other cause but for the Lunges; for al thinges *that haue no Lungs, haue neither necke nor voyce, except 16 fishe. And you shal vnderstande, that the necke is all that is conteyned betweene the head and the shoulders, and betweene the chinne and the brest. It is compounde and made of foure thinges, that is to say, of 20 *Spondillus*,[1] of *Seruicibus*,[2] of *Gula*, and of *Gutture*, the which shal be declared more playnely hereafter; and through these, passe the waye of the meate and of the ayre; but they be not of the substance of the necke.

24 The Spondelles of the necke be seuen: The fyrst is ioyned vnto the lower parte of the head, called Paxillarie or Bazillarie,[3] and in the same wise are ioyned euery Spondel with other, and the laste of the seuen 28 with the fyrst of the Backe or Ridge: and the Lygamentes that keepe these Spondels together, are not so hard and tough as those of the backe: for why? those of the necke be more feebler and subtiller: The cause

The Neck is made for the Lungs.

* sign. G. i, bk.

It consists of 4 things:

1. Seven Spondels or Vertebræ,

---

[1] *Vertebre:* f. A turning ioynt, or ioynt wherein the bones meet so as they may turne; as in the huckle-bone, &c.—1611. Cotgrave.
[2] The Fax-wax or *Pax-wax*, the *Ligamentum Nuchæ.* See next page.
[3] See p. 44, note 2, above.

is this, for it is necessary otherwhile that the head moue without the necke, and the necke without the head, the whiche might not well haue beene done if they had beene strong and boystrous. Of these afore- 4 sayd seuen *Spondels of the necke, there springeth seuen payre of Senewes, the whiche be diuided into the head and into the Vysage, to the shoulders, and to the armes. From the hole of the first spondel springeth 8 the fyrst payre of senewes, betwene the fyrst spondel and the seconde; and so foorth of al the rest in like maner as of these. Also these senewes receyue subtil wylle of the senews of the Brayne; of the which wylle, 12 and senewes, and fleshe, with a pannikle, make the composition of Muskels, Lazartes, and Brawnes, the which three thinges be al one, and be the instrumentes of voluntarie mouing of euery member. 16

The Muskles of the neck, after Galen, are numbred to be .xx. mouing the head and the necke. Likewise it is to be noted, that there be three maner of fleshes in the necke: the first is called *Pixwex*[1] or Seruisis, and 20 it is called of Chylder 'Golde heire, or yellowe heire,' the whiche are certayne longitudinales lying on the sides of the Spondels from the head downe to the latter Spondel. And they are ordeyned for this cause, that 24 when the Senewes be weery of euer muche labour with mouing and tra*uayle, that they might rest vpon them as vpon a bedde. The second fleshe is musculus, from whome springeth the Tendons and cordes that moue 28 the head and the necke, whiche be numbred twentie, as is afore declared: The thirde fleshe replenisheth the voyde places, &c. The thirde parte of the necke is

---

[1] Called also Fax-wax, Fick-fack, Fig-fag, Fix-fax, Pack-wack, Pease-wease, and Tax-wax. '*Pax-wax*, synewe,' ab. 1440. *Promptorium Parvulorum*, ed. Way. The tough strong elastic ligament running along the spinal vertebræ into the occipital bone.—1866. Wheatley. *Dict. of Reduplicated Words. Philolog. Soc. Trans.*, 1865, p. 67. A.Sax. *feax* is hair.

Ch. VI.] *The Neck, its Throat-boll and Gullet.* 47

called *Guttture*,[1] and it is the standing out of the throte **3. Guttur, or Throat-boll.**
boll. The fourth part is called *Gula*,[2] and the hinder **4. Gula, or gulic,**
parte *Ceruix*,[3] and hath that name of the Philosophers,
4 because of the marowe comming to the Ridge bones.
It is so called, because it is (as it were) a seruaunt to
the brayne: For the necke receiueth and taketh of the
brayne, influence of vertue of mouing, and sendeth it
8 by senewes to other parts of the body downwardes,
and to al members of the body. Heere you shal vnder-
stand, that the way of the meate, & Mire or Isofagus,[4] **or Æsophagus,**
is al one thing; and it is to be noted, that it stretcheth **running from the mouth to the**
12 from the mouth to the stomache, by the hinder part of **Stomach.**
the necke inwardly, fastned to the spondels of the
necke, vntyl he come to the fyfthe spondel, and there
he leaueth the spondel, and stretcheth tyl he come to
16 the for*most part of the brest, & passeth through Dia- *** sign. G. iii.**
fragina[5] tyl it come to the mouth of the stomacke, and
there he is ended. Furthermore it is to be noted, that
this Wesande is compounde and made of two Tunikles **The Gullet is made of 2 Coats,**
20 or Cotes, that is to say, of the inner and of the vtter.
The vtter tunikle is but simple, for he needeth no
retention but onely for his owne nourishing: but the
inner Tunikle is compounde and made of Musculus **the inner one of longitudinal mus-**
24 Longitudinal Wyl, by which he may drawe the meate **cular fibre.**
from the mouth into the stomack, as it shal be more
playnely declared in the Anatomie of the stomacke.[6]
Furthermore, *Caua*[7] *pulmonis via, trachia arteria*,[8] al

---

[1] *Guttur* (*Lat.*), the Throat, or Head of the Windpipe. *Guttural Cartilage* (in *Anat.*), a Gristle, which, with others, makes up the Larynx, or top of the Throat.—Kersey.
[2] *Gula* (*Lat.*), the Gullet, or upper part of the Throat.—Kersey.
[3] *Cervix*, the hinder part of the Neck. *Cervical*, belonging to the Neck, as the *Cervical* or *Vertebral Vessels*, a Term us'd by Anatomists for the Arteries and Veins that pass thro' the *Vertebræ*, or Turning-Joints and Muscles of the Neck up to the Scull.—1706. Kersey.
[4] *Oesophagus*, the Gullet or Weasand-pipe, the Conduit or Funnel that conveys the Meat and Drink from the Mouth to the Stomach.—1706. Kersey.
[5] for 'Diafragma.' [6] Chapter VIII, p. 60, below. [7] orig. *Cana*.
[8] See *Arteria trachea*, p. 44, note 4, in Chapter V, above.

48　The Throat-Boll.　The Shoulder-Bones.　[Ch. VII.

The Throat-boll or Epiglottis is within the neck, next the Gullet, and made of gristle.

these be one thing, that is to say, the throte boll;[1] and it is set within the necke besides yᵉ Wesande towardes *Gula*, and is compounde of the grystle knytte eache with other. And that pannikle that is meane be- 4 tweene the Wesand and the throte bol, is called *Isinon*.[2] Also ye shal vnderstand, that the great Veines which ramefie by the sydes of the necke to the vpper part of the head, is of some men called *Gwidege*,[3] & of others 8 *Vena organices*, the incision of whom is perillous.

[It. Guidegi; see p. 86.]

* sign. G. lii, bk.　*And thus it is to be considered, that the Necke of man is compounde and made of skinny fleshe, Ligamentes, and bones. And this suffiseth for the necke 12 and the throte.

[CHAPTER VII.]

¶ *The Anatomie of the*
Shoulders and Armes.　16

The Shoulder has 2 bones,

AND fyrst to speake of the bones: It is to be noted, that in the shoulder there be two bones, that is to say, the Shoulder bone, and the Cannel bone;[4] and also the adiutor bone[5] of the arme 20 are ioyned with yᵉ shoulder bones, but they are not numbred among them, but amongst the bones of the armes. In the composition of the shoulder, the fyrst bone is *Os spatula*, or shoulder blade, whose hinder 24 part is declined towards yᵉ chinne, & in that ende it is broade, & thin, and in the vpper part it is round, in whose roundnes is a concauitie, which is called yᵉ boxe or coope of the shoulder, into which entereth the 28 Adiutor bones; and they haue a bynding togethers

the Shoulderblade

---

[1] '*epyglotum : anglice*, the throtebolle.' 15th-century Glossary in Wülker's *Old English Vocabularies*, i. 580/21. '*Epiglottis* or *Sublinguium* (in *Anat.*), the fifth Cartilage or Gristle of the *Larynx*, the cover or flap of the Wind-pipe.' —1706. Kersey.　[2] Arab. *isa* is fat matter.—N.
[3] 'A corruption for Arab. *'irek*, vein, says Dr. Neubauer; the '*i = ain*, is the strongest guttural, written by *gŵ;* the confusion of *r* and *d* is common in Arabic texts.'　[4] The Clavicle.　[5] The Humerus or upper bone of the arm.

Ch. VII.] *The Clavicle & the Bones of the Arm.* 49

with strong flexible Senewes, and are conteyned faste with the bone called *Clauicula,* *or the Cannel bone :[1] and this Cannel bone stretcheth to bothe the shoulders, one ende to the one shoulder, and another to the other; and there they make the composition of the shoulders. The bones of the great arme, that is to say, from the shoulder to the fingers endes, be .xxx. The first is the Adiutor bone, whose vpper ende entreth into the concauitie or boxe of the shoulder bone : It is but one bone, hauing no felowe, and it is hollowe, and ful of marowe; and it is also crooked, because it shoulde be the more habler to grype thinges; and it is hollowe, because it should be lighter and more obedient to the steering or mouing of the Brawnes. Furthermore, this bone hath two emynences, or two knobs in his nether extremitie, or in the iuncture of the Elbowe, of the which the one is more rising then the other, and are made lyke vnto a Polly[2] to drawe water with; and the endes of these bones enter into a concauitie proportioned in the vppermoste endes of the two Focel bones ;[3] of whiche two bones, the lesse goeth from the Elbowe to the Thombe, by the vppermoste part of the arme, and the greater is the *nether bone from the Elbowe to the little finger. And these two bones be conteyned with the Adiutor bone,[4] and bee bounde with strong Ligamentes, and in like maner with the bones of the hande. The whiche bones be numbred .viij.; the .iiij.

sign. G. iv. and the Clavicle.

The great Arm has 30 bones :

1. the Adjutor, or *Humerus,*

with 2 knobs at the elbow,

made like a bucket-pulley, and fitting into the top ends of Ulna and Radius.

*sign. G. iv, bk.

The arm-bones are bound to the 8 hand-bones (or carpal-bones).

---

[1] *Claviculæ* (in *Anat.*), the Clavicles, or Channel-bones : two small Bones which fasten the Shoulder-bones and Breast-bone, as it were a Key, being situated at the Basis, or bottom of the Neck above the Breast, on each Side one. —1706. Kersey.

[2] Fr. *Poulie :* f. A pullie.—1611. Cotgrave.

[3] The *Ulna* and *Radius,* or lower bones of the Arm. See *Focile minus,* from Kersey, p. 52, note 3, below. Fr. *Focile :* m. The arme from the elbow to the wrist ; the leg, or shanke, from the knee to the ankle ; each consisting of two bones : *Focile grand,* Th' upper of these two bones, being the longer and greater ; *Focile mineur,* ou, *petit focile,* The vndermost, and lesse of them.—1611. Cotgrave. [4] The Humerus, or upper bone of the Arm.

VICARY. E

50 *Bones of the Hand. Sinews & Muscles.* [Ch. VII.

	vppermost be ioyned with the .iiij. nethermost towardes the handes: and in the thirde warde of bones, be .v. and they are called *Ossa patinis*, and they are in the
To the 5 metacarpal bones are joined the bones of the fingers and thumb;	palme of the hande. And to them be ioyned the bones 4 of the Fingers and the Thombes, as thus; in euery fynger .iii. bones, and in the thombe two bones: that is to say, in the fingers and thombe of euery hand
14 finger bones,	.xiiij. called *Ossa digito*ru*m*; in the palm of the hand 8
5 metacarpal,	.v. called *Patinis*;[1] and betweene the hande and the
8 carpal,	wryste, viij. called *Rucete*;[2] and from the wryst to the
3 in arms:	shoulder .iij. bones: al which beeing accompted together,
30 in all.	ye shal finde thirtie bones in eche hand and arme. 12
	To speake of Senewes, Lygamentes, Cordes and Brawnes: Here fyrst ye shal vnderstand that there
Through the Vertebræ run 4 Sinews,	commeth from Mynuca,[3] thorowe the Spondels of the necke, foure senewes, which most playnly do appeare in 16
* sign. H. j.	sight, as thus: one commeth into the *vpper parte of the arme, another into the nether parte, and one into the inner side, and another into the vtter side of y*e* arme; and they bring from the brayne and from 20 Minuca, both feeling and mouing into the armes, as thus: The senewes that come from the Brayne and from the marow of the backe that is called Minuca, when they come to the iuncture of the shoulder, there 24
which, with the Ligaments of the Shoulder,	they are mixed with the Lygamentes of the selfe shoulder, and there the Lygamentes receyue both feeling and mouing of them; and also in their medling
form a Tendon,	together, they are made a Corde or a Tendo*n*. Three 28

[1] *Metacarpus*, or *Metacarpium* (in *Anat.*), the Back of the Hand, consisting of four small and somewhat long Bones, which stretch out the Palm of the Hand, and are call'd *Post-brachialia*. *Metacarpus* is also a Bone of the Arm, made up of four Bones, which are ioyn'd to the Fingers; that which bears up the Fore-finger, being the biggest and longest.—1706. Kersey.

[2] ? meaning. Of these carpal bones, Cotgrave (A.D. 1611) has, ' *Os sesamoïdes*. Certaine little flat bones wherewith the ioynts of the fingers and toes are filled, setled, and strengthened: their number is vncertaine, and their name they haue of the oylie graine *Sesame*, the which they somewhat resemble.'

[3] 'the marow of the backe,' spinal cord.

Ch. VII.] *Sinews. How a Tendon becomes a Muscle.* 51

causes I find why the senewes were medled with the
Lygamentes: The first cause is, that the excellent feeling of the senewes, whiche many waies be made weery
4 by their continual mouing, should be repressed by the   which relieves
insenciblenesse of the Lygaments: The seconde is, that
the littlenesse of the Senewes shoulde be fulfilled
through the quantitie of the Lygaments. The third is,
8 the feeblenesse of the senewe, that is insufficiente and
too feeble to vse his offices, but by the strength and   and strengthens the Sinews.
hardnes of the Lygamentes.

Nowe to declare *what a Corde is, what a Ligament,   * sign. H. j. bk.
12 and what a Muskle or a Brawne, it is ynough rehearsed
in the Chapter of the Simple members.[1] But if you
wyl, thorough the commaundement of the Wyl or of
the Soule, drawe the arme to the hinder part of the   When you draw back the arm,
16 body, then the vtter Brawne is drawen together, and   the outer Muscle contracts, and the
the inner is inlarged: And likewise inwards, when the   inner expands.
one brawne dothe drawe inwardes, the other doth
stretche; & when the arme is stretched in length, then
20 the Cordes be lengthened: but when they passe the
iuncture of the shoulder and of the Elbowe, by three
fingers breadth or thereabout, then it is deuided by   The Tendon is mixed with flesh,
subtill wyl, and medled with the simple fleshe: and   and called a Brawn.
24 that whiche is made of it is called a Brawne. And
three causes I finde why that the simple fleshe is
medled with the Corde in the composition of the
brawne: The fyrst is, that the aforesayde Wylle
28 might drawe in quiet through the temperaunce of the
fleshe. The seconde is, that they temper and abate
the drought of the cord with his moystnes, the which
drought he getteth through his manifold mouing. The
32 thirde is, that the forme *of the brawny members   * sign. H. ii.
shoulde be the more fayre, and of better shape: wherfore God and nature haue clothed it with a Panikle,
that it might the better be kept. And it is called of

[1] Ch. ii. p. 17, above.

52 Muscles. Veins & Arteries of the Arm. [Ch. VII.

or Muscle, that is, a little mouse.

yᵉ Philosophers 'Musculus,' because it hath a forme like vnto a Mouse. And when these Brawnes come neere a Ioynt, then the Cordes spring foorth of them, and are medled with the Lygaments agayne, and so moueth that Ioynt.[1] And so ye shall vnderstande, that

Between every 2 ioints is a Brawn;

alwayes betweene euery two Ioyntes, is ingendred a Brawne, proportioned to the same member and place, vnto the last extremitie of the fingers, so that aswell the least iuncture hath a proper feeling and mouing when it needeth, as hath the greatest. And after

and there are 14 Brawns or Muscles in the arm and hand.

Guydo, there be numbred .xiiij. in the arme and hande, as thus : .iiij. in the Adiutor,[2] mouing the vpper part of the arme : and .iiij. in the Focels,[3] mouing that part of the arme : and fiue in the hande, mouing the fyngers.

Of Veins and Arteries. From the Vena Cava one branch runs to each arm-pit, ✱ sign. H. ii, bk. when it divides into 2:

Now to speake somewhat of the Veynes and Arteirs of the arme : It is to be vnderstoode, that from *Vena-kelis*[4] there commeth two braunches, the one commeth to the one arme pyt, and the other ✱commeth to the other. And nowe marke the spreading; for as it is of the one, so it is of the other, as thus : when the braunch is in the arme pyt, there he is deuyded into

1. at the bend of the arm called Bazilica,

two braunches; the one braunche goeth along in the inner side of the arme vntil it come to the bought of the arme, and there it is called *Bazilica* or *Epatica*,[5]

---

[1] *Ligament* (Lat. in *Anat.*), a Band or String partaking of the Quality of a Cartilage and a Membrane, design'd by Nature for ioyning together of Parts, especially Bones, in order to the better performing of their Motions.—1706. Kersey.

[2] Humerus. *Adjutory*, aiding or helping; as the *Adjutory Bones*, two Bones that reach from the Shoulders to the Elbows, and are so call'd by some Anatomists. —1706. Kersey. *Adiutor.* A helper; a furtherer.—1578. Cooper.

[3] *Focile Majus*, the greater bone of the Arm, peculiarly call'd *Vlna*, or the greater Bone of the Leg, nam'd *Tibia*. *Focile Minus*, the lesser Bone of the Arm, known by the Name of *Radius*, or the lesser Bone of the Leg, termed *Fibula*.—1706. Kersey's *Phillips*. See the extract from Cotgrave, p. 49, note 3, above.

[4] *Vena Cava*, the largest vein of the body. See p. 57, note 4.

[5] *Basilica*, or the *Basilick Vein*, the inner Vein of the Arm, otherwise call'd

Ch. VII.] *The Veins Salvatella and Cephalica.* 53

and so goeth downe the arme til it come to the wryst, and there it is turned to the backe of the hand, and is found betwene the little fynger and the next, and there
4 it is called *Saluatella*.[1] Nowe to the other braunche that is in the arme hole, which spreadeth to the vtter side of the shoulder, and there he deuideth in two: ye one goeth spreading vp into ye carnous parte of the
8 head, and after discendeth through the bone into the Brayne, as is declared in the Anatomie of the head[2]: The other braunche goeth on the outward side of the arme, and there he is deuided in two also; the one
12 parte is ended at the hande, and the other part is folded about the arme, tyl it appeare in the bought of the arme, and there it is called *Sephalica*:[3] from thence it goeth to the backe of the hande, & appeareth be-
16 tweene the thombe *and the formost fynger; and there it is called *Sephalica occularis*. The two braunches that I spake of, whiche be diuided in the hinder part of the shoulders; from eche of these two (I say) springeth
20 one; and those two meete together, and make one veyne,

and at the back of the hand,

Salvatella.

2. the other runs into the head.

The other branch runs outside the arm,

and at its bend is called Cephalica;

* sign. H. iii. the other runs between the thumb and 1st finger.

Hepatica, being the lower Branch of the *Axillaris*, divided into three Branches, under the *Musculus Pectoralis*.—1706. Kersey. '*Veine basilique*. (Called by our Anatomists) the liuer veine: issues from the *Sousclaviere*, and is diuided into two branches, a deepe and a superficiall one; the later whereof, being neere the inwarde processe of th' arme, and verie neere the skinne, is diuided into other two; *viz*. a lesse, which runnes into the head veine, and together with it, makes the *Mediane*; whilest the greater passes by th' elbow vnto the hand, & there makes the *Salvatelle*.'—1611. Cotgrave. '*Basilica*. The liuer vaine. *Hepaticus*. Of the liver.'—1578. Cooper.

[1] *Salvatella* (in *Anat*.), a Vein which takes its rise from the Liver, and runs thro' the Arm and Wrist into the Little Finger.—1706. Kersey. See Cotgrave's definition, p. 54, note 3.

[2] See p. 24, above, and p. 57, note 4, below.

[3] *Veine cephalique*. The head veine; or, a third branch of *la Sousclaviere*; passes betweene the muskle *Deltoïde*, and that of the breast, and goes vnto the bought of the elbow, where it diuides it selfe into two branches; the lower, and lesse, going along th' inner part of the arme, ioynes with a branch of *la basilique*, and together with it, makes the *Mediane*; the higher, and greater, seated in the outside of the elbow, yeelds on both sides many branches, the greatest whereof meets with *la Basilique*, and together with it, makes *la Salvatelle*.—1611. Cotgrave.

54 The 5 chief Veins. Arteries. The Chest. [Ch. VII.

The 5 chief Veins, which appeareth in the bought of the arme, and there it is called *Mediana*, or *Cordialis*, or *Commine*.[1] And thus it is to be vnderstoode, that of *vena Sephalica* springeth *vena occularis;* and of *vena Bazilica*[2] springeth *vena Saluatella;*[3] and of the two veynes that meete, springeth *vena Mediana;* and in ramefying from these fyue principal Veines springeth innumerable, of the whiche a Chirurgion hath no great charge, for it suffiseth vs to knowe the principals.

with many less ones, which a Surgeon needn't trouble about.

To speake of Arteirs, you shall vnderstande, that wheresoeuer there is founde a Veine, there is an Arteire vnder him: and if there be founde a great Veine, there is found a great Arteir; and where as is a little Veine there is a little Arteir: For whersoeuer there goeth a veine to geue nutriment, there goeth an Arteir to bring the spirite of lyfe. Wherfore it is to be noted, that the Arteirs lye *more deeper in the flesh then the Veines doo: for they cary and kepe in them more precious blood than doth the Veine; and therefore he hath neede to be further from daungers outwardly: and therefore God and Nature haue ordeyned for him to be closed in two cotes, where the Veine hath but one, &c.

Under every Vein is an Artery,

carrying the spirit of life.

* *sign. H. lit, bk.*

Arteries have two coats.

The Breast is the Chest of the Spiritual Members.

The Brest or Thorax is the Arke or Chest of the spiritual members of man, as sayth the Philosopher: where it is to be noted, that there be foure thinges conteyning, and eyght conteyned, as thus: The foure conteyning are, the Skinne, Musculus fleshe, the

---

[1] *Veine mediane.* The middle, common, or black veine, compounded of the two lesse branches of the liuer and head veines, and running along the middle of the arme almost vnto the wrist, where it passes in the form of an *Y* into the hand: there is likewise another of this name vnder th' Instup. — 1611. Cotgrave.

[2] *Basilica,* the liuer vaine.—Cooper. 1578. See p. 52, note 5.

[3] *Salvatelle;* f. Th' outward branch of the shoulder veine, falling down, ouer the wrist, vnto the partition between the ring finger and the little one. —1611. Cotgrave.

Ch. VII.] *The Chest, its Paps, Muscles, & Bones.* 55

Pappes, and the Bones: The partes conteyned are, the Hart, the Lunges, Panikles, Ligamentes, Nerues, Veines, Arteirs, Mire or Isofagus. Nowe the skinne and the fleshe are knowen in their Anatomie. It is to be noted, that the fleshe of the Pappes differeth from the other fleshe of the body, for it is white, glandulus, & spongeous: and there is in them, both Nerues, Veines and Arteirs; and by them they haue Coliganes[1] with the hart, the lyuer, the brayne, and the generatiue members. Also there is in the brest, as old Authors *make mention, lxxx. or .xc. Muskles; for some of them be common to the necke, some to the shoulders, and to the spades, some to Diafragma[2] or yᵉ Mydriffe, some to the Ribbes, some to the Backe, & some to the brest it selfe. But I fynde certayne profitablenes in the creation of yᵉ Paps, aswel in man as in woman: for in man it defendeth the spirituals from annoyannce outwardly: and another, by their thicknes they comfort the natural heate in defience of the spirites. And in women there is the generation of milke: for in women there commeth from the Matrix into their Brestes manye Veines which bring into them menstrual blood, the whiche is turned (through the digestiue vertue) from red colour into white, like the colour of the Pappes, euen as Chylley comming from the stomocke to the Lyuer is turned into the colour of the Lyuer.

Nowe to speake of the bones of the Brest: They be sayde to be triple or threefolde; and they be numbred to be seuen in the Brest before; and their length is according to the breadth of the brest; and their extremities or endes be grystlie, as the ribbes be.

The Breast contains 8 parts: Heart, Lungs, &c.

In it are 80 or 90 Muscles.
* sign. H. iv.

The Paps protect the spiritual members,

and in women make milk.

The breast has 7 bones,

with gristly ends.

---

[1] *Colligance:* f. A binding, tying, or knitting together.—1611. Cotgrave.
[2] *Diaphragme.* The Midriffe: a long and round muscle, whereby the vitall parts are separated from the naturall; and the heart and lights from the stomack, and nether bowels.—1611. Cotgrave.

56 Ensiform Cartilage. Spine, Ribs, Heart. [Ch. VII.

*sign. H. iv, bk.

At the lower end of the Thorax is a gristle, *Ensiforme*.

And in the *vpper ende of Thorax is an hole or a concauitie, in which is set the foote of the Furklebone,[1] or Canel bone; and in the nether ende of Thorax, agaynst y⁰ mouth of the stomacke, hangeth a gristle called *Ensiforme*[2]: and this grystle was ordeyned for two causes: One is, that it shuld defende the stomacke from hurte outwardly: The seconde is, that in time of fulnes it should geue place to the stomacke in time of neede when it desireth, &c.

The Spine has 12 Spondels or Vertebræ,

and 12 Ribs,

7 true, 5 false.

Nowe to speake of the parts of the backe behindefoorth: There be .xij. Spondels,[3] through whom passeth Mynuca, of whom springeth .xij. payre of Nerues, br[i]nging both feeling and mouing to the Muscles of the Brest aforesayde. And here it is to be noted, that in euery syde there be .xij. Rybbes; that is to say, .vij. true, and .v. false, because these .v. be not so long as the other .vij. be, and therefore be called false Rybbes, as it may be perceiued by the sighte of the Eye.

The Heart is King of all members;

*sign. I. i.

Likewise of the partes that be inwardly; and fyrst of the Hart, because he is the principal of al other members, and the beginning of life: he is set in the middest of the brest seuerally by him selfe, as Lord and King of *al members. And as a Lorde or King ought to be serued of his subiectes that haue their liuing of him, So are al other members of the body subiectes to the Hart, for they receyue their liuing of him, and they doo seruice many wayes vnto him agayne. The substaunce of the Hart is, as it were,

---

[1] *Furcale Os, Furcule Superior*, or *Furcella* (in *Anat.*), the upper Bone of the *Sternum*, or breast-bone, otherwise call'd *Jugulum*.—1706. Kersey.

[2] *Ensiformis Cartilago* or *Mucronata* (*Lat.* in *Anat.*), is the lowest part of the *Sternum*, or Breast-bone, so nam'd from its sharp-pointed Triangular Shape, resembling the Edge of a Sword.—1706. Kersey.

[3] *Spondylus*, A rounde thyng of stone, or leade, put on a spindle: a wherue. A ioynt or knot of the backe bone.—1578. Cooper.

Ch. VII.] *The Heart, its Work, & its 2 Ventricles.* 57

Lazartus[1] fleshe; but it is spermatike, and an official[2] member, and the beginning of life; and he geueth to euery member of the body both blood of life, and *and gives each, life-blood and heat.*
4 spirite of breath and heate: for if the Hart were of Lazartus fleshe, his mouing and steering should be voluntarie, and not natural; but the contrarie is true, for it were impossible that the Hart should be ruled
8 by Wyl onely, and not by nature. The Harte hath the shape and forme of a Pyneapple; and the brode ende thereof is vpwardes, and the sharpe ende is downewardes, depending a little towardes the left *The Heart is like a Pine-apple.*
12 side. And here it is to be noted, that the Hart hath blood in his substaunce, whereas al other members haue it but in their Veines & Arteirs: also the hart is bounde with certayne Ligamentes to the backepart of
16 the brest, but these Lygamentes touche *not the substaunce of the Hart, but in the ouerpart they spring foorth of him, and is fastened, as is aforesayde. * sign. I. l, bk.
Furthermore, the Hart hath two Ventrikles[3] or con- *It has 2 Ventricles.*
20 cauities, and the left is hyer then the right; and the cause of this holownesse is this, for to keepe the bloud for his nourishing, and the ayre to abate and temper the great heate that he is in, the which is kept in his
24 concauities. Nowe here it is to be noted, that to the right Ventrikle of the harte commeth a veyne from the great veyne called *Venakelis*,[4] that receyueth al the *Into the right one, comes a vein from the Vena Cava,*

[1] *Lacertus*, the sinewy part of the arm, between the shoulder and the elbow. *Lacertosus*, brawny, sinewy, musculous, nervous, strong. Bailey's Forcellini. *Lacertosus*, Hauing great brawnes and strong sinewes: *Coloni lacertosi.* Husbandmen that be strong brawned.—1578. Cooper.

[2] That which fills an office, has a function.

[3] *Ventriculi Cordis*, the Ventricles of the Heart, which are two large Holes, one on the Right, and the other on the Left Side of the Heart: The former receiving the Blood from the *Vena Cava*, a great hollow Vein, sends it to the Lungs; whilst the other receives the Blood from the Lungs, and distributes it thro' the whole Body by the *Aorta*, or great Artery, and its branches.—1706. Kersey's *Phillips.*

[4] *Vena Cava*, the largest Vein in the Body, so named from its great Cavity or hollow Space, into which, as into a common Channel, all the lesser Veins

58 Ventricles of the Heart, & their Functions. [Ch. VII.

substaunce of the blood from the Lyuer. And this veine that commeth from Venakelis, entreth into the hart at the right Ventrikle, as I sayde before; and in *and brings the Heart some of the thickest blood;* him is brought a great portion of the thickest blood to nourishe the Hart with; & the residue that is left of this, is made subtil through the vertue of the hart; and then this blood is sent into a concauitie or pytte in the myddest of the Harte betweene the two Ventrikles, and therein it is made hote and pured; and then it *the rest is refined in the left Ventricle,* passeth into the left Ventrikle,[1] and there is ingendred in it a spirit that is clearer, brighter, and subtiller then ** sign. I. ii.* any corporal *or bodely thing that is ingendred of the *into a clear spirit, between body and soul.* foure Elementes; For it is a thing that is a meane betweene the body and the soule. Wherfore it is likened of the Philosophers, to be more liker heauenly thinges then earthly thinges. Also it is to be noted, that from *From the left Ventricle springs one Artery taking blood to the Lungs;* the left Ventrikle[2] of the Hart springeth two Arteirs: The one hauing but one cote, and therefore it is called *Arterea venalis:* and this Arteir carieth blood from the Hart to the Lungs,[3] the which blood is vaporous, that is tried and left of the Harte, and is brought by this Artery to the Lunges, to geue hym nutriment: and there he receyueth of the Lunges ayre, and bringeth it to the hart to refreshe him with. Wherefore Galen

4

8

12

16

20

24

---

except the *Pulmonaris* empty themselves; being divided into two thick Branches called the *Ascending and Descending Trunks.* This Vein receiving the Blood from the Liver and other Parts, carries it to the Right Ventricle of the Heart, that it may be there a-new improved and inspirited.—1706. Kersey's *Phillips.*

[1] *Pulmonaria Vena,* or *Arteria Venosa,* a Vessel which, after having accompany'd the Wind-pipe and Pulmonary Artery in all their Branchings in the Lungs, and by its small Twigs receiv'd the Bloud out of that Artery, it discharges itself thro' the left Auricle of the Heart into the Ventricle of the same Side.—1706. Kersey.

[2] The Pulmonary Artery springs from the *right* Ventricle.

[3] *Pulmonaria Arteria,* or *Vena Arteriosa* (in *Anat.*), a Vessel in the Breast, that springs immediately out of the right Ventricle of the Heart, from whence it conveys the Bloud to the Lungs, having a double Coat like that of the Arteries. —1706. Kersey.

Ch. VII.] *Aorta or Beating Vein. Heart's Valves.* 59

sayth, that he fyndeth that mans harte is natural and frendly to the Lungs, for he geueth him of his owne nutrimental to nourishe him with; and the Lunges rewarde him with ayre to refreshe him with agayne, &c. The other Arterye that hath two cotes, is called *vena Arterialis*, or the great Artery that ascendeth and dissendeth;[1] and of him springeth al the other Arteirs that spreade to euery member of the body, for by him is vnified *and quickneth al the members of the body. For the spirite that is reteyned in them, is the instrument or treasure of al the vertue of the soule. And thus it passeth vntil it come to the Brayne; & there he is turned into a further digestion, and there he taketh another spirite, and so is made animal, and at the Lyuer nutrimental, and at the Testikles generatiue; and thus it is made a spirite of euery kinde, so that he, beeing meane of al maner of operations and workinges, taketh effect. Two causes I fynde why these Arteirs haue two cotes: One is, that one cote is not sufficient nor able to withstande the violent mouing and steering of the spirite of lyfe that is caryed in them: The seconde cause is, that the thing that is caried about from place to place, is of so precious a treasure that it had the more neede of good keeping. And of some Doctors this Arteir is called Pulsatiue veyne, or the beating veyne: for by him is perceyued the power & might of the Hart, &c. Wherfore God and Nature haue ordeyned that the Arteirs should haue two cotes.

Also there is in the Harte three Pelikels, opening and *closing the gooing in of the Harte blood and spirite in conuenient time. Also the Hart hath two

margin notes:
and another, the Aorta,
from which all other Arteries proceed.
* sign. I. ii, bk.
The Aorta goes to the Brain,
and is there made animal; in the Liver nutrimental, and in the Testicles, generative.
This Artery is sometimes called the Beating Vein.
The Heart has 3 Valves.
* sign. I. iii.

---

[1] *Arteria Aorta* or *Magna*, the great Artery, a Vessel consisting of four Coats, and continually beating, which carries the spirituous Blood from the left Ventricle of the Heart, by its Branches, to all parts of the Body.—1706. Kersey.

little Eares, by whome commeth in and passeth out the ayre that is prepared for the Lunges.[1] Also there is founde in the Hart a Cartilaginus auditament, to helpe and strength the selfe Harte. Also the Harte is couered with a strong Pannikle, which is called of some, *Capsula cordeo,* or *Pericordium,*[2] the whiche is a strong case, vnto whome commeth Nerues, as to other inwarde members. And this Panicle *Pericordium* springeth of the vpper Panicle of the Midriffe. And of him springeth another Panikle, called *Mediastinum*,[3] the which departeth the Brest in the middest, and keepeth that the Lunges fal not ouer the Hart. Also there is an other Pannikle that couereth the Ribbes inwardly, that is called *Plura,*[4] of whom the Midriffe taketh his beginning. And it is sayde of many Doctors, that Duramater is the originall of all the Pannicles within the body: and thus one taketh of another, &c.

[CHAPTER VIII.]

¶ The Anatomie of the Lunges.

THE Lunges is a member spermatike of his fyrst creation; and his natural complexion is colde and dry; and in his accidental complexion he is colde

---

[1] *Auriculæ Cordis* (in *Anat.*), the two Auricles, or Bosoms of the Heart, which are seated at its *Basis* over the *Ventricles,* and so call'd from their somewhat resembling the Ears of a Man's Head: Their Use is, to receive the Venal Blood from the *Vena Cava* and *Pulmonaris,* and (as it were) to measure it into the Ventricles.—1706. Kersey.

[2] *Pericardium* (in *Anat.*), a double Membrane, Skin or Bag, which surrounds the whole Substance of the Heart, containing a Liquor to moisten, make slippery, and (as some say) to cool that noble Part.—1706. Kersey.

[3] *Mediastinum,* the double Skin, or folding of the *Pleura,* which proceeds from the *Vertebra's* or Turning-joynts of the Back, and divides the whole Breast from the Throat to the Midriff into two hollow Bosoms.—1706. Kersey.

[4] *Pleura* (*Gr.* in *Anat.*), the Membrane or Skin which covers the In-side of the Chest, sticking to the Ribs.—1706. Kersey.

Ch. VIII.] *The Lungs, their Lobes, and Functions.* 61

and moyste, lapped in a Nerueous Pannikle, bicause <small>are wrapt in a nervous membrane.</small>
it should gather togeather the softer substaunce of
the Lunges, and that the Lunges might feele by the
4 meanes of the Pannicle, that whiche he might not feele
in himselfe. Nowe to proue the Lunges to be colde
and drye of kinde, it appeareth by hys swift steering,
for he lyeth euer wauing ouer the hart, and about the <small>They lie waving about the heart.</small>
8 harte. And that he is colde and moyst in rewarde, it
appeareth wel, that he receyueth of the brayne many
cold matters, as Cataries,[1] and Rumes, whose substaunce
is thinne. Also I fynde in the Lunges, three kinds of
12 substaunce: One is a Veyne comming from the Liuer,[2]
bringing with him the Crude or rawe parte of the
Chylle[3] to feede the Lunges: Another is *Arterea venalis*,[4]
comming from the hart, bringing *with him the spirite <small>* sign. I. iv.</small>
16 of lyfe to nourishe him with: The third is *Trachia*[5]
arteria, that bringeth in ayre to the Lunges; and it
passeth through al the left part of them to doo his
office. The Lunges is deuided into fiue Lobbes[6] or <small>The Lungs are divided into 5 Lobes,</small>
20 Pellikels, or fiue portions, that is to say, three in the
right side, and two in the left side. And it was done
for this cause, that if there fel any hurt in the one part,
the others shoulde serue and doo their office. And
24 three causes I finde why the Lunges were principally
ordeyned: First, that they should drawe colde winde, <small>to refresh the Heart with cool air</small>
and refreshe the hart: The seconde, that they shoulde

---

[1] *Catarrhus*, A rewme or stilling downe of humors from ye hed.—1578. Cooper.

[2] *Vena Portæ*, the Portal Vein, according to Vicary.

[3] *Chyle*, a white Juice in the Stomach and Bowels, proceeding from a light dissolution and fermentation of the Victuals; which Juice, mingling and fermenting with the Gall and Pancreatick Juice, passes the Lacteal Veins, &c., and at last is embodied with the Bloud.—1706. Kersey.

[4] The Pulmonary Artery: see note 1, p. 58, above.

[5] *Trachea, siue Trachia.* The wesin or pipe of the lungs: the winde pipe. —1578. Cooper. See *Arteria Trachea*, note 4, p. 44, above.

[6] Lobes (*Gr.* in *Anat.*), the several Lappets or Divisions of the Lungs, or Liver.—1706. Kersey.

62 The Œsophagus, Diaphragm, & Belly. [Ch. VIII.

and to purify the air.

chaunge and alter, and purifie the ayre before it come to y*e* hart, least the hart were hurte and noyed with the quantitie of the ayre: The thirde cause is, that they shoulde receyue from the harte the fumous superfluities 4 that he putteth foorth with hys breathing, &c.

Behind the Lungs is the Œsophagus.

Behinde the Lunges, towarde the Spondels, passeth Mire or Isofagus, of whom it is spoken of in the Anatomie of the necke.[1] And also there passeth both 8 Veynes and Arteirs; and al these with *Trachia arteria*

* *sign. l. iv, bk.*

doo make a Stoke, replete vnto the Gullet with * Pannikles, and strong Lygaments, and Glandulus fleshe to fulfil the voyde places. And last of al is the 12

The Midriff or Diaphragm

Midriffe;[2] and it is an official member, made of two Pannikles, and Lazartus flesh; and his place is in the middest of the body ouerthwart or in bredth vnder the region of the spirituall members, departing them 16 from the matrix. And three causes I finde why the

divides the spiritual organs from the nutritive ones.

Midriffe was ordeyned: First, that it should diuide the spirituals from the nutrates: The seconde, that it should keepe the vital colour or heate to dissende 20 downe to the nutrates: The last is, that the malicious fumes reared vp from the nutrates, should not noye the spirituals or vytals, &c.

The Wombe or Belly

The wombe is the region or the citie of al the 24 Intrils; the whiche reacheth from the Midriffe downe vnto the share inwardly, and outwardly from the Reynes or Kydnes, downe to the bone Pecten, about

is made of Syfac (Peritoneum) and Myrac (Epigastrium).

the priuie partes. And thys wombe is compounde and 28 made of two thinges, that is to say, of *Syfac* and *Myrac*.[3] *Syfac*[4] is a Pannicle, and a member spermatike, official,

---

[1] p. 45 above.

[2] *Midriff*, a Membrane, or Skin, which separates the Heart and Lungs from the Lower Bowels. See *Diaphragm*. Kersey. And p. 55, note 2, above.

[3] *Arab. Sifāc*, the peritoneum; *Marāc*, the soft parts of the belly.—See Forewords, § 13, 'Vicary's *Anatomie*.' Also p. 63, note 2.

[4] *Siphack* (Arabick), the inner rim of the belly, which is joyned to the cawl,

Ch. VIII.] *The Belly, its 2 Parts, and its Muscles.* 63

sensible, senowy, compound of subtil Wyl, and in complexion colde * and drye, hauing his beginning at the   * sign. K. i.
inner Pannicle of the Midriffe. And it was ordeyned
4 because it shoulde conteyne and bind together al the   *Syfac* binds all
Intrals, and that he defende the Musculus so that he   the entrails together.
oppresse not the natural members. And that he is
strong and tough; it is because he should not be lightly
8 broken, and that those thinges that are conteyned goo
not foorth, as it happeneth to them that are broken,[1] &c.
Myrac[2] is compound and made of foure things, that
is, of skin outwardly, of fatnes, of a carnous pannicle,
12 and of Musculus fleshe. And that it is to be vnder-
standed, that all the whole from Sifac outwarde, is called   Everything outside the Peritoneum is *Myrac*:
Myrac, it appeareth wel by the wordes of Galen, where
he commaundeth, that in al woundes of the wombe, to
16 sewe the Sifac with the Myrac; and by that it proueth,
that there is nothing without the Sifac, but Myrac.
And in this Myrac or vtter parte of the wombe,[2] there   in it are 8 Muscles;
is noted eyght Muscles, two Longitudinals, proceeding   2 longitudinal,
20 from the sheelde of the Stomache vnto Os Pecten:[3] two   2 latitudinal,
Latitudinales comming from the backe-wardes to the   * sign. K. i, bk.
wombe: and foure Tran*uerse, of the which, two of them   4 transverse.'
spring from the Ribbes on the right side, and go to the   2 going from the ribs on the left, over the belly on the right.
24 left side, to the bones of the Haunches, or of Pecten: and
the other two spring from the Ribbes on the left syde,
and come ouer the wombe to the righte partes, as the
other before doth. Heere it is to be noted, that by the

where the intrals are covered.—1681. (1st ed. 1656) T. Blount, *Glossographia.*
(Blount died in 1679.)

[1] Ruptured, having hernia.

[2] *Myrach*, an Arabick Word of the same Signification with 'Epigastrium,
the Fore-part of the *Abdomen* or lower Belly, whose upper part is call'd
Hypochondrium; the middle part, *Vmbilicalis;* and the Lowermost *Hypogastrium.*'—1706. Kersey's *Phillips.*

[3] *Pectinis Os,* or *Pubis Os,* the Share-bone, which is the lower and inner, or
the fore-part of the *Os Innominatum.* The upper Part of this Bone is call'd its
Spine, into which the Muscles of the lower Belly are inserted.—1706. Kersey's *Phillips.*

64 The Belly's Muscles. Omentum or Caul. [Ch. VIII.

These Muscles have 3 powers

vertue of the subtyl wyl that is in the Musculus longitudinal, is made perfect the vertue attractiue: and by the musculus Tranuerse is made the vertue retentiue: and by the musculus Latitudinale is made the vertue expulsiue. It is thus to be vnderstoode, that by the

and 3 functions.

vertue attractiue, is drawen downe into the Intrals al superfluities, both water, wynde, and dyrt: By the vertue retentiue, all thinges are withholden and kept, vntil nature haue wrought his kinde: And by the vertue expulsiue, is put foorth al thinges when Nature prouoketh any thing to be done. Galen sayth that woundes or incisions be more perillous in the middest of the wombe then about the sides, for there the partes be more tractable, then any other partes be. Also he sayth, that in wounds persing the womb there shal not

* sign. K. ii.

be made * good incarnation, except Sifac be sewed with Myrac.

Nowe to come to ye parts conteyned within: Fyrst,

The Caul lines the Peritoneum,

that which appeareth next vnder the Sifac is *Omentum*,[1] or *Zirbus*,[2] the which is a pannicle couering the stomacke and the Intrals, implanted with many Veynes and Arteirs, and not a little fatnes ordeyned to keepe moyst the inwarde partes. This Zirbus is an official member, and is compound of a veyne and an Arteir, the which entreth and maketh a line of the vtter tunikle of the stomacke, vnto whiche tunikle hangeth the Zirbus,

protects the nutritive organs,

and couereth al the guttes downe to the shayre. Two causes I finde why they were ordeyned: one is, that they shuld defend ye nutratiues outwardly: the seconde is, that through his owne power & vertue he should

and helps Digestion.

strength and comfort the digestion of al the Nutrates, because they are more feebler then other members be, bicause they haue but a thin wombe or skinne, &c.

---

[1] *Omentum.* The call or sewet wherein the bowels are lapt. The rim or thin skinne wrapping the braine called *Pia mater.*--1578. Cooper.

[2] Zirbus (*Arab.*), the Caul that covers the Bowels.—1706. Kersey.

Ch. VIII.] *The Guts, Duodenum, Jejunium, Ileum.* 65

Next Zirbus, appeareth the Intrals or guttes, of which Galen saith, that the Guttes were ordeined in the fyrst creation to conuey the drosse of the meate and drinke, & to clense the body of their superfluities. And here it is to be noted that there be sixe portions *of one whole Gutte, which both in man and beast beginneth at the nether mouth of the stomacke, and so continueth foorth to the end of the Fundament. Neuerthelesse he hath diuers shapes and formes, and diuers operations in the body; and therfore he hath diuers names. And here-vpon the Philosophers say, that y⁵ lower wombe of a man is like vnto the wombe of a swine. And lyke as the stomack hath two tunikles, in like maner haue al the Guttes two tunikles. The fyrst portion of the Guttes is called *Duodenum*, for he is .xij. ynches of length, and couereth the nether parte of the stomacke,¹ and receyeth al the drosse of y⁵ stomacke: The second portion of the Guttes is called *Ieiunium*,² for he is euermore emptie; for to him lyeth euermore the chest of the Gal, beating him sore, and draweth forth of him al the drosse, and clenseth him clene: The .iij. portion of gutte is called *Yleon*,³ or final gutte, and is in length .xv. or .xvj. Cubites. In this gutte oftentimes falleth a disease called *Yleaca passio*.⁴ The .iiij. gut is called *Monoculus*, or blind

The Entrails or Guts
carry off the refuse of food.
A Gut has 6 parts,
* sign. K. ii, bk.
from the mouth of the Stomach, to the Fundament:

1. *Duodenum*,

2. *Jejunium*,

3. *Ileum* (these 3 make the Small Intestine),

---

¹ *Duodenum* (in *Anat.*), the first of the thin Guts, about Twelve Fingers breadth long, which is continu'd to the *Pylorus*, or lower Orifice of the Stomach, and ends at the first of the Windings under the *Colon:* This Gut differs from the *Jejunum* and *Ileum*, in that it is straighter, and its Coats thicker.—1706. Kersey.
² *Ieiunium*. The vppermost gutte next the bottome of the stomacke. *Ieunium.* Fasting. 1578.—Cooper. *Iejunum* or *Iejunum Intestinum*, the second of the small Guts, which is about eight Foot long in Men, and so call'd from its being often found empty.—1706. Kersey.
³ *Ileum*, or *Ileon*, the third of the small Guts, so call'd by reason of its great turnings, and being about 21 Hands-breadth in Length: it begins where the Gut *Iejunium* ends, and ends itself at the *Cæcum*.—Kersey.
⁴ *Iliack Passion*, a painful wringing or twisting of those Guts, when they are stopt up, or full of Wind, or troubled with sharp Humours, or when the upper
VICARY. F

66 *The Guts, Cæcum, Colon, Rectum.* [Ch. VIII.

4. *Monoculus* (*Cæcum*, or *caput Coli*),
* sign. K. iii.
5. *Colon*,

6. *Rectum* (these 3 make the Large Intestine).

The Mesentery is a texture of mesenteric Veins, protected by membranes, &c.

gut;[1] and it seemeth to haue but one hole or mouth; but it hath two, one neere vnto the other; for by the one al thinges go in, *and by the other they goo out agayne: The fyfth[2] is called *Colon*,[3] and receyeth al the 4 drosse depriued from al profitablenesse; and therefore there commeth not to him any veynes Miseraices, as to the other: The syxte and last is called *Rectum*[4] or *Longaon*,[5] and he is ended in the Fundament, and hath 8 in his nether end foure Muscles, to holde, to open, to shutte, and to put out, &c. Next is to be noted of *Mesenterium*,[6] the which is nothing else but a texture of innumerable veynes Miseraices,[7] ramefied of one 12 veyne called *Porta epates*,[8] couered and defended of Pannicles and Ligamentes comming to the Intrals, with the backe ful of fatnes and glandulus fleshe, &c.

The stomacke[9] is a member compound and sper- 16 part of any Entrail sinks or falls in with the lower: It is also call'd *Chordapeus* and *Volvulus*.—1706. Kersey.

[1] *Cæcum Intestinum* (in *Anat.*), the blind Gut, so nam'd, because one end of it is shut up, insomuch that the Ordure, and the Humour call'd *Chyle*, both come in and go out at the same Orifice.—1706. Kersey.

[2] *orig.* fyrst.

[3] *Colon* is one of the thick Guts, and the largest of all, being about 8 or 9 Hands-breadths long, and full of little Cells, which are sometimes stuff'd with Wind and other Matters that cause the Pains of the *Colick*.—Kersey.

[4] *Rectum Intestinum* (in *Anat.*), the straight Gut, which begins at the first *Vertebra* or a Turning-joynt of the *Os Sacrum*, and goes directly downward to the end of the Rump, or the utmost end of the Backbone.—1706. Kersey.

[5] *Longanon*. The arse gutte.—1578. Cooper.

[6] *Mesenterion*. The double skinnes that fasten the bowels to the backe, and eche to other, and also wrappeth and incloseth a number of veynes being branches of the gret carrying veyne by which both the guttes are nourished, and the iuice of meate concocted is conueyed to the liuer to bee made bloude.—1578. Cooper. *Mesenterium* or *Mesentery*, the double Skin in the middle of the Belly, which fastens the Bowels to the Back, and one to another; being enrich'd with Glandules or Kernels, Nerves, Arteries, Veins, and Vessels, that carry the Juices call'd *Chyle* and *Lympha*.—1706. Kersey.

[7] *Mesaraick Veins* (in *Anat.*), Branches of the *Venæ Portæ*, that arise from, or rather are enclos'd in the *Mesaræum* or *Mesentery*.—Kersey.

[8] *Mesenterick Vein*, is the Right Branch of the *Venæ Portæ*, which spreads it self over the Guts, *Iejunum*, *Ileum*, *Cæcum*, and *Colon*.—1706. Kersey.

[9] *Ventricle* (i. e. a little Belly), the Stomach, a skinny Bowel seated in the

Ch. VIII.] *The Stomach, its Tunicles & Helpers.* 67

matike, senowy and sensible; and therein is made perfect the fyrst digestion of Chile. This is a necessarie member to al the body; for if it fayle in his working, al the 4 members of the body shal corrupte. Wherefore Galen sayth, that the stomacke was ordeyned principally for two causes: The first, that it shoulde be to al the members of the body, as ye earth is to al that are 8 ingendred of the earth, that is, that it shoulde desire sufficient *meate for al the whole body: The seconde is, that the stomacke should be a sacke or chest to al the bodie for ye meate, and as a Cooke to al the 12 members of the body. The stomacke is made of two pannicles, of which the inner is Nerueous, and the vtter Carneous. This inner pannicle hath musculus longitudinales that stretcheth along from the stomacke 16 to the mouth, by the which he draweth to him meate and drinke, as it were handes. Also he hath Tranuers wyl,[1] for to withholde or make retention. And also the vtter pannicle hath Latitudinal wyl, to expulse and 20 put out; and that by his heate he shoulde helpe the digestiue vertue of the Stomacke, and by other heates geuen by his neighbours, as thus: It hath the lyuer on the right side, chafing & heating him with his lobes 24 or figures: & the Splen[2] on the left syde, with his fatnes, and veynes sending to him melancolie, to exercise his appetites: and aboue him is the Harte, quickening him with his Arteries: Also the brayne, send to 28 him a braunche of Nerues to geue him feeling. And

The Stomach

sign. K. iii, bk.

is the body's foodsack and cook,

and is made of 2 Membranes,

with longitudinal and latitudinal fibres.

The Liver is on its right,

the Spleen on its left,

the Heart above it.

---

lower Belly, under the Midriff, between the Liver and the Spleen: It consists of four Tunicks, or Coats, *viz.* a Nervous, Fibrous, Glandulous, and Membranous one; and its Office is, to ferment or digest the Meat.—1706. Kersey. *Stomachus* (in *Anat.*), is properly the left or upper Orifice of the *Ventricle* or Stomach, by which Meats are received into it; and not the whole Stomach, which is termed *Ventriculus.*—*Ibid.* [1] On Will, see p. 20.

[2] *Splen* or Lien, the Spleen or Milt, a Bowel under the left Short Ribs over against the Liver, being a Receptacle for the Salt and earthy Dregs of the Blood; where, by the help of the Animal Spirits, they are refin'd, and returning to the Blood, promote its further Fermentation.—1706. Kersey.

sign. K. iv.
The Stomach is long, like a gourd.

It holds about 2 pitchers of water, and is liable to many diseases.

he hath on the hinder parte, dissending of the partes of the backe, many Lygamentes, with the which he is *bounde to the Spondels of the backe. The forme or figure of the Stomacke is long, in likenes of a gowrde, crooked: and that both holes be in the vpper part of the body of it, is because there should be no going out of it vnaduisedly of those thinges that are receyued into it. The quantitie of the stomacke commonly holdeth two pitchers of water, and it maye suffer many passions; and the nether mouth of the stomacke is narrower then the vpper, and that for three causes: the first cause is, that the vpper receyueth meate great and boystrous in substaunce, that there beeying made subtile, it might passe into the nether: The second is, for by him passeth al the meates, with their chilositie, from the Stomacke to the Lyuer: The thirde is, for that through him passeth al the drosse of the Stomacke to the guttes. And this suffiseth for the Stomacke, &c.

The Liver

does the second Digestion.

sign. K. iv, bk.

It has curdled blood,

The Lyuer[1] is a principal member, and official; and of his first creation, spermatike; complete in quantitie of blood, of him self insencible, but by accidence he is sencible, and in him is made the seconde digestion, and is lapped in a Senowy pannicle. And that he is a principall *member, it appeareth openly by the Philosophers, by Auicen and Galen. And it is official, as is the stomacke; and it is of spermatike matter, and senowy, of the which is ingendred his Veynes. And because it was little in quantitie, nature hath added to it cruded blood, to the accomplishment of sufficient quantity, and is lapped in a senowy pannicle. And why the Lyuer is cruded, is, because ye Chile[2] which

---

[1] *Liver*, one of the noble Parts of the Body, and the thickest of all the Bowels; its office being to purify the Mass of Bloud by straining.—1706. Kersey.

[2] *Chyle*, a white Juice in the Stomach and Bowels, proceeding from a light dissolution and fermentation of the Victuals; which Juice, mingling and

Ch. VIII.] *Of the Liver; & of the Five Humours.* 69

commeth from the stomacke to the Lyuer, should be turned into the colour of blood. And why the Lyuer was ordeined, was, because that al the nutrimental 4 blood shoulde be ingendred in him. The proper place of the Lyuer is vnder the false Ribbes in the righte side. The forme of the lyuer is gibbous[1] or bunchy on the back side, & it is somewhat hollow, like the insyde 8 of an hande. And why it is so shapen, is, that it should be plycable to the stomacke, like as a hande dothe to an apple, to comforte her digestion; for his heate is to the stomacke as the heate of the fyre is to 12 the Potte or Cauldron that hangeth ouer it. Also the Lyuer is bounde with his pellikles to the Diafragma,[2] and with strong Lygamentes. And also he hath Colyganes[3] with the *Stomack and the Intrals, and with 16 the Hart and the Raines, the Testicles, and other members. And there are in hym fiue Pellikles like fiue fingers. Galen calleth the Lyuer *Massasanguinaria*, conteyning in it selfe foure substances, Natural and Nutri- 20 mental. The naturals is sent with the blood to all partes of the body to be ingendred and nourished. And the nutrimentals be sequestrate, and sent to places ordeyned for some helpinges. These are the places of 24 the humors: the blood in the Lyuer, Choler in the chest of gal, Melancolie to the Splen, Flegme to the Lunges and the Iunctures, the watery superfluities to the Reynes and the Vesike.[4] And they goo with y^e

marginalia: and makes nutritive blood. It is bent so as to fit the stomach, and comfort it. The Liver is bound to the Diaphragm and to the Stomach, &c. * sign. L. i. It has 5 pellicles like fingers. The 5 Humours lie in places in the body.

---

fermenting with the Gall and Pancreatick Juice, passes the Lacteal Veins, &c., and at last is embodied with the Bloud.—1706. Kersey.

[1] *Gibbous*, hunch-back'd, crump-shouldered, bossed, bunchy.—Kersey.

[2] *Diaphragm* (q. d. a Fence or Hedge set between), a Term us'd by Anatomists, for the Midriff, a large double Muscle which passes a-cross the Body, and separates the Chest, or middle Cavity, from the Belly or lower one: It is also sometimes call'd *Septum Transversum* and *Disseptum*.—1706. Kersey.

[3] Attachment, binding together. See p. 55, note 1.

[4] *Vesica Urinaria*, the Urine-Bladder, a Vessel shap'd like a Pen, which is appointed to receive the Urine separated in the Kidneys, and brought to it from the Ureters.—1706. Kersey.

blood, and sometime they putrifie and make Feuers; and some be put out to the skin, and be resolued by sweat, or by skab, by Pushes, or by Impostumes. And these foure natural humours, that is to say, Sanguin, Choler,[1] Melancoly, & Fleme, be ingendred and distributed in this maner: First ye shal vnderstande, that from the Spermatike matter of the Liuer inwardly, there is ingendred two greate veynes, of the whiche the first and the greatest is called *Porta*, and commeth *from the concauitie of the Lyuer, of whom springeth al the smal veynes *Miseraices*;[2] and these *Miseraices* be to *Vena porta* as the braunches of a tree be to the stocke or tree. For some of them be conteyned with the botome of the stomacke, some wyth *Duodenum*, some with *Ieiunium*, some with *Yleon*, & some with *Monoculus* or *Saccus*.[3] And from al these guttes they bring to *Vena porta* the succozitie of Chiley gooing from the stomacke, & distribute it into the substaunce of the Lyuer. And these veynes *Miseraices* be innumerable. And in these vaynes is begon the seconde digestion, and ended in the Lyuer, like as is in the Stomacke the fyrst digestion. So it proueth that *Vena porta* and *vena Miseraices* serue to bring al the succozitie of the meate and drinke that passeth the Stomacke, to the Liuer, and they spreade them selues thorough the substaunce of the Liuer inwardly; and al they stretche towards the gibbos or bowing part of the Liuer, and there they meete and goe al into one vnitie, & make the seconde great veine called *Venakelis*, or *Concaua*, or *Vena ramosa*, al is one;[4] and he with his rootes draweth *out al the blood ingendred from the

---

[1] *Choler*, a hot and dry yellow Humour, contain'd in the Gall-bladder, which is of great use for the Fermentation of the Juice nam'd *Chyle*, and bringing it to Perfection: In a Figurative Sense, it is taken for Passion, Anger, or Wrath. —1706. Kersey.
[2] See notes 6, 7, 8, p. 66.   [3] See p. 65.   [4] See note 4, p. 52.

Ch. VIII.] *Of the Gall, and the Spleen or Milt.* 71

Lyuer, and with his braunches ramefying vpwardes and downewardes, carieth and conueyeth it to al other members of the body to be nourished with, where is
4 made perfect the thirde digestion. And also there goeth from the Lyuer, veines bearing the superfluities of the thirde digestion to their proper places, as it shal be declared hereafter.

The Vena Cava carries nutritive blood to all parts of the body.

8   Nowe to speake of the Gal, or the chest of the Gal: it is an official member, and it is spermatike and senowy, and hath in it a subtil wyl; and it is as a purse or a pannicular vesike[1] in the holownesse of the Lyuer,
12 about the middle pericle or lobe, ordeyned to receyue the Cholerike superfluities which are ingendred in the Lyuer. The which purse or bagge hath three holes or neckes: by the fyrste, he draweth to him from the Lyuer
16 the choler, that the blood be not hurt by the choler: by the seconde necke, he sendeth to the bottome of the stomacke, Choler to further the digestion of the stomacke: And by the third neck, he sendeth the
20 choler regularly from one gutte to another, to clense them of their superfluities and drosse: and the quantitie of the purse may *conteyne in it halfe a pinte, &c.

The Gall is like a purse in the hollow of the Liver.

This purse has 3 holes or necks, running to the Liver,

the Stomach,

and the Guts.

* sign. L. ii, bk.

And next is the Splen or the Milte,[2] the whiche is
24 a spermatike member, as are other members, and official, and is the receptory of the melancolious superfluities that are ingendred in the Liuer. And his place is on the lefte side transuerslye lincked to the stomacke, and his
28 substaunce is thinne. And two causes I finde why he was ordeyned there: The first is, that by the melancolious superfluities that are ingendred of the Lyuer which he draweth to him, he is nourished with: The
32 seconde cause is, that the nutritiue blood should by

The Spleen or Milt receives the melancholy superfluities of the Liver.

---

[1] *Vesica*, a Bladder: In *Anatomy*, a membranous or skinny Part, in which any Liquor or Humour is contained; as the Urine, Gall, Seed, &c.—1706. Kersey.

[2] See note 2, p. 67.

him be made the more purer and cleane, from the drosse and thicking of melancolie, &c.

The Reins and Kidneys.

Kidneys cleanse the blood.

And next of the Reynes and Kidnes :[1] It is to be vnderstoode, that within the region of the Nutrites backwardes, are ordeyned the Kidnes, to clense the blood from the waterie superfluities, And they haue ech of them two passages, or holes, or neckes : by the one is drawen the water from Venakelis by two veynes, whiche are called *vena emulgentes*,[2] the length of a fynger of a man, and issueth from the Liuer : and by the other is sente the same water to the Bladder, and is called **Poros vrithides*.[3] The substaunce of the Kidnes is Lazartus[4] fleshe, hauing Longitudinal wyl.[5] And their place is behinde on eache side of the Spondles, and they are two in number; and the righte Kidney lyeth somewhat hyer then the lefte, and is bounde fast to the backe with Lygamentes. The Philosopher sayth, that mans kidneys are like to the kidnes of a Cowe, ful of harde knottes, hauing in him many harde concauities, and therefore the sores of them be harde to cure. Also they are more harder in substaunce then any other fleshy member, and that for two causes : one is, that he bee not muche hurt of the sharpnesse of the vrin : The other is, that the same vrin that passeth from him might the better be altered and clensed throughe the

* sign. L. iii.

The Kidneys are brawny,

full of knots,

and cleanse the urine.

---

[1] *Reins* or *Kidneys*, certain Bowels of a fleshy Substance, whose Office is to strain the Urine into the *Pelves* or Beasons in the middle of their Body, and to cause it to run thro' the Vessels call'd Ureters into the Bladder.—1706. Kersey.

[2] *Emulgent* (*i. e.* milking out, or stroking), a Term in *Anatomy*, as the *Emulgent Vessels*, i. e. two large Arteries and Veins, which arise, the former from the descending Trunk of the *Aorta*, or great Artery, the latter from the *Vena Cava* : They are both inserted to the Kidneys ; so that the *Emulgent Arteries* carry the Blood with the Humour call'd *Scrum*, to them, and the *Emulgent Veins* bring it back again, after the *Scrum* is separated from it by the Kidneys.—1706. Kersey.

[3] *Ureters* (*Gr.* in *Anat.*), two Conduits or Pipes that proceed from the Reins, and convey the Urine thence to the Bladder. *Celsus* calls them *The White Veins.*—1706. Kersey. See p. 76, below.

[4] *Lacertosus*, brawny, muscleful.   [5] See p. 47.

Ch. VIII, IX.] *Kidneys, Vena Cava, & Hanches.* 73

same. Also there commeth from the harte to eche of the kydnes an Arteir, that bringeth with him blood, heate, spirite, and lyfe. And in the same maner there
4 commeth a veyne from the Lyuer, that bringeth blood to nourishe the kydnes, called 'blood nutrimental.' The grease of these kydnes or fatnes is as of other inwarde members, but it is an official member, made of thinne
8 blood, congeled & cruded *through colde: and there is ordeyned the greater quantitie in his place, because it should receyue and temper the heate of the kydnes, which they haue of the biting sharpnesse of the water.
12 Nowe by the kydnes vpon the Spondels passeth *Venakelis*, or *venecaua*,[1] which is a veyne of a great substaunce, for he receyueth al the nutrimental blood from the Lyuer: and from him passeth many smal
16 pypes on euery side; and at the Spondel betweene the Shoulders, he deuideth him selfe whole into two great braunches; the one goeth into the one arme, and the other into the other, and there they deuide them selues
20 into many veynes and branches: as it is declared in the armes (p. 52).

The Kidneys are fed by an Artery from the Heart,

and a vein from the Liver.

* sign. L. iii, bk.

Vena Cava gets the nutritive blood from the Liver,

and branches between the Shoulders into each arm.

[CHAPTER IX.]
¶ *The Anatomie of the*
24   Hanches and their parts.

THE Hanches are the lower parte of the wombe, ioyning to the Thies, and the secret members. And three thinges there are to be noted thereof: the
28 first is of the partes conteining: the seconde is of the partes conteyned: and the thirde is of the partes *proceeding outwardes. The partes conteyning outwardly be *Myrac* and *Sifac*,[2] the *Zirbus*, and the bones.
32 The partes conteyned inwardly, are the *Vezike*, or bladder, the spermatike vessels, the Matrix in women,

The Hanches

enclose the entrails,

* sign. L. iv.

the Bladder, the Womb in women, &c.

[1] See note 4, p. 57.   [2] See pp. 62, 63.

74   *The Hanches. Man has 30 Vertebræ.* [Ch. IX.

Longaon, Nerues, Veynes, and Arteirs dissending downwards. The partes proceeding outwardes, are The Buttockes, and the Muscles dissending to the Thies, of whiche it is to be spoken of in order.   4

And first of the partes conteyning, as of Myrac, Sifac, and Zirbus, there is ynough spoken of in the Anatomie of the wombe.[1] But as for the bones of the Hanches, There be of the partes of the backe, three Spondels 8 of *Ossa sacri*,[2] or of the Hanches, and three *cartilaginis* spondels of *Ossa caude*,[3] called The tayle bonne. And thus it is proued, that there is in euery man, woman and childe .xxx. spondels; and thus they are to be 12 numbered : In the Necke .vij., in the Ridge .xij., in the Reynes .v., and in the Hanches .vj. And it is to be noted, that euery spondel is hollowe in the middest, through which holownesse passeth Nuca from the 16 Brayne, or the marowe of the backe. And some Authors say, that Mynuca is of the same substaunce that the *Brayne is of, for it is like in substaunce, and in it selfe geueth to the Nerues both the vertue of 20 mouing and feeling. And also euery Spondel is holed on euery side, through the which holes both Arteirs and veynes doo bring from the hart and the Lyuer both lyfe & nourishment, like as they doo to the 24 brayne : and from the pannicle of Minuca or the marowe of the backe, through the holes of the sides of the spondels, springeth forth Nerues motiues; and there they intermedle them selues with the strong Lygamentes 28

Side notes:
- The Hanches haue 3 real and 3 false Vertebræ.
- Everybody has 30 Vertebræ,
- hollow in the middle, through which the Spinal Cord passes.
- * sign. L. iv, bk.
- Every Vertebra has holes to let Arteries and Veins through.

---

[1] p. 62.
[2] *Os sacré.* The great bone whereupon the ridge-bone resteth.—1611. Cotgrave. ' *Sacrum Os* (in *Anat.*), the broadest of all the Bones of the Back, which bears up all the other *Vertebræ* or Turning-joints, and in shape somewhat resembles a Triangle : It consists of Five or Six Bones, which are plainly distinguishable in Infants, but cannot be so well discern'd in grown Persons.'— 1706. Kersey.
Os de la hanche. The third part of *Os Ilium;* it selfe consisting also of three parts.—1611. Cotgrave.
[3] *Cauda (Lat.)*, the Tail of a Beast, a Rump.—Kersey.

Ch. IX.] *Spondels or Vertebræ. The Whirlbone.* 75

that be insencible; and so the Lygamentes receyue that
feeling of the Nerue which the Nerue taketh of Mynuca.
And by this reason many Autors proue, that Mynuca *The Spinal Cord*
4 is of the same substaunce that the Brayne is of, and *is of the same substance as the*
the panicles of the Nuca¹ is of the substaunce of the *Brain.*
pannikles of the Brayne, &c. And eche of these
spondels be bounde fast one with another, so that one
8 of them maye not wel be moued without another.
And so al these spondels together, conteined one by *The Spondels*
another, are called yᵉ Ridgbone, which is the fundation *form the Back-bone.*
of al the shape of the body. They, with the laste
12 spondel, be conteyned or ioyned to the bones of the
*Hanches, and they be the vpholders of al the spondles. * sign. M. i.
And these bones be smal towardes the tayle bone, and
broade towardes the Hanches, and before they ioyne
16 and make *Os pectinis*.² And so they be brode in the
partes of the Iles,³ and therefore some Authors calleth
it *Ilea*. And ech of these two bones toward the lyuer
hath a great rounde hole, into the whiche is receyued
20 the bone called *Vertebra*,⁴ or The whorle bone; Also *The Vertebra or Whirlbone (see page 84).*
besides that place there is a great hole or way, through
the which passeth from aboue Musculus, veynes and
Arteirs, and go into the Thees. And thus it is to be
24 noted, that of this bone *Pecten*, and the bone *Vertebra*,
is made the iuncture of the Thye.⁵

---

¹ *Nucha*, the hinder Part, or Nape of the Neck, otherwise call'd *Cervix*.—Kersey.

² *Pectinis Os*, or *Pubis Os*, the Share-bone, which is the lower and inner, or the fore-part of the *Os Innominatum*. The upper Part of this Bone is call'd its *Spine*, into which the Muscles of the lower Belly are inserted.—1706. Kersey.

³ Fr. *Iles:* m. The flankes; or the sides of the lower part of the bellie (so tearmed by Anatomists). *Os des Iles*. Is ioyned to the transuerse passages of the sacred bone; and diuided by Anatomists into three parts; the first whereof (being the highest, and broadest) retaines this name; th' other two are [*Os barré* and *Os de la hanche*: see opposite].—1611. Cotgrave.

⁴ *Vertebra*. A ioynt in the body, where the bones so meete that they may turne, as in the backe bone or chine.—1570. Cooper. See p. 85, note 3, below.
Uertebra, whyrlebone, 632/6. *Scia*, the whyrlebon, 610/11. 15th cent. Glossaries, in Wülker's *Gloss*.   ⁵ See *Ginglymus*, note 1, p. 85.

Of the Bladder and its Vessels. [Ch. IX.

The Bladder Now to speake of the parts conteined: The first thing that commeth to sight is the Bladder, the which is an official member, compounde of two Neruous *has a neck,* Pannicles, in complexion colde and dry, whose necke is carnous, and hath Muscles to withholde, and to let *longer in men than in women.* go: And in men it is long, and is conteyned with the yard, passing through *Peritoneum;* but in women it is shorter, and is conteyned within the *Vulua*.[1] The ** sign. M. i, bk.* place of the bladder is *betweene the bone of the Share and the tayle gutte called *Longaon;*[2] and in women it is betweene the foresayde Bone and the Matrix. And *Two long vessels bring it Urine from the Kidneys.* in it is implanted two long vessels comming from the kidnes, whose names be *Torri vrichides*,[3] bringing with them the Urin or water from the kidnes to the bladder, whiche priuily entreth into the holes of the pannicles of the bladder by a natural mouing betweene tunicle and tunicle; and there the vrin fyndeth the hole of the nether tunikle, and there it entreth priuily into the concauitie of the bladder; and the more that the bladder is filled with vrin, the straighter be the two tunicles comprised togeather; for the holes of the tunicles be not euen, one agaynst another; and therefore if the bladder be never so ful, there may none goe backe *The Bladder is round, and holds a pitcher ful.* agayne. The forme of it is rounde; the quantitie is a pitcher full; in some, more; & in some, lesse, &c. *The spermatic vessels come from Venakelis or Vena cava.* Also there is founde two other vessels, called *vasa seminaria,* or the spermatike vessels.[4] And they come from Venakelis, bringing blood to the Testikles, as wel in man as in woman, in the which, by his further digestion, it is made sparme or nature in men. They ** sign. M. ii.* *be put outwarde, for the Testikles be without; but

4
8
12
16
20
24
28

---

[1] *Vulva,* the Matrice, Mother, or Womb; also the Womb-passage, or Neck of the Womb; a Woman's Privities.—1706. Kersey.
[2] The Rectum: see p. 66.   [3] Ureters. See note 3, p. 72.
[4] *Vasa seminalia,* or *Vasa spermatica,* those Arteries and Veins which pass to the Testicles.—1706. Kersey.

Ch. IX]. *The Womb, its Neck, and Hymen.*

in women it abydeth within, for their Testicles stande within, as it shal be declared hereafter.

Next foloweth the Matrix in women: The Matrix in woman is an official member, compounde and Nerueous, and in complexion colde and dry. And it is the felde of mans generation; and it is an instrument susceptiue, that is to say, a thing receying or taking: and her proper place is betweene the bladder and the gutte *Longaon:* the likenes of it is as it were a yarde reuersed or turned inwarde, hauing testikles likewise, as aforesaid: also the Matrix hath two cauities or selles, and no more; but al beasts haue as many selles as they haue pappes heades. Also it hath a long necke lyke an vrinal; & in euery necke it hath a mouth, that is to saye, one within, and an other without. The inner in the tyme of conception is shutte, and the vtter parte is open, as it was before: and it hath in the middest a Lazartus pannicle, whiche is called in Laten *Tentigo*.[1] And in the creation of this Pannicle is founde two vtilities: The first is, that by it goeth forth the vrin, or els it should be shed through*out al the Vulua: The seconde is, that when a woman doth set hir thies abrode, it altereth the ayre that commeth to the Matrix for to temper the heat. Furthermore, the necke that is betweene these two foresayde mouthes, in her concauitie hath many inuolutions and pleates, ioyned together in the maner of Rose leaues before they be fully spread or ripe, and so they be shut togeather as a Purse mouth, so that nothing may passe foorth but vrin, vntil the time of chylding. Also about the middle of this necke be certain veynes in Maydens, the which in tyme of deflouring be corrupt & broken.[2] Furthermore, in the

Side notes:
The Matrix in women
has 2 cells only,
and a long Neck, with 2 mouths.
* sign. M. ii, bk.
This Neck has many folds or pleats.
In the middle of the Neck is the Hymen.

---

[1] *Tentigo.* A stiffenesse.—1578. Cooper.

[2] *Hymen*, a fabulous Heathen Deity, presiding over Marriage: In *Anatomy* a folding of the inner Skin of the Neck of the Womb, which is commonly taken

78   *Of the Engendering of the Embryo.*   [Ch. IX.

sides of the vtter mouth of the mouth are two testicles or stones, and also two vessels of sparme, shorter than mans vessels; and in time of coyt the womans sperme is shed downe into the bottome of the Matrix. Also 4 *Veins come from the Liver to nourish the fœtus,* from the Lyuer there commeth to the Matrix many veynes, bringing to the childe nourishing at the time of a womans beeing with childe: and those veynes, at suche time as the Matrix is voyde, bring therto super- 8 *or to produce menses.* fluities from certayne members of the body, whereof are ingendred womans flowres, &c.

* sign. M. iii. *And forasmuche as it hath pleased almightie God to geue the knowledge of these his mysteries and 12 workes vnto his creatures in this present worlde, Here *The Embryo* I purpose to declare what thing Embreon¹ is, and his creation. The noble Philosophers, as Galen, Auicen, Bartholomeus, and diuers other writing vpon this 16 *is engenderd of the seed of man and woman.* matter, say, That Embreon is a thing ingendred in the mothers wombe, the original wherof is yᵉ sparme of the man and the woman, of the which is made, by the might and power of God, in the mothers wombe a 20 chylde, as hereafter more at large shal be declared. *In the Matrix* First, the feelde of generation called the Matrix, or the mother, is knowen in the anatomie, whose place is properly betwixt the Bladder and Longaon in the 24 *is sown a seed by natural heat,* woman, in which place is sowen, by the tillage of man, a couenable matter of kindly heate; for kindly heate is cause efficeens² bothe of dooing and working, and spirite that geueth vertue to the body, and gouerneth 28 and ruleth that vertue: the which seede of generation commeth from al the partes of the body, both of the

for a mark of Virginity, and whose Fibres drawn together make the *Myrtiform Glandules.*—1706. Kersey.

¹ *Embryo, (Gr.)* a Child in the Mother's Womb, after its Members come to be formed, but before it has its perfect Shape.—1706. Kersey. See the curious account of the formation of the Embryo in the *Legendary or Lives of Saints*, in the E. E. Text Soc.'s volume for 1887, p. 319; and earlier in the late Thos. Wright's thin *Popular Treatises on Science*, 1841, p. 138-40.   ² L. Causa efficiens.

Ch. IX.] *Of the Embryo, and the Seed creating it.* 79

man and the woman, with consent & wyl of al
members, and is shed into the place of concey*uing,   *sign. M. iii, bk.
where, through the vertue of Nature, it is gathered   which is gatherd
4 together in the selles of the matrix or the mother, in   into the womb-cells,
whom—by the way of the working of mans seede,
and by the way of suffering of the womans seede
mixte together, so that eche of them worketh in other,
8 and suffereth in other—there is ingendred Embreon.   and engenders the Embryo.
And further it is to be noted, that this sparme that
commeth both of man and woman, is made & gathered   The seed is made of the purest blood
of the most best and purest drops of blood in all the   in the body.
12 body; and by the labour and chafing of the testikles
or stones, this blood is turned into another kinde, and
is made sparme. And in man it is hotte, white, &
thicke, wherfore it may not spread nor runne abroade
16 of it selfe, but runneth and taketh temperaunce of the
womans sparme, which hath contrarie qualities; for the
womans sparme is thinner, colder, and feebler. And as
some Authors holde opinion, when this matter is gathered
20 into the right side of the matrix, then it happeneth a
male kinde; and likewise on the lefte, the female;
and where the vertue is most, there it sauoureth most.
And further it is to be noted, that lyke as the Renet[1]
24 of the Cheese hath by him selfe the *way or vertue of   *sign. M. iv.
working, so hath the mylke by way of suffering: and   And as rennet curdles milk, and
as the Renet and mylke make the cheese, so doth the   the two make cheese, so does
sparme of man and woman make the generation of   the seed of man and woman make
28 Embreon, of the which thing springeth, by the vertue   the Embryo,
of kindly heate, a certayne skinne or caule, into the   which is wrapt in a Caul.
which it lappeth it selfe in, wherewith afterwardes it
is tyed to the mothers wombe, the whiche couering
32 commeth foorth with the byrth of the childe; and if   And if this Placenta comes not
it happen that any of the skinne remaine after the byrth   away with the child, the mother's
of the childe, then is the woman in peril of her lyfe.   life is in danger.

[1] *Rennet*, or *Runnet*, the Maw of a Calf, commonly us'd to turn Milk for Cheese-curds.—1706. Kersey.

Furthermore it is sayde, that of this Embreon is ingendred the Hart, the Lyuer, the Brayne, Nerues, veynes, Arteirs, Cordes, Lygamentes, Skinnes, Gristles, & Bones, receyuing to them by kindly vertue the menstrual blood, of whiche is ingendred both fleshe and fatnes. And as wryters say, the fyrst thing that is shapen be the principals, as is the Harte, Lyuer, and Brayne. For of the Hart springeth the Arteirs, of the Lyuer the Veynes, and of the Brayne the Nerues: and when these are made, Nature maketh & shapeth Bones and grystles to keepe & saue them, as the bones of the head for the *Brayne, the Brest-bones and the Ribbes for the Harte and the Lyuer. And after these springeth al other member, one after another. And thus is the childe bred foorth in four degrees, as thus: The first is, when the sayde sparme or seede is at the fyrst as it were mylke: The seconde is, when it is turned from that kinde into another kinde, [it] is yet but as a lumpe of blood; and this is called of Ypocras, *Fettus*:[1] The thirde degree is, when the principals be shapen, as the Hart, lyuer, and Brayne: The fourth and laste, as when al the other members be perfectly shapen, then it receyeth the soule wyth life and breath; and then it beginneth to moue it-selfe alone. Nowe in these foure degrees aforesayde, in the fyrst, as milke, it continueth vij. dayes: in the seconde, as Feetus, ix. dayes: in the thirde, as a lumpe of fleshe ingendring the principals, the space of ix. dayes: and the fourth, vnto the tyme of ful perfection of al the whole members, is the space of xviij. dayes: So is there xlvj. dayes from the day of conception vnto the day of ful perfection and receyuing of the soule, as God best knoweth.

---

[1] *Fœtus*. All things brought forth by the generation of man, beast, fish, &c.—1578. Cooper. '*Fœtus*, the Young of all kind of Creatures, especially Humane; but in a stricter Sense, it is a Young Child, whose Parts are perfectly formed in the Womb.'—1706. Kersey.

Ch. IX.] *Of the Rectum, Anus, and the Yard.* 81

Now to come agayne to the Anatomie *of the Hanches : Then come we to *Longaon*,¹ otherwise called, The tayle gutte, whose substance is panniculer, as of al the other bowels; the length of it is of a spanne long, stretching nigh to the Raynes ; his nether parte is called *Annis*, that is to say, The towel.² And about him is found two Muscles, the one to open, the other to shutte. Also there is founde in him fiue veynes or braunches of veynes, called *vena emoraidales*,³ and they haue coliganes⁴ with the bladder ; wherefore they are partners in their greeues. And when this Longaon is raysed vp, then ye may see the veynes and arteirs, and senowes, howe they be braunched and bounde downe to the nether partes. The partes proceeding outwardly are, *Didimus*,⁵ *Peritonium*, the Yarde, the Testikles, and the Buttocks.

* sign. N.
The Hanches.
The tail-gut's end is the Anus.
It has 5 Veins.

And fyrst it shal be spoken of the yarde, or of mans generatiue members, the which dureth vnto that parte that is called Peritoneum, the which place is from the Coddes vnto the Fundament, wherevpon is a seame. Wherfore sayth the Philosopher, Mans yard is in the ende and terme of the share.⁶ The yarde is an official mem*ber, and the tyller of mans generation, compounde and made of skinne, brawnes, Tendons, veynes, arteirs, senewes, and great Lygaments : and it hath in it two passages or principal issues, that is to say, one for the sparme, and another for the vrin. And as the Philosophers say, the quantitie of a common yard is viij. or

Of man's generative members.

The Yard
* sign. N. i, bk.
has 2 passages,
a. for sperm ;
b. for urine ;
and is 8 or 9 inches long.

---

¹ *Longanon*. The arse gutte.—1578. Cooper. See p. 66, above.
² *Anus*, tuel, fundament. *Tuel*, (among Hunters) the Fundament of a Beast.—1706. Kersey.
³ *Hæmorrhoidal Veins* are twofold, *viz.* either Inward or outward : The former being Branches of the Mesenterick Vein, pass to the Gut *Rectum*, and thence to the Fundament ; But the other arise from the Hypogastrick Vein, and sometimes from a double Branch of it, spreading about the *Sphincter* of the *Anus*.—1706. Kersey.   ⁴ Attachment, binding together : p. 55, n. 1.
⁵ A Membrane : not *Didymi*, Twins ; also the Testicles of a Man.—Kersey.
⁶ *Share* . . a Man's Yard or Groin.—Kersey.
VICARY.   G

ix. ynches, with mesurable bignes proportioned to the quantitie of the matrix. This member hath, as sayth Auicen, three holes; through one passeth incensible polissions[1] and wynde, that causeth the yard to ryse: the other two holes be declared before. Also the yard hath a skinne; and about the head thereof it is double; and that men call *Prepusium*;[2] and this skinne is mouable, for through his consecration the spermatike matter is the better and sooner gathered together, and sooner cast foorth from the Testicles: for by him is had the more delectation in the dooing. And the formost part of the head of the yard before is made of a subtil brawny fleshe, the whiche, if it be once lost, it is neuer restored agayne, but it may wel be skinned, &c.

The Coddes is a compounde mem*ber and an officiall; and though it be counted amongst the generatiue members, yet it is called a principal member, because of generation. This purse was ordeyned for the custodie & comfort of the testikles and other spermatike vesselles. And it is also made of two partes, of the inner and the vtter. The vtter is compound, and made of skinne and lazartus, longitudinal and transuersal, in like maner as is the Myrac.[3] The inner parte of the Coddes is of the substaunce of the Sifac,[3] and are in similitude as two pockets drawen together by themselues, and they differ not from the Sifac: and there be two, bicause if there fal any hurt to the one, the other should serue. The Testikles or stones be two, made of glandulus fleshe or curnelly fleshe. And furthermore, through the *Didimus* commeth to the Testicles, from the Brayne, Senowes; and from the Hart, Arteirs; and from the Lyuer, veynes, bringing vnto them both

The Yard has a Prepuce.

* sign. N. il.
The Cods.

The Testikles are in a purse.

They are of muscular flesh.

---

[1] Pollutions, discharges of seed.
[2] *Præputium*, (in *Anat.*) the Fore-skin that covers the Nut or Head of a Man's Yard; also the fore-part of the *Clitoris* in Women.—1706. Kersey.
[3] See ch. viii, p. 63, above.

Ch. IX, X.] *Groin, Hips. The Leg & its 3 Parts.* 83

feeling and steering, lyfe and spirite, and nutrimental blood, and the most purest blood of al other members of the body, whereof is made the sparme by the labour 4 of the Testicles, the which is * put foorth in due tyme, as is before rehearsed. — They make Sperm of the purest blood. * sign. N. ii, bk.

The Groynes be knowen: they be the emy(?) iunctures or purging place vnto the Lyuer; and they haue 8 curnelly fleshe in the plying or bowing of the Thyes. — The Groins.

The Hippes haue great brawny fleshe on them; and from thence dissende downwards brawnes, cordes, and lygaments, mouing and bynding together the thies with 12 the Haunches themselues. — The Hips.

[CHAPTER X.]
¶ *The Anatomie of the Thyes, Legges, and Feet.*

16 THE Legge reacheth from the Ioynt of the Thie vnto the extremitie of the Toes; and I wyl diuide it in partes, as the armes were deuided. One parte is called *Coxa*, or Thigh, and that is al that is 20 conteyned from the ioynt of the Haunche vnto the knee: The seconde part is called *Tibia;* and that reacheth from the Knee to the Ankle: The thirde is the little foote, and that is from the Ankle vnto the 24 end of the Toes. And here it is to be noted, * that the Thigh, Legge, and Foote are compounde and made as the great arme or hande, with skinne, fleshe, veynes, arteirs, senewes, brawnes, tendons, and bones, whereof 28 they are to be spoken of in order. Of the skinne and fleshe there is ynough spoken of before. And as of veynes and arteirs, in their discending downewardes, at the laste spondels they be deuided into two partes, 32 whereof the one parte goeth into the right Thye, and the other into the lefte; and when they come to the Thye, they be deuided into other two great braunches:

— The Leg is of 3 Parts.
1. *Coxa* (or Femur), Thigh.
2. *Tibia* (with the Fibula),
3. Foot,
* sign. N. iii.
all made of skin, flesh, &c.
The Veins and Arteries divide and run down each leg,
in 2 branches,

G 2

84 *The Legs, their Veins and Sinews.* [Ch. X.

<small>one inside, and one outside, to the ankles, forming 4 veins used for bleeding.</small>
the one of them spreadeth into the inner side of the Legge, and the other spreadeth into the vtter syde, and so braunching, dissende downe the Legge to the ankles & feete, and be brought into foure veynes, which be 4 commonly vsed in letting blood, as hereafter foloweth. One of them is vnder the inner ankle towarde the heele, called *Soffena;*[1] and another vnder the vtter ankle, called *Siattica;*[2] and another vnder the hamme, called 8 *Poplitica*,[3] the fourth betweene the little Toe and the next, called *Kenalis*.[4] And it is to be noted of these foure great veynes in the legges, of the manyfolde
<small>* sign. N. lii, bk.</small> daungers that mighte * fal of them, as oft it happeneth. 12 There be many other braunches which a Chirurgion nedeth not much to passe vpon.

<small>The Sinews</small>
The Senewes spring of the last spondel, and of *Os sacrum*, and passeth through the hole of the bone of 16
<small>run downwards</small>
the Hippe, and dissendeth to the Brawnes, and moueth the Knee & the ham; and these dissende downe to
<small>and move the foot.</small>
the ankle, and moue the foote: and the brawnes of the feete moue the Toes, in lyke maner as is declared in the 20 bones of the handes. The first is called *Coxa*, that is, The thye bone, and he is without a fellowe,[5] and is ful
<small>The top or head of the thigh-bone, or *Femur*, is called *Vertebrum*;</small>
of marow, and is round at eyther ende: The roundnes that is at the vpper ende is called *Vertebrum*, or Whurle 24 bone, and boweth inwardes, and is receyued into the boxe or hole of the hanche bone: And at the lower

---

[1] *Saphœna* (*Gr.* in *Anat.*), the Crural Vein, a Vein that goes down under the Skin of the Thigh and Leg, and turns towards the upper part of the Foot, where it sends forth several Branches, some of which go to the great Toe.—Kersey.

[2] *Sciatick-Vein*, a Vein seated above the outward part of the Ancle.—1706. Kersey.

[3] *Poplitick*, belonging to the Ham, as *The Poplitick Vein* or *Muscle*. *Poplitica Vena*, the Vein of the Ham, which takes rise from the *Iliacal* Branches of the *Vena Cava*, and sometimes reaches down the back of the Leg, even to the Heel.—1706. Kersey. *Poples, poplitis*, The hamme of ones leg behynde the knee.—1578. Cooper.   [4] *Chenalis?*

[5] The Femur is a single bone, and also the biggest in the body.

Ch. X.]   *The Bones of the Leg and Foot.*   85

[end] & towards the Knee, there it hath two roundes, which be receyued into the concauities[1] of the bone of the legge at the knee, called the great Fossels.[2] There is also at the knee a rounde bone called The knee panne.[3] Then followeth the legge, wherin is two bones called *Focile maior* and *Focile minor ;*[4] the bygger of them passeth before, making the shape of the shinne, and it is called the shinne bone, and passeth *downe, making the inner ankle. The lesse passeth from the knee backwardes, dissending downe to the vtter ankle, and there formeth that ankle, &c.

The bones of the foote are xxvj., as thus : Fyrst, next the ankle bone is one, called in Laten *Orobalistus :*[5] next vnder that towardes the Heele is one called *Calcany :*[6] and betweene them is another bone called *Os nauculare.*[7] In the seconde warde there be foure bones, called *Raceti,*[8] as be in the handes. In the thirde and fourth warde be xiiij., called *Digitori,*[9] and .v. called *Pectens,*[10] at the extremities of the Toes next to the

Margin notes: Its 2 lower ends (or condyles) fit into the knee Fossels. Here, too, is the Knee-pan or Patella. The Leg has 2 bones, Tibia and Fibula. * sign. N. iv. The Foot has 26 bones; 7 in the Tarsus, and 19 in the Metatarsus :

---

[1] *Ginglymus,* (*Gr.*) a joyning of Bones, when the Head of one is receiv'd into the Cavity or Hollow of another, and again the Head of the latter into the Cavity of the other ; as the joynting of the Thigh-bone with the *Tibia,* and of the Shoulder-bone with the *Ulna.*—1706. Kersey.   [2] See note 3, p. 52.

[3] *Patella,* (*Lat.*) a deep Dish, with broad Brims : Among *Anatomists,* the round, broad Bone at the joynting of the Thigh and Leg ; the Whirle bone of the Knee.—1706. Kersey.

[4] *Tibia,* (*Lat.*) a Pipe, Flute, or Flagelet . . . In *Anatomy* the Leg, or Part betwixt the Knee and the Ancle, consisting of two Bones, one outward, nam'd *Focile minus* [Fibula], another inward and larger, which has usurp'd the Name of the whole, and is termed *Tibia,* but others call it *Focile Majus,* and *Canna Major.*—1706. Kersey.

[5] *Astralagus,* the Huckle-Bone : Also the first principal Bone of the Foot, which with other little ones, makes up that Part which immediately succeeds the Leg, and is call'd the Pastern in Beasts.—1706. Kersey.

[6] *Calcaneus,* or *Os Calcis,* (in *Anat.*) the Heel-bone, or bone of the *Tarsus,* which lies under the *Astragali,* and is united to them by the Joynting call'd *Ginglymus.*—1706. Kersey.

[7] *Naviculare Os,* (in *Anat.*) otherwise call'd *Cymbiforme,* the third Bone of each Foot, in that part of it which immediately succeeds the Leg.—Kersey.

[8] The Cuboid bone, and the Internal, Middle and External Cuneiform bones.

[9] The Metatarsal bones,   [10] The Phalanges.

nayles. And thus be there in the foote xxvj. bones, with the Legge from the Ankle to y⁰ Knee, two in the Knee, and one rounde and flat bone, and in the Thye one. And thus shal you finde in the whole Legge and Foote .xxx. bones. And this sufficeth for young Practitioners.

with 3 in the Leg,

30 bones.

4

8

FINIS.

¶ Imprinted at London
by Henry Bamforde,
1577.

12

---

p. 20, 67. *Wyl.* Fr. *Fibres:* f. The small strings, or haire-like threads of roots; also, the fibers, threads, or strings of muscles, & veines; in Lincolneshire they are tearmed Cheyres.—1611. Cotgrave.

p. 36. *Secondyna* is probably the choroid or pigmented coat of the Eye.—D'Arcy Power.

p. 36. *Unia* (for *Uvea*). Culpepper's translation of Riolanus, 1671, Lond. p. 138, says: 'The *Uvea* or Grape skin Coat, and its open hole, which makes the Pupilla or sight of the Eye: the external Face or Circle of the Pupilla is termed Iris, or the Rain bow.'—D'Arcy Power.

p. 43. *Cansales.* ? Ital. *Causale*, casuall, subiect to chance. *Casuale*, casuall, by fortune.—1598. Florio.

p. 48. *Gwidege.* It. *Guidegi*, the names of certeine veines in the throte. 1598. Florio. *A Worlde of Wordes. Gwidege* must be the jugular vein.—D'Arcy Power.

p. 48. *Isinon* is the *Isthmus*, of which Kersey's *Phillips* says: 'In Anatomy it is taken by some for that part which is between the Mouth and the Gullet; also the Ridge that separates the Nostrils.

p. 48. *Vena organices:* ? Vena carotidis, the Carotid Artery.—D'A. P.

p. 62, 64, 74. '*Zirbus* or *Omentum.* Ital. *Zirbo*, the Caule wherein the bowels are lapt, as *Omento*, a fat pannicle, caule, sewet, rim or couering, which, being inserted with manie veines springing in branches from *Vena porta*, representeth the forme of a net. Properly the caule or sewet, rim or kell, wherein the bowels are lapt. Also the rinde or thin skin inwrapping the braine, called *Pia Matre*, or *Matre pia*.' 1598. Florio. *A Worlde of Wordes.*

p. 75, &c. *Spondels.* It. *Spondili*, any small bones, namely,[1] the spondils, the knuckles or turning ioints of the back-bone or chine. Also spindle-wherues. Also a kind of Serpent. Also a kind of great Oyster like an Asses-hoofe. 1611. Florio, condenst from ed. 1598; as for 'spindle-wherues,' ed. 1598 has 'a wherue of wood or stone to put on a spindle.'

p. 85. *Os nauculare.* Ital. *Osso nauiforme*, a bone next to the ankle in the inside, called in English the 'ship, or betelike bone.' 1611. Florio. *Queen Anna's New World of Words.* Not in ed. 1598.

[1] especially.

APPENDIX OF DOCUMENTS, EXTRACTS, ETC.

I. Grants to Vicary by Kings and Queens:—

a. Posts of Serjeant of his Surgeons, and Chief Surgeon to Himself, by Henry VIII (29 April, 1530) 89
b. 21 Years' Lease, from 25 March 1539, of the Tithes and Glebe of the Rectory of Boxley, and the House and 10 pieces of Land in Boxley, Kent; by Henry VIII 91
c. Bailiffship of Boxley Manor, &c., and 2 Annuities of £10 each, for the joint and separate lives of Thos. Vicary and his son William, by Henry VIII (5 Oct. 1542) 93
d. Arrears of his Annuity under (*a*), by Queen Mary (20 Oct. 1553) 95
e. Bailiffship of Boxley Manor, &c., and 2 Annuities of £10 each for his Life, by Philip and Mary (23 Jan. 1555) 96

II. Payments to Vicary and other Surgeons, &c. by Kings and Queens:—

f. £20 Quarterly, by Hen. VIII (1528-31, from the Record Office) ... 99
g. £20 Quarterly, by Hen. VIII (1538-41, from Arundel MS. 97) ... 103
h. £20 Quarterly, by Hen. VIII (1543-4, from Phillipps MS. 3852)... 110
i. 40 Marks (£26 13*s.* 4*d.*) by Hen. VIII, 1536-8 ; by Ed. VI, 1551 ; by Eliz., 1560-1 (from the Record Office) 111
j. £20 Yearly : last payments to M. de la More ; first and other to Vicary (from the Record Office) 113
k. £20 Quarterly, by Edw. VI (1547, 1552-4, from the Record Office) 117
l. £20 Yearly, by Elizabeth (1559-61, from the Record Office) ... 121

III. Extracts from the City Repertories, &c. at Guildhall:—

1. As to the Foundation of Bartholomew's Hospital, and to Vicary 123
2. Supplementary Extracts as to Bartholomew's 147
3. As to Barbers, Unlicenst Surgeons, and the Plague 155
4. London Street-Scenes in Vicary's Days 168

IV. Vicary's Bailiff's Accounts of Boxley Manor 179

V. Thos. Dunkyn's £100 Mortgage to Vicary of Watsole, and 60 acres in Boxley, and Wyldes, 18 acres, in Stowting, Kent (March 1557-8) 181

VI. Vicary's Will (27 January 1560-1) 187

Appendix of Documents, Extracts, &c.

		PAGE
VII.	Henry VIII's Statutes relating to Surgeons	195
	m. 3 Hen. VIII, ch. 11, requiring their Examination and Licensing, and forbidding unlicenst folk to practise	197
	n. 5 Hen. VIII, ch. 6, discharging them from serving as Constables, &c.	198
	o. 22 Hen. VIII, ch. 13, declaring Alien Surgeons not to be Handicraftsmen	200
	p. 32 Hen. VIII, ch. 40, empowering Physicians to practise Surgery	202
	q. 32 Hen. VIII, ch. 42 (A.D. 1540), uniting the Barbers and Surgeons	202
	r. 34 & 35 Hen. VIII, ch. 8, empowering unlicenst folk to treat simple diseases, without liability under (*m*), 3 Hen. VIII, ch. 11	208
VIII.	Supplement to the Statutes; from the Guildhall Records	
	City Orders as to the Exemption from Watch, &c. of Surgeons and Physicians (not Barbers): and the Compromise with the Barber-Surgeons as to Service on Inquests, &c.	210
IX.	Ten Recipes by Henry VIII and his Physicians, with a Poem 'What Veins to Bleed in'	220
X.	Payments by Henry VIII and Princess Mary to Surgeons other than Thomas Vicary, in 1529-43	231
XI.	Pay of Army and Navy Surgeons to Henry VIII	236
XII.	Some of Henry VIII's payments to Holbein, and to Players	238
XIII.	The 185 Freemen of the Barber-Surgeons' Company in 1537, with the numbers of the other City Companies	243
XIV.	Ordinances of the Barber-Surgeons of London, A.D. 1529	247
	The Wardens of the Surgeons and Barber-Surgeons, 1488-91	260
	The Barber-Surgeons' right to the 17th Place in the City Companies	261
XV.	Ordinances of the Barber-Surgeons of York, A.D. 1592	269
	The Fresh Ordinances of 1614, as to the Master of Anatomy, &c.	279
	Later Ordinances, 1676-1768	283
XVI.	The Corporation of London's Order for the Government of St. Bartholomew's Hospital, A.D. 1552	289

I.

GRANTS TO VICARY BY KINGS AND QUEENS.

a. 29 April, 1530. Grant by K. Henry VIII to Thomas Vicary, his Surgeon, (for his past and future services) of the post of Serjeant of the King's Surgeons, and Chief Surgeon to the King, with its pay of 40 Marks a year, after the death or resignation of Marcellus de la More [1536], appointed Serjeant, 6 Aug. 1513.

Patent Roll. 22 Henry the Eighth. Part 22. membrane (23) 13.

Domino Thoma Vicarie.   Rex Omnibus ad quos & cetera,[1] salutem.

Cum nos, per literas nostras patentes datas apud Westmonasterium, sexto die Augusti Anno regni nostri quinto, dederimus & concesserimus dilecto seruienti nostro Marcello de la More, Principali Cirurgico nostro, Officium seruientis Cirurgicorum nostrorum, Habendum, occupandum officium predictum dicto seruienti nostro durante vita sua, cum feodis & vadiis, tam de Hospicio nostro, quam aliter ab antiquo debitis & consuetis Habendum & percipiendum modo & forma ante tunc visitatis, simul cum omnibus allocationibus, tam le bouge le Courte, quam vini, ceri, & aliorum requisitionum pro curis, & cum omnimodis preemu[2]nenciis, auctoritatibus, proficiis, commoditatibus & auantagiis dicto officio pertinentibus siue spectantibus, in tam largo & amplo modo & forma, prout aliqua alia persona, tempore celibris [sic] memorie Edwardi, nuper Regis Angliæ quarti, aut aliorum progenitoroum nostrorum, dictum officium perantea habens, habuit & percepit in & pro exercitacione eiusdem, prout in eisdem literis patentibus plenius continetur. Cum que eciam nos, per alias literas nostras patentes, datas apud Westmonasterium Tercio die Nouembris, Anno regni nostri septimo, dederimus & concesserimus prefato Marcello de la More, per

As We, on Aug. 6, 1513,

made Marcellus de la More, Serjeant of our Surgeons, for his life,

with customary fees

and Bouge of Court,

as in Edw. IV's time;

And as We, on Nov. 3, 1515,

gave the said M. de la More

---

[1] &c. = hae literæ nostræ pervenerint.
[2] *u* is used for *i* in *proficuis* below.

90 App. I. a. 1530. Vicary, Serjeant of Surgeons, &c.

an Annuity of 40 marks,

payable at Michaelmas and Easter,

now We (for his good service to Us)

make Thomas Vicary

Serjeant of our Surgeons,

and also our Chief Surgeon, for his life,

with wages, bouge of Court, wine, wax, and requisites for cures,

as soon as M. de la More shall die, or resign or forfeit his post,

and with all fees and benefits

nomen 'dilecti Magistri Marcelli de la More, seruientis nostri & principalis Cirurgici nostri,' quandam annuitatem, siue quendam annualem redditum, quadraginta marcarum sterlingorum, Habendum & annuatim percipiendum eidem Marcello, a festo Pasche Anno regni nostri sexto, durante vita eiusdem Marcelli, ad duos anni terminos, videlicet, ad festa sancti Michaelis Archangeli, & Pasche, per equales porciones, ad Receptam Scaccarij nostri, per manus Thesaurarii & Camerarii nostrorum ibidem pro tempore existentium, prout in eisdem literis patentibus plenius continetur, Sciatis, quod nos, de gratia nostra speciali, ac ex certa sciencia & mero motu nostris, ac in consideracione boni & fidelis ac diutini seruicij nobis, per dilectum seruientem nostrum, Thomam Vicarie, ante hec tempora impensi, & imposterum impendendi, dedimus & concessimus, ac per presentes damus & concedimus, prefato Thome Vicarie, predictum officium seruientis Cirurgicorum nostrorum, ac officium Principalis Cirurgici nostri, Necnon ipsum Thomam, seruientem Cirurgicorum nostrorum ac Principalem Cirurgicum nostrum, constituimus, ordinauimus, deputauimus, fecimus & nominauimus, ac per presentes constituimus, ordinamus, deputamus, facimus, & nominamus; ac vadia, feoda, regarda & allocaciones, tam le bouge the Courte de Hospicio nostro, quam vini, ceri, & aliorum requisitorum pro curis, cum omnibus & omnimodis proficiis, commoditatibus, preeminenciis, auctoritatibus, & auantagiis, dicto officio qualitercumque pertinentibus, siue spectantibus, damus & concedimus per presentes, immediate & quamcito officia predicta, per mortem dicti Marcelli de la More, sursum reddicionem literarum patentium predictarum, cessionem, forisfacturam, seu quouis alio modo vacare contigerint, Habendum, occupandum, & gaudendum, dictum officium seruientis Cirurgicorum nostrorum, ac officium Principalis Cirurgici nostri, prefato Thome, durante vita sua, immediate & quamcito dictum officium seruientis Cirurgicorum nostrorum ac officium Principalis Cirurgici nostri, per mortem prefati Marcelli de la more, sursum reddicionem literarum patentium predictarum, cessionem, forisfacturam, seu quouis alio modo vacare contigerint, vel in manibus nostris quouismodo extiterint, cum omnibus & omnimodis vadiis, feodis, regardis, allocacionibus, proficuis, commoditatibus, preeminenciis, auctoritatibus, & auantagiis, dictis officiis, seu eorum altero ab antiquo debitis & consuetis, & in tam amplis modo & forma prout aliquis alius, siue aliqui alij, officia

App. I. a. 1530. Vicary's Chief Surgeoncy; b. Lease. 91

predicta, seu eorum alterum perantea habens, occupans, siue exercens, aut habentes, occupantes, vel exercentes, habuerunt vel perceperunt, de & pro occupacione & exercicione eorundem, vel eorum vtriusque. Et vlterius, de vberiori gratia nostra predicta, dedimus, & concessimus, ac per presentes damus & concedimus, prefato Thome Vicarie, durante vita sua, predictam quandam annu[i]tatem, siue quendam annualem redditum quadraginta marcarum sterlingorum, immediate & quamcito dictum officium seruientis Cirurgicorum nostrorum ac officium Principalis Cirurgici nostri, per mortem prenominati Marcelli de la More, sursum reddicionem literarum patencium predictarum, cessionem, forisfacturam, aut quouis alio modo in forma predicta vacare contigerint, vel in manibus nostris aliter aliquo modo deuenire extiterit, ac habendum & annuatim percipiendum dictam annuitatem, siue annualem redditum quadraginta marcarum sterlingorum, eidem Thome Vicarie, immediate & quamcito officia predicta vacare contigerint, in forma predicta, durante vita ipsius Thome Vicarie, ad festa Pasche & sancti Michaelis Archangeli, equis porcionibus, ad Receptam Scaccarij nostri predicti Soluendam, per manus Thesaurarii Camerarii eiusdem Scaccarij nostri pro tempore ibidem existentibus, Absque compoto, vel aliquo alio, inde nobis vel heredibus nostris reddendo, soluendo, seu faciendo. Eo quod expressa mencio de vero valore annuo, aut de certitudine premissorum, vel de aliis donis siue concessionibus per nos prefato Thome Vicarie ante hec tempora factis in presenti minime factis existit, aut aliquo statuto, ordinacione, prouisione, siue restriccione, inde incontrarium factis, editis, ordinatis, prouisis siue restrictis, aut aliqua alia re, causa, vel materia quacumque non obstante. In cuius & cetera. Teste Regis apud Westmonsterium, xxix die Aprilis.
per breve de priuato sigillo, & de dato, & cetera.

that have ever been held with the said posts.

And further, of our more abounding grace, We grant to Thomas Vicary, for his life, that Annuity of 40 marks,— so soon as M. de la More shall die, or resign or forfeit it,—

to be paid to him at Easter and Michaelmas,

without deduction.

Witness the King, at Westminster, 29 April 1530.

---

b. A.D. 1539. Henry VIII's 21-years' Lease to Vicary, of the Tithes, Glebe, and House of the Rectory of Boxley, Kent, with 10 pieces of Land there.

(Court of Augmentations, Inrolments of Leases, Vol. 210, f. 71. 30 Henry VIII.)

Hec Indentura facta inter excellentissimum Principem et Dominum, Dominum Henricum Octauum, Dei gracia, etc', ex vna parte, et Thomam Vycary, vnum Chirurgi-

This Indenture made between Henry VIII and Thos. Vicary

92 App. I. b. 1539. *Vicary's Lease of Boxley Land, &c.*

[witnesses that the King leases to]

corum dicti Domini Regis, ex altera parte, Testatur, quod idem Dominus Rex, per aduisamentum et consensum Consilij Curie Augmentacionum reuencionum Corone sue, tradidit, concessit, et ad firmam dimisit,

[Vicary the Tithes of Grain, Glebe-lands, and chief House, of the Rectory of Boxley, Kent;]

prefato Thome, omnes et omnimodas decimas granorum et terras glebas Rectorie de Boxley in Comitatu Kancie, nuper Monasterio de Boxley in eodem Comitatu, modo dissoluto, spectantes et pertinentes; Ac totum Capitale Mesuagium, ac omnia orrea, stabula, domos, et edificia, dicte Rectorie spectantia et pertinentia; Necnon omnes

[Also 10 pieces of land, Boxley Field, Squires and Carters crofts, Herpole, Wheat park, Blackland, the Hale, Rishett, and Hoyton Meadow.]

illas decem pecias terre arabilis, prati, et pasture vocatas Boxley felde, Squyers croft, Carters croft, grete Herpole, lyttell Harpole, le Whete parke, Blackeland, le Hale, Rysshett, et Hoyton medoo, cum pertinenciis, in Boxley predicta, dicto nuper Monasterio spectantes et pertinentes: Exceptis tamen premissorum,[1] et dicto Domino Regi, heredibus, et successoribus suis omnino

[(Except all big trees and woods, and the Advowson of Boxley parish Church.)]

reseruatis, omnibus grossis arboribus et boscis premissorum, ac aduocacione vicarie ecclesie parochialis de Boxley predicta: Habendum et tenendum omnia et singula premissa cum pertinenciis, exceptis preexceptis,

[To hold the same to Thos. Vicary for 21 years from 25 March 1539,]

prefato Thome et assignatis suis a festo Anunciacionis beate Marie Virginis vltimo preterito vsque ad finem termini et per terminum viginti et vnius annorum extunc proximo sequencium et plenarie complendorum:

[at the rent of £40, that is,]

Reddendo inde annuatim dicto Domino Regi, heredibus, et successoribus suis quadraginta libras legalis monete Anglie; videlicet, pro predictis decimis granorum et terris

[for the Tithes and Glebe £26 13s. 4d.,]

glebis dicte Rectorie viginti sex libras, tresdecim solidos, et quatuor denarios; Et pro predicto Mesuagio, orreis,

[and for the Rectory House and 10 pieces of land, £13 6s. 8d.,]

stabulis, domibus, et edificijs, ac predictis decem pecijs terre arabilis, prati, et pasture, tresdecim libras, sex solidos, et octo denarios; ad festa Sancti Michaelis

[half yearly, at Michaelmas and Lady Day.]

Archangeli et Anunciacionis beate Marie Virginis, vel infra vnum mensem post vtrumque festum festorum illorum, ad Curiam predictam per equales porciones soluendos durante termino predicto. Et predictus Do-

[The King covenants that Vicary shall hold the premises free from other charges.]

minus Rex vult et per presentes concedit, quod ipse, heredes, et successores sui dictum Thomam et assignatos suos de omnibus redditibus, pencionibus, porcionibus, et denariorum summis quibuscumque de premissis seu de aliqua inde parcella exeuntibus seu soluendis, preterquam de redditu superius reseruato, versus quascumque personas de tempore in tempus,[2] [exonerabunt acquietabunt et defendent, ac omnia domos et edificia pre-

---

[1] Sic.
[2] There is an obvious omission of several words here; the omitted words are supplied from similar leases in the same volume.

App. I. c. 1542. *Vicary's Boxley Bailiwick.* 93

missor*um*, tam in maeremijs qu*a*m in cooperturis tegular*um* et 'slate', de tempore in tempus] tociens quociens necesse *et* oportunu*m* fuerit, bene *et* sufficienter reparari, sustentari, *et* manuteneri facient dura*nte* termino [predicto]. Et pred*ic*t*us* Thomas concedit p*er* presentes, q*uo*d ipse *et* assign*ati* sui coopertur*am* straminis ac om*n*es alias necessar*ias* reparaciones reparac*iones* [1] premissor*um*, p*re*ter maerem*ium*, tegulas, *et* 'slate' predict*a*, de tempore in tempus tociens quociens necesse *et* oportunu*m* fue*r*it, bene *et* sufficient*er* reparabunt, sustentabunt, *et* manutenebunt dura*nte* termino p*re*dic*to*. Et p*re*d*ic*t*us* Dom*inus* Rex vlt*er*ius vult, *et* p*er* presentes concedit, q*uo*d bene licebit p*re*fato Thome *et* assigu*ati*s suis de tempore in tempus cape*re*, percipe*re*, et habe*re* de, in, *et* sup*er* p*re*missis competen*s* *et* sufficien*s* hedgebote, fyrebote, ploughbote, *et* cartebote, ibidem *et* non alibi annuatim expendend*um* *et* occupand*um*, durante te*r*mi*ne* p*re*dic*to*. In cuius rei testimonium vni parti *et* cet*er*a, alteri vero parti *et* cet*er*a. Data apud Westmo*n*asterium, ———————[2] Anno regni di*c*ti Domini Regis —————.[2]

Vicary covenants that he'll keep the buildings in good repair, and will thatch them with straw, but not shingle, tiles, or slates.

The King grants to Vicary sufficient wood for hedges, firing, and repair of ploughs and carts.

Date blank.

---

c. 5 Oct., A.D. 1542. Henry VIII's Grant to Vicary and his son William, of the post of Bailiff of Boxley Manor, with 2 Annuities of £10 each.

(Court of Augmentations; Inrolments of Leases; Vol. 235, f. 98. 34 Henry VIII.)

Rex, Omn*ibus* ad quos *et* cet*er*a, salut*em*. Sciatis q*uo*d nos, in consideraci*one* boni, v*er*i, *et* fidelis seruicij quod dilec*ti* seruientes n*ost*ri, Thomas Vycary, Chirurgicus n*oste*r, *et* Will*elmus* Vacary,[1] filius ipsius Thome, ante hec tempora nob*is* fecerunt, de grac*ia* n*ost*ra speciali, ac ex certa sciencia, *et* mero motu n*ost*ris, dedim*us* et concessim*us*, ac p*er* p*re*sentes dam*us et* concedim*us* eisdem Thome *et* Will*elm*o, officium Balliuatus Man*er*ii n*ost*ri de Boxley, in Comitatu n*ost*ro Kanc*ie*, Ac omnium Maneriorum, terrar*um*, tenement*orum et* hereditamentorum n*ost*rorum quor*um*cumque, cum pertinenc*iis* tam in Boxley *et* alibi vbicumq*ue* in d*i*cto Com*itatu* Kanc*ie*, quam alibi vbicumq*ue* infra regnu*m* n*ost*ru*m* Angl*ie*, que nuper Monast*er*io de Boxley in eodem Com*itatu* nostro Kanc*ie*, quam alibi vbicu*m*q*ue*, dic*t*o nuper Monast*er*io spectab*ant* siue pertineb*an*t.[3] Ac ipsos Thomam *et* Will*el*mum Balliuos Maner*iorum*, terrarum, tenem*entorum*, possessionum, *et* hereditamen-

For the good service done to Us by Our Surgeons, Thos. Vicary, and his son William,

We grant them the office of Bailiff of Our Manor of Boxley in Kent,

and all other Manors

late belonging to Monastery of Boxley;

[1] Sic. [2] Blank.
[3] "spectan et pertinen" is an error for "spectabant et pertinebant."—R. Kirk.

94 App. I. c. 1542. *Vicary's Boxley-Manor Bailiwick.*

And we make Thos. and Wm. Vicary, Keepers of Our woods; to hold and exercise the said offices personally or by deputy,

from 25 March 1542,

for the life of the longest liver of them.

And We grant the said Thomas and Wm. Vicary as fee £10 a year

out of the said Manor, &c.,

from March 25, 1542, for the life of the longest liver of them,

payable at Michaelmas and Lady Day.

And Further, We grant to Thos. and Wm. Vicary

a 2nd Annuity of £10 out of the said Manor, &c.,

for the life of the longest liver of them,

payable at Michaelmas and Lady Day.

Witness, Sir Richard Riche, at Westminster Oct. 5, 1542.

torum predictorum, Ac Custodes boscorum predictorum,[1] facimus, ordinamus, et constituimus per presentes : Habendum, exercendum, et gaudendum officia predicta, ac eorum vtrumque, prefatis Thome et Willelmo, tam per se quam per sufficientem deputatum siue deputatos suos sufficientes, a festo Annunciacionis beate Marie Virginis vltimo preterito, ad terminum et pro termino vite ipsorum Thome et Willelmi et eorum alterius diucius viuentis. Et vlterius, de vberiori gracia nostra, damus et per presentes concedimus prefatis Thome et Willelmo pro exercicio officiorum predictorum, quoddam annuale feodum siue vadia decem librarum sterlingorum, exeuncium et exiturarum de Manerijs, terris, et tenementis predictis : Habendum, gaudendum, et annuatim percipiendum easdem decem libras eisdem Thome et Willelmo, a dicto festo Annunciacionis beate Marie Virginis vltimo preterito ad terminum et pro termino vite predictorum Thome et Willelmi, et eorum alterius, vt prefertur, diucius viuentis, de exitibus et reuencionibus et proficuis Maneriorum predictorum et ceterorum premissorum, tam per manus suas proprias, quam per manus Receptorum, firmariorum, tenencium, siue occupatorum ea[2]rundem pro tempore existencium, ad festa Sancti Michaelis Archangeli et Annunciacionis beate Marie Virginis per equales porciones soluendas. Et vlterius, de vberiori gracia ac pro consideracione predicta, per presentes concedimus prefato Thome et Willelmo quandam aliam annuitatem siue annualem redditum decem librarum sterlingorum, annuatim exeuncium et exiturarum de Maneriis, terris, et tenementis predictis : Habendum, gaudendum, et annuatim percipiendum easdem decem libras prefatis Thome et Willelmo, et eorum assignatis, ad terminum vite predictorum Thome et Willelmi, et eorum alterius diucius viuentis, tam per manus suas proprias quam per manus Receptorum, tenencium, firmariorum, seu aliorum occupatorum dictorum Maneriorum, terrarum, tenementorum, et hereditamentorum predictorum pro tempore existencium, de exitibus et reuencionibus eorundem, ad festa Sancti Michaelis Archangeli et Annunciacionis beate Marie Virginis per equales porciones soluendas. Eo quod expressa mencio et cetera. In cuius rei et cetera. Teste Ricardo Riche, Milite, apud Westmonasterium, quinto die Octobris, Anno regni nostri tricesimo quarto.
per breue de priuato Sigillo,
virtute Warranti regij.

[1] This shows that something has been omitted above. Compare with Patent Roll, 1 & 2 Philip and Mary.—R. Kirk.   [2] Sic.

App. I. d. 1553. *Arrears of Vicary's 1530 Annuity.* 95

d. 20 Oct. 1553. Queen Mary's Order that Thomas Vicary shall be paid the arrears of his Annuity of 20 Marks since the death of Marcellus de la More, under Henry VIII's Grant of 29 April, 1530 (p. 89).

Patent Roll, 1 Mary, part 14, membrane 19 (25).

Regina etc' Thesaurario et Camerarijs suis qui nunc sunt, et qui pro tempore erunt, salutem. Cum Dominus Henricus, nuper Rex Anglie octauus, pater noster, per literas suas patentes, gerentes datam vicesimo nono die Aprilis anno regni sui vicesimo secundo, dederit et concesserit dilecto seruienti suo Thome Vicarie officium Seruientis Cirurgicorum suorum, ac officium principalis Cirurgici sui, necnon ipsum Thomam Seruientem Cirurgicorum suorum ac principalem Cirurgicum suum constituerit, ordinauerit, deputauerit, fecerit, et nominauerit per literas suas predictas, ac vadia, feoda, regarda, et allocaciones, tam le bouge the Courte de Hospicio suo, quam vini, ceri, et aliorum requisitorum pro curis, cum omnibus et omnimodis proficuis, commoditatibus, preeminencijs, auctoritatibus, et aduantagijs dicto officio qualitercumque pertinentibus siue spectantibus, dederit et concesserit per literas predictas, immediate et quam cito officia predicta, per mortem Marcelli de la More (tunc habentis officia predicta), sursum reddicionem literarum patencium eidem Marcello de la More inde antea confectarum, cessionem, forisfacturam, seu quouis alio modo vacare contingerent: Habendum, occupandum et gaudendum dictum officium Seruientis Cirurgicorum suorum, ac officium principalis Cirurgici sui, prefato Thome durante vita sua, immedietate¹ et quamcito dictum officium Seruientis Cirurgicorum suorum, ac officium principalis Cirurgici sui, per mortem prefati Marcelli de la More, sursum reddicionem literarum patencium predictarum, cessionem, forisfacturam, seu quouis alio modo, vacare contingerent, vel in manibus dicti patris nostri quouismodo existerent, cum omnibus et omnimodis vadiis, feodis, regardis, allocacionibus, proficuis, commoditatibus, preeminencijs, auctoritatibus, et aduantagiis dictis officijs, seu eorum altero, ab antiquo debitis et consuetis; Et vlterius dederit et concesserit, per literas predictas, prefato Thome Vicarie, durante vita sua predicta, quandam annuitatem, siue quendam annualem redditum, quadraginta marcarum sterlingorum, immediate et quamcito dictum officium Seruientis

Pro Thoma Vicarie, de liberate.

As Henry VIII, by Patent of 29 April, 1530,

made Thos. Vicary Serjeant of his Surgeons, and Chief Surgeon to Himself,

and gave him the wages, Bouge of Court, wine, wax, and requisites for cures,

pertaining to these posts,

so soon as Marcellus de la More (who then held them) should surrender or vacate them, or die,

(To hold the said posts to the said Thos. Vicary, with all their profits, after the same became vacant);

And as Henry VIII also gave to Thos. Vicary during his life an Annuity of 40 Marks,

¹ Sic.

96 App. I. *d. Arrears of Annuity. e. Bailiwick* (2).

so soon as the said Posts should be vacated by the said Marcellus de la More,	Cirurgicor*um* suor*um* ac officium principalis Cirurgici sui, p*er* mortem p*re*nomi*n*ati Marcelli de la More, sursum reddicionem lit*er*ar*um* patencium p*re*dic*t*ar*um*, cessionem, forisfac*t*uram, aut quouis alio modo in forma pre*dict*a, vacare continge*re*nt, vel in manib*us* ejusdem pa*t*ris no*s*tri, aut aliquo alio modo deuenire
To hold and take the said Annuity to the said Thos. Vicary	existe*re*nt; Ac hab*e*ndum et annuatim p*er*cipiendu*m* dictam annuitat**e**m siue annualem redditum quadraginta marcar*um* sterlingor*um* eidem Thome Vicarie, im-
(as soon as it became payable)	mediate et q*u*amcito officia p*re*dic*t*a vacare continge*re*nt in forma pre*d*ic*t*a, durante vita ipsius Thome Vicarie,
by equal half-yearly payments at Ea*s*ter and Michaelmas,	ad festa Pasche et S*anct*i Michae*l*is Arch*a*ngeli, equis porc*i*onib*us*, ad receptam Sc*a*cc*a*rij sui soluend*arum*, p*er* manus Thesa*u*rar*ij* et Cam*er*ariorum ejusdem Scaccarij sui p*ro* tempore ibidem existen*cium*, absq*ue* com-
free from all deductions;	poto vel aliquo alio inde dic*t*o pat*r*i no*s*tro, vel heredi*bus* suis, reddendo, soluendo, seu faciendo; prout in eisdem
And as Marcellus de la More is dead	li*t*eris plenius continet*ur*: Et quia p*re*dic*t*us Marcellus de la More diem clausit extremum, vt p*ro* c*er*to intel-
We bid you, our Treasurer and Chamberer, to pay Thos. Vicary all arrears of his said Annuity of 40 marks,	lexe*ri*m*us*: Vob*is* mandam*us*, quod eidem Thome id quod ei aretro est de p*re*dic*t*a annuitate siue annuali redditu quadraginta marcar*um*, a die mortis p*re*dic*t*i Marcelli, et eandem annuitatem siue annualem reddi*tum* quadraginta marcar*um* exnunc singulis annis,
and also all future payments of it during his life, half-yearly,	durante vita ipsius Marcelli[1] Vicarie, ad festa p*re*dic*t*a, de T*h*esa*u*ro no*s*tro ad receptam p*re*dictam, de tempore in tempus soluatis, iuxta tenorem li*t*erar*um* p*re*dic*t*ar*um*, recipientes a p*re*fato Thoma, de tempore in tempus,
you taking his receipts for the same.	li*t*eras suas acquietancie de tempore in tempus huiusmodi soluc*i*ones v*e*stras testificantes, que p*ro* nobis sufficientes fue*ri*nt in hac parte. Teste R*egi*na apud
20 Oct. 1553.	Westm*on*a*s*terium, xx die Octob*ri*s. [1553.]

e. 28 January, 1555. Grant by Philip and Mary, to Thomas Vicary for Life, of the post of Bailiff of Boxley Manor, &c.; and of Two Annuities of £10 each.

Patent Roll, 1 & 2 Philip and Mary, part 11, m. 5 (23).

De conc*essione* p*ro* Thoma Vicarye, ad vitam.	Rex et Regina, Omn*ibus* ad quos, etc'.,[2] sal*u*tem. Sciatis q*uo*d nos, in considerac*i*o*n*e boni, veri et fidelis seruicij quod dilec*t*us seruiens no*s*ter, Thomas Vycarye,
For Thos. Vicary's faithful service to Hen. VIII and Edw. VI,	seruie*n*s, siue senior et principalis Chirurgus no*s*ter, tam p*re*charissimis Principib*us*, Henrico Octauo et Edwardo Sexto, nup*er* Regib*us* Angl*ie*, quam nob*is*, ante hec

[1] So, by mistake for 'Thome.'
[2] etc. = hae literae nostrae pervenerint.

App. I. e. 1555. *Vicary's Bailiwick of Boxley* (2). 97

tempora impendit, de gracia nostra speciali, ac ex certa
sciencia, et mero motu nostris, dedimus et concessimus,
ac per presentes, pro nobis, heredibus et successoribus   We grant to the
nostris, damus et concedimus eidem Thome, officium   said Thos. Vicary
Balliuatus Manerij nostri de Boxley in Comitatu nostro   the post of Bailiff
Kancie, ac omnium Maneriorum, terrarum, tenemen-   of Our Manor of
torum et hereditamentorum nostrorum quorumcumque,   Boxley in Kent,
cum pertinenciis, tam in Boxley, et alibi vbicumque in   and all other
dicto Comitatu Kancie, quam alibi vbicumque infra   Manors belonging
regnum nostrum Anglie, que nuper Monasterio de Box-   to the dissolvd
ley, in eodem Comitatu nostro Kancie, modo dissoluto,   Boxley Abbey;
ludum spectabant et pertinebant, ac parcelle terrarum,
tenementorum, et possessionum inde existebant; Ac offi-   and the post of
cium Custodis omnium boscorum nostrorum tam in dicto   Keeper of our
Comitatu nostro Kancie, quam alibi vbicumque, dicto   Woods there,
nuper Monasterio spectancium siue pertinencium; Ac
ipsum Thomam Balliuum Maneriorum, terrarum, tenementorum, possessionum, et hereditamentorum predictorum, Ac Custodem boscorum predictorum, facimus,
ordinamus, et constituimus per presentes; Habendum,   to hold and exer-
exercendum, et gaudendum officia predicta, et eorum   cise the said posts,
vtrumque, prefato Thome, tam per se quam per suffi-   to the said Thos.
cientem deputatum siue deputatos suos sufficientes, ad   Vicary,
terminum et pro termino vite ipsius Thome. Et vlter-   personally or by
ius, de vberiori gracia nostra, damus, et per presentes,   deputy, during
pro nobis, heredibus, et successoribus nostris, concedimus his life.
prefato Thome Vicarye, pro exercicio officiorum predic-   And further we
torum, quoddam annuale feodum, siue vadia, decem   give the said
librarum sterlingorum, exeuncium et exiturarum de   Thos. Vicary, for
Maneriis, terris, et tenementis predictis: Habendum,   his said posts,
gaudendum, et annuatim percipiendum easdem decem
libras eidem Thome, a Festo Sancti Michaelis Arch-   one Annuity of
angeli, Anno regni nostri dicte Regine primo, a quo   £10,
tempore officia predicta et eorum vtrumque exercuit, ad   from Michaelmas,
terminum et pro termino vite naturalis ipsius Thome, de   1553, (since
exitibus, reuencionibus, et proficuis Maneriorum predicto- when he has fild
rum, et ceterorum premissorum, per manus suas proprias,   the said posts,)
vel per manus receptorum, firmariorum, tenementorum,   out of the profits
siue occupatorum eorundem, siue de Thesauro nostro ad   of the said
receptam Scaccarij nostri Westmonasterii, heredum, et   Manors,
successorum nostrorum, per manus Thesaurarii et Came-   or from our
rariorum nostrorum, heredum, et successorum nostrorum,   Treasurer,
ibidem pro tempore existencium, ad festa Annuncia-
cionis beate Marie Virginis et Sancti Michaelis Arch-   at Lady Day and
angeli, per equales porciones soluendas. Et vlterius, de   Michaelmas.
vberiori gracia nostra, ac pro consideracione predicta,   And further
pro nobis, heredibus, et successoribus nostris, per pre-   We grant to the
VICARY. H

8 *

98 App. I. e. *Vicary's Bailiwick of Boxley Manor* (2).

said Thos. Vicary another Annuity of £10

out of the said Manors,

from Michaelmas, 1553, for his life,

either from the Receivers of the said Manors,

or our Treasurer.

Witness the King and Queen at Westminster, Jan. 28, 1555.

sentes concedimus prefato Thome Vycarye quandam aliam annuitatem, siue annualem redditum, decem librarum sterlingorum, annuatim exeuncium et exiturarum de Maneriis, terris, et tenementis predictis : Habendum, gaudendum, et annuatim percipiendum easdem decem libras prefato Thome Vicarie, a dicto Festo Sancti Michaelis Archangeli, Anno regni nostri dicte Regine primo, ad terminum vite sue, per manus suas proprias, vel per manus receptorum, tenencium, firmariorum, seu aliorum occupatorum dictorum Maneriorum, terrarum, tenementorum et hereditamentorum predictorum, de exitibus et revencionibus eorundem Maneriorum et ceterorum premissorum, siue de Thesauro nostro, ad receptam Scaccarij nostri Westmonasterij, heredum, et successorum nostrorum, per manus Thesaurarii et Camerariorum nostrorum, heredum, et successorum nostrorum, ibidem pro tempore existencium, ad dicta festa Annunciacionis beate Marie Virginis et Sancti Michaelis Archangeli, per equales porciones soluendas. Eo quod expressa mencio etc'. In cuius rei etc'. Testibus Rege et Regina apud Westmonasterium xxviij die Januarij.

per breve de priuato sigillo.

---

p. 93. *hedgebote*, &c. Hedgebote, Is necessary Stuff to make *Hedges*, which the Lessee for Years &c. may, of common Right, take in his ground leased.— Jacob, *Law Dict.*

Firebote, Fuel for *Firing* for necessary Use, allowed by Law to Tenants out of the Lands &c. granted them. See *Estovers* (Fr. *Estover*, from the Verb *Estoffer*). It signifies to supply with Necessaries ; and is generally used in the Law for Allowances of Wood made to Tenants, comprehending *House-bote*, *Hedge-bote* and *Plough-bote* for Repairs &c.—Jacob.

Plow-bote, a Right of Tenants to take Wood to repair *Ploughs*, Carts and Harrows ; and for making Rakes, Forks, &c.—Jacob, *Law Dict.*

II.

PAYMENTS TO VICARY AND OTHER SURGEONS, &c., BY KINGS AND QUEENS.

Payments by Henry VIII to his Physicians, Surgeons, Apothecaries, Barber, &c., from Christmas, 1528 to Lady Day, 1531.

(From Bryan Tuke's MS. Accounts presented to the Record Office by Sir W. C. Trevelyan.)

Quarter Wages due at Cristmas anno xx^{mo} [A.D. 1528].

(lf. 8, bk.) Item, for Anthony Chabo, surgion, fee	... ... x li
Item, for Doctour Bentley,[1] phisicion, fee	... ... x li
Item, for Doctour Buttes, phisicion, fee	... ... xxv li
*(lf. 10) Item, for Thomas Vicary, surgion, wages	... ... v li

(lf. 13, bk.) Rewardes geuen on Wedenesdaye, Newyeres day, at Grenewich, anno xx^{mo} [1529].

Item, to Iohn Penn, Barbour, in Rewarde	... ... xl s
(lf 15, bk.) Item, to Doctour Bentley seruaunte	... vj s viij d
Item, to Doctour Chambre seruaunte[2] ...	... ... xiij s iiij d

(lf. 25, bk.) Yet quarter wages due at our Lady day (aº xxº, A.D. 1529).

Item, for Doctour Bentley, phesicion, fee	... ... x li
Item, for Doctour Buttes, phisicion, fee	... ... xxv li
(lf. 26, bk.) Item, for Iohn Penn, Barbour [3]	... ... lxvj s viij d
*(lf. 27) Item, for Thomas Vicary, surgion), wages	... ... v li

(lf. 28, bk.) Yet halfe yeres Wages due at our Lady [day, 25 March, an. 21mo. 1529].

Item, for Iohn Clemente, phesicion, fee	... ... x li
Item, for Nicholas Simpson),[3] fee	... ... l s

[1] For Bentley, Buttes, Chambre, Harman, Penn, Simpson, &c., see the cut from Holbein's Picture in the Forewords.
[2] ? Divines:
Item, to Doctour Stokeleies seruaunte ... ... xiij s iiij d
Item, to Doctour Rawson) seruaunte ... ... xiij s iiij d
[3] For the liveries of damask, budge, velvet, cotton cloth, fustian, canvas, &c., for the robes of John Penn, Nicholas Simpson, and Edmund Harman, in 27 Hen. VIII, see Sir Andrew Windsor's account in the Miscellaneous Books, Augmentation Office, No. 455, leaf 31 back. (We see none in No. 456.) Also for Jn. Penn's liveries under the Warrant of Nov. 22, 1526 (an. xviij^{mo}), see Wardrobe Accounts, Exch. of Receipts, Parcel 1, a. 11, shelf 298, leaf 9.
For Henry VIII's books, pictures, clothes, utensils, &c., see the excellent

100 App. II. *Henry VIII's Payments to Vicary, &c.*

(lf. 39, bk.) Quarter waigis due at Midsomer aº xxj^{mo} [A.D. 1529].

(lf. 40) Item, for Anthony Skabo, surgion), fee ... ... x ti
Item, for Doctour Bentley, phesicion, fee ... ... x ti
Item, for Doctor Buttes, phesicion, fee[1] ... ... xxv ti
(lf. 41) Item, for Iohn) Pen, Barbour ... ... lxvj s viij d
*(lf. 41, bk.) Item, for Thomas Vicary, surgion), wagis[2] ... C s

(lf. 53, bk.) Quarter Wagis due at Michelmas, anno xxj^{mo} [A.D. 1529].

Item, for Anthony Shabo, surgion, fee ... ... x ti
(lf. 54) Item, for Doctour Bentley, phesicion, fee ... ... x ti
Item, for Doctour Buttes, phesicion, fee ... ... xxv ti
*(lf. 55) Item, for Thomas Vicary, surgion, fee ... ... v ti

(lf. 56) Half yeres Wagis due at Michelmas, aº xxj^{mo} [A.D. 1529].

(lf. 56, bk.) Item, for Iohn Clement, phesicion, fee ... ... x ti
Item, for Nicholas Sampson, fee[3] ... ... l s

(lf. 67) Quarter Wagis due at Cristmas, aº xxj^{mo} [A.D. 1529].

Item, for Anthony Schobo, surgion, fee ... ... x ti
Item, for Doctour Bentley, phesicion, fee ... ... x ti
Item, for Doctour Buttes, phesicion, fee[4] ... ... xvv ti
(lf. 68) Item, for Iohn Pen, Barbour ... ... lxvj s viij d
*(lf. 68, bk.) Item, for Thomas Vicary, surgion, fee ... ... v ti

(lf. 72) Rewardes geuen on Saterday, Newyeres daye, as folowith, at Grenewiche, anno xxj^{mo} [A.D. 1530] as hath byn accustummyde.

Item, to Iohn Penn, Barbour ... ... xl s
(lf. 73) Item, to Doctor Bentleys seruaunte ... ... vj s viij d
Item, to Doctour Chambers seruaunte ... ... xiij s iiij d

MS., signed by Henry on leaf 1, *Royal Household Book*, temp. Hen. VIII and Edw. VI, *Miscellaneous Books*, Augmentation Office, No. 160 : a MS. which ought to be printed. We sadly want a Record-Office-Document printing Society, ¤ọ́ to say half-a-dozen of them.

[1] (lf. 40, bk., and 54, bk.) Item, for Barnardyne de bolla, myllyner, wages ... ... vj li xx d
(lf. 41) Item, for Mr Whittington), scolmaster to thenxmen) ... v li
On lf. 44 bk., 60 bk. Dr Sampson, Dean of the King's Chapel, occurs.
[2] Item, for bastard Falconbridge, fee [*occurs elsewhere*] ... ... x li
Item, for Lodwicus Vives, [author] fee ... ... x li
[4] Item for Piro, the frenche coke, fee [*and elsewhere*] ... lxvj s viij d

App. II. *Henry VIII's Payments to Vicary, &c.* 101

(lf. 84, bk.) Quarter Wag*is* due at o*ur* Lady Day [25 March, an. 21, 1530].

(lf. 85) Item, for Anthony Skabo, surgion, fee ... ... x li
Item, for Doct*our* Bentley, phisicion, fee ... ... x li
Item, for Doct*our* Butt*es*, phisic*io*n, fee ... ... xxv li
(lf. 86) ¹Item, for Iohn Penn, barbour ... ... lxvj s viij d
*(lf. 86, bk.) Item, for Thomas Vicarie, surgion, fee ... ... v li

(lf 87, bk.) Halue yeres wag*is* due at o*ur* Lade day [an. 21, A.D. 1530].

(lf. 88) ²Item, for Iohn Clement, Phesicon, fee ... ... x li
Item, for Nicholas Sampson, fee ... ... 1 s

(lf. 99) Yet payment*es* in Maye, anno xxij^{do} [A.D. 1530].

(lf. 99, bk.) Item, more paid the said x^{th} day of maye to ⎫
Anthony Chabo, the king*is* Surgion, by the kingis war- ⎪
raunte datid at Windesour, xviij° Aprill, anno xxj°, xl li ⎪
ster̃ling, vpon an obligacon takin of the same Anthony to ⎪
repaye the said xl li to the Treasourer of the chamber for ⎬ xl li
the tyme being, to the king*is* vse, in mane*r* & forme folowing, ⎪
that is to say, at Ester next cumyng, x li, and so yerly after ⎪
at the said feast of Easter, x li, till the said sum*m*e of xl li ⎪
be paid.³ ⎭

(lf. 113) Quarter Wag*is* due at Mydesme*r* [an. 22, A.D. 1530]

Item, for Anthony Skabo, Surgion ... ... x li
Item, for Doct*our* Bentley, phisic*io*n, fee ... ... x li
Item, for Doct*our* Butt*es*, phisic*io*n, fee⁴ ... ... xxv li
(lf. 107, bk.) Item, for Iohn Penn, barbour ... ... lxvj s viij d
Item, for Thomas Vicary, surgion, fee ... ... C s

(lf. 124) Yet Quarter wagis [Michelmas] anno xxij^{do} [A.D. 1530].

(lf. 123, bk.) Item, for Anthony Scabo, surgion, fee ... ... x li
Item, for Doct*our* Bentley, phisic*io*n, fee ... ... v li
Item, for Doct*our* Buttes, phisicon, fee ... ... xxv li

[1] Item, for Anthony Annesley, tenesplay-keper ... ... vj s viij d
Item, for M*aste*r Whitington, scolmaster of Th*en*xm*e*n ... ... C s
* These payments are repeated elsewhere in the MS.
[2] Item, for Iodowicus Vives, annuite ... ... x li
[3] On leaf 103 Anthony Toto and Barthilmewe Pe*n*ne, paynters of Florence, get a quarterly payment (£18 15s.) of their wages of £25 a year each, during the King's pleasure. On lf. 145 they get £12 10s.
[4] Item, for Piro, the frenche coke, fee ... ... lxvj s viij d

102 App. II. *Henry VIII's Payments to Vicary, &c.*

(lf. 124, bk.) Item, for Iohn Penne, barbour ... ... lxvj s viij d
*(lf. 125) Item, for Thomas Vicary, surgion) ... ... v li

(lf. 126) Halue yeres wagis due at Michelmas a° xxij^{do} [1530].
(lf. 126, bk.) Item, for Iohn Clement, phisicon, fee ... ... 1 s

(lf. 143) Quarter Wages due in December [an. 22, A.D. 1530].
Item, for Anthony Scabo, Surgion, fee ... ... x li
Item, for Doctor Bentley, phisicion, fee ... ... x li
Item, for Doctour Buttes, phisicion, fee ... ... xxv li
(lf. 144, bk.) Item, for Iohn) Peyn, barbour ... ... lxvj s viij d
*Item, for Thomas Vicary, chirurgen, fee ... ... v li

(lf. 146, bk.) Paymentes in Januari, Anno Regni Regis Henrici octaui xxij^{do} [A.D. 1531].

Rewardes geuen on Sonday, Newcyeres day at Grenewiche, as hathe ben accustomde[1]
(lf. 147, bk.) Item, to Doctor Bentlis seruaunt ... vj s viij d
Item, to Doctor Chambers seruaunt[2] ... ... xiij s iiij d

(lf. 159) Quarter wagis due at our Lades Annunciacon [25 March, an. 22, A.D. 1531].
Item, for Antony Skabo, Surgion, fee ... ... x li
(lf. 159, bk.) Item, for Doctour Bentley, phisicon, fee ... ... x li
Item, for Doctour Buttes, phisicion, fee ... ... xxv li
*(lf. 160, bk.) Item, for Thomas Vicarie, Surgion, fee ... ... v li

(lf. 161, bk.) Halfe yeres wages due at our lades annunciacon [25 March, 1531].
Item, for Iohn) Clement, phisicion, fee ... ... x li
Item, for Nicholas Sampson, fee ... ... 1 s

In a thin volume of scraps of Wages of Hen. VIIIs household, Record Office, B. v. 4, the only entry on our subject is in *An.* 12,
'Item, for Doctour Farnande, þ^e quenes fysician, xxxiij li vj s vij d —*per* annum, lxvj li xiij s iiij d.

[1] Item to Master Crane, for playing before the Kinges grace with the childerne of the Kinges chapell ... ... vj .: xiij s iiij d
(lf. 149) Item to the Kinges plaiers, for plaing befor his grace v} !.. xiij s iiij d
Item to the princesse plaiers, for plaieng befor his grace ... iiij li·
Item to one that gave the king a nightingall singing ... ... xx s
Item to the gardyner of Wansted that gave the King two hechcokes } v s
[? hedgehogs or heathcocks?] ... ...
Other payments to Players occur; and the Musicians get monthly wages, &c.
[2] Item, to Doctor Wolman seruaunt ... ... xiij s iiij d

App. II. *Henry VIII's Payments to Vicary, &c.* 103

1538-41. Henry VIII's Quarterly and other Payments to his Surgeons (including Thos. Vicary), Physicians, Apothecaries, and Barbers, from the Arundel MS 97, in the British Museum.

Payments in March 1538 (a°. 29).

(leaf 6, back), Item, paide to Thom*as* Ashe, poti-cary, . . . for certain medicines, by doctour Cromer and other phesic*i*ons, and by the poticarye employed for the releif and conse*r*uac*i*on of the helth of lady Marget Douglas,[1] during the tyme of her beinge in the Towre of London, & also sins the same ... ...   xiiij ti iiij d

(leaf 9) Quarte*r* Wag*es* at o*ur* Lady day, a*nn*o vt sup*ra* (March 25, 1538)

Item, for Anthony Chabo, surgion [2] ... ...   x ti
Item, for Docto*ur* But*es*, phesicon, fee ... ...   xxv ti
Item, for Docto*ur* Bentley, phesic*i*on [3] ... ...   x ti
(lf 9, bk) Item, for Ioh̄n Penn, Barbour   ...   lxvj s viij d
*(leaf 10) Item, for Thom*as* Vicary, Surgion ... ...   C s
(lf 10, bk) Item, for Docto*ur* Mighel, phesic*i*on   xvj ti xiij s iiij d
Item, for Ioh̄n Sodo, poticary to the lady Mary   vj ti xiij s iiij d
(lf 11) Item, for Ioh̄n Alif, Surgion ... ...   C s
(leaf 11, back) Item, for Austen de Augustyns,[4] phesic*i*on   xxv ti

(leaf 24, back) Quarter Wag*es* at Midsom*er*, a*nn*o xxx° (1538).

Item, for Anthony Chabo, surgion   ... ...   x ti
Item, for Docto*ur* Buttes, phesicon   ... ...   xxv ti
Item, for Docto*ur* Bentley, phesic*i*on [5] ... ...   x ti
Item, for Ioh̄n Penn, Barbour ... ...   lxvj s viij d
*Item, for Thom*as* Vicary, Surgion ... ...   C s
(leaf 26, back) Item, for Docto*ur* Mighel de la se [? Delasco], p*h*esic*i*on to the lady Marye   xvj ti. xiij s. iiij d
Item, for Joh̄n de Sodo, poticary to y*e* lady Mary   vj li. xiij s. iiij d
Item, for Ioh̄n Aylif, Surgion, fee   ... ...   v ti

---

[1] She gets £20 for necessaries on Oct. 1, 1539, an. 30; leaf 40, back.
[2] See the extract from Brewer's Calendar in the Forewords.
[3] (leaf 9, back) Item, for M*aste*r Whitington, scholemaster to thenxmen   v li.
See Holbein's Picture for Penn, Butts, Bentley, Ayliff, &c.
[4] Agostino degli Agostini, Physician to Cardinal Wolsey.
[5] (leaf 25) Item, for M*aste*r Whitington, scholemaster to thenxme*n*   v li

104 App. II. *Henry VIII's Payments to Vicary, &c.*

(leaf 36, back) Quarter Wages at Mighelmas, anno vt supra (1538).

Item, for Anthony Chabo, surgion	x li
Item, for Doctour Buttes, phesicon, fee	xxv li
Item, for Doctour Bentley, phesicion	x li
(leaf 37, back) Item, for Iohn Penn, barbour, wagis	lxvj s viij d
*Item, for Thomas Vicary, Surgion, fe	v li
(leaf 38, back) Item, for Doctour Mighel, phesicion to my lady Mary...	xvj li xiij s iiij d
Item, for Sodo, poticary to the saide lady Mary	vj li xiij s iiij d

(leaf 39, back) Yet half yeres wages at Michelmas, Anno xxx° (1538).

Item, for Iohn Clement, phesicion, fee	x li
Item, for Austyn de Augustyns, phesicion	xxv li

(leaf 47, back) Yet paymentes in December, anno xxx° (1538).

Item, payde to Anthony Chabo, the kinges surgion, vpon his obligacion of his half yeres wages beforehande, after the rate of xl li by yere, which half yere is accompted to begyn at Christmas nowe, and shall ende and be fully ronne at Midsommer next commynge, the somme of ... } xx li

Item, paid to Augustyne de Augustinis, phesicion, in advauncement of his half yeres wages, which shalbe fully ronne at the Anunciacion of our Lady next, after the rate of L li by yere ... ... } xxv li

(leaf 49, back) Quarter Wages at Christemas, Anno vt supra (1538).

Item, for Anthony Chabo, surgion, fee	x li
Item, for Doctour Buttes, Phesicion, fee	xxv li
Item, for Doctour Bentley, phesicion, fee[1]	x li
Item, for Iohn Penn, Barbour, wagis	lxvj s. viij d
*(leaf 51, back) Item, for Thomas Vycary, Surgion ...	v li
(leaf 52, back) Item, for doctour Mighell de la so, phesicion to ye lady Mary ...	xvj li. xiij s. iiij d
Item, for Iohn de Sodo, poticary to the sayde lady	vj li. xiij s. iiij d
Item, for Iohn Aylif, Surgion, fee	v li

(leaf 53) Rewardes geuen on Wensday, Newyeres day, at Grenewiche, anno vt supra (xxx° : A.D. 1539).

Item, to Iohn Penn, Barbour, in rewarde	xl s
Item, to Edmund[2], Barbour, in rewarde	xl s

[1] (leaf 51) Item, for Master Whitington, scholemaster to thenxmen v li
[2] ? Edmund Harman.

App. II. *Henry VIII's Payments to Vicary, &c.* 105

(lf 53, bk) Item, to Doctour Augustin, phesicion, seruaunt[1]   x s
Item, to doctour Bentley seruaunt   ...   ...   vj s viij d
Item, to doctour Chambre seruaunt[2]   ...   ...   xiij s iiij d

(leaf 66, back) Yet quarter Wages at our Lady day, anno xxx°
(1539).

Item, for Iohn Penn, Barbour ...   .   lxvj s viij d
*Item, for Thomas Vicary, Surgion   ...   .   v li
(leaf 67) Item, for Doctor Tragonnell, fee   ...   ...   x li
(leaf 67, back) Item, for doctour Mighell, phi- } xvj li xiij s iiij d
  scicon to y® lady Mary   ...   ...
Item, for Iohn de Sodo, poticary to the lady Mary   vj li xiij s iiij d

(leaf 68, back) Yet half-yeres Wages at our Ladyday, anno
xxx° (1539).

Item, for Austen de Augustins, phe- } nil, quia prius in decembre
  sicion ...   .

(leaf 78) Yet paymentes in June, Anno xxxj° (1539).

Item, to Thomas Bill, doctour of phisicke, by the kingis
  Warraunte, dated primo Aprilis, anno xxx° (1538) for the
  yerly payment to him of his yerly annuitie of x li by yere,
  to be yerely paide to him from the feast of the natiuitie of } C s
  our lorde last, quarterly, by even porcions, v li for twoo
  quarters fully ronne at the feast of the Natiuite of sainct
  Iohn Baptist, anno tricesimo primo ...   .
Item, paid to Robert Huicke, Doctour of phisicke, by
  warraunte dated primo Aprilis, anno xxx domini Regis
  nunc, for his yerely annutie of x li by yere, to be paide } C s
  vnto him from the feast of Christimas last, quarterly, by
  even porcions, the somme of v li, for ij quarters fully ronne
  at the natiuitie of saint Iohn Baptiste, Anno xxxj° ...

(leaf 79, back) Quarter wages in June, Anno ut supra, (1539).
Item, for Anthony Chabo, surgion   ...   ...   m t [= nil[3]]
Item, for Doctour Buttes, phesicion   ...   .   xxv li
Item, for Doctour Bentley, phesicion   ...   .   x li
* (leaf 80, back) Item, for Thomas Vicarie, Surgion   ...   v li
(leaf 81, back) Item, for doctour Mighel de la } xvj li. xiij s. iiij d
  Soo, phesicion...   ...   .
Item, for Iohn de Sodo, poticary   ...   ...   vj li. xiij s. iiij d

---
[1] (leaf 54) Item, to doctour Lupton [a divine] seruaunt   ...   xiij s iiij d
[2] (leaf 55, back) Item, to Bastard Falconbridge seruaunt   ...   vj s viij d
  Item, to Bartlet [Berthelet], the kinges printer seruaunt, }
  that broght the king a boke couered with crimosen saten } vj s viij d
  embradred ...   .   }
[3] See leaf 91 back, 89, 82, 68 back, &c.

106 App. II. *Henry VIII's Payments to Vicary, &c.*

(leaf 91, back) Quarter Wages, Anno vt supra (Sept. 1539).

Item, for Anthony Chabo, surgion	nil
Item, for Doctour Buttes, phesicion, fee	xxv ƚi
Item, for doctour Bentley, phesicion	x ƚi
(leaf 92) Item, for Iohn Penn, Barbour, wagis ...	lxvj s. viij d
*(leaf 92, back) Item, for Thomas Vicary, Surgion ...	v ƚi
(leaf 93, back) Item, for Doctour Mighell de la Soo, phesicion	} xvj ƚi. xiij s iiij d
Item, for Iohn de Sodo, poticary	vj ƚi xiij s iiij d
Item, for Iohn Aylif, Surgion, wagis	v ƚi

(leaf 102) Yet paymentes in Decembre, Anno xxxj° (1539).

Item to Doctour Augustyne, in aduauncement of his half [yeres] wagis beforehand; which half yere is accompted to begynne primo Octobris, Anno xxxj° [1539], and shall ende vltimo Marcij then next folowinge ...	} xxv ƚi
Item, prested [advanced] to Anthony Chobo, the kingis Surgion, in aduauncement of his half yeres wagis beforehande; which half yere is accompted to begynne primo Ianuarij, Anno xxxj° [1540], and shall ende vltimo Iunij then next followinge	} xx ƚi
Item, payde to Nicholas Alcoke, Surgion, by the kingis warraunt, dated the xx day of Novembre, Anno xxxj° [1539], for the yerely payment to him of x ƚi by yere, quarterly, by even porcions, from Mychaelmas dicto Anno xxxj° duringe his lyf, the somme of 1 s for one quarter fully ronne vltimo Decembris dicto Anno xxxj°	} 1 s

(leaf 104, back) Quarter Wagis a Cristumas, anno vt supra (1539).

Item, for Anthony Chabo, surgion	x ƚi
(leaf 105) Item, for Doctour Buttes, phesicion	xxv ƚi
Item, for doctour Bentley, phesicion	x ƚi
(leaf 105, back) Item, for Iohn Penn, barbour ...	lxvj s. viij d
*Item, for Thomas Vycary, surgion	v ƚi
(leaf 106, back) Item, for Doctour Mighell de la Soo, phesicion	} xvj ƚi xiij s. iiij d
Item, for Iohn de Sodo, poticarye	vj ƚi xiij s iiij d
(leaf 107. Item, for Hans Holbyn, paynter ...	... vij ƚi x s)
Item, for Iohn Aylif, Surgion	v ƚi
Item, for doctour Hyll, phesicion	1 s

(leaf 108) Rewardes geuen on Thursday, Newyeres day, at Grenewiche, as hathe be accustumed. Anno tricesimo primo (1540).

Item, to Iohn Penn, Barbour, in rewarde ... ... xl s

App. II. *Henry VIII's Payments to Vicary, &c.* 107

Item, to Edmunde[1], barbo*u*r, in rewarde[2] ... ... xl s
Item, to docto*u*r Bentleis ser*u*aunt, in rewarde ... ... vj s viij d
Item, to docto*u*r Chambre ser*u*aunt, in rewarde xiij s iiij d
Item, to docto*u*r Augustine ser*u*aunt[3] ... ... x s
(leaf 109) Item, to docto*u*r Cromer ser*u*a*u*nt ... ... vj s viij d

(leaf 123, back) Quarter wag*is* at o*u*r Lady day, A*nn*o vt supra (xxxj°, A.D. 1540).

Item, for Anthony Chobo, surgion ... ..., n*i*l
Item, for Docto*u*r Butte*s*, phesic*i*on ... ... xxv ti
Item, for Docto*u*r Bentley, phesic*i*on ... ... x ti
*(leaf 124) Item, for Thomas Vicary, Surgion ... ... v ti
(leaf 125) Item, for Nicholas Alcocke, surgion ... ... 1 s
Item, for Iohn Aylif, surgion ... ... v ti
(leaf 125, back) Item, for Thom*a*s Bi*l*l, phesicion ... 1 s
Item, for doc*tou*r Huie, phesicion ... ... 1 s
Item, for Docto*u*r Augustyn } n*i*l, qu*i*a sol*v*itur pr*i*mam di*e*m iij Dece*m*bris vltim*o*

(leaf 136) Quarter wag*is* a Midsom*er*, A*nn*o xxxij° (1540).

Item, for Anthony Chabo, surgion ... ... nil, q*ui*a pr*i*us
Item, for Docto*u*r Butte*s*, phesicon, fee[4] ... ... xxv ti
Item, for Docto*u*r Bentley, phesic*i*on ... ... x ti
*(leaf 136, back) Item, for Thomas Vycary, surgion ... v ti
(leaf 137, back) Item, for Docto*u*r Mighell de la Soo, phesic*i*on ... ... } xvj ti. xiij s. iiij d
Item, for Iohn Sodo, poticary ... ... vj ti. xiij s. iiij d
Item, for Nicholas Alcocke, surgion ... ... 1 s
Item, for Iohn Alif, Surgion ... ... v ti
Item, for Thom*a*s Bill, phisic*i*on ... ... 1 s
Item, for Docto*u*r Huie, phesic*i*on ... ... 1 s

(leaf 149, back) Quarte*r* wag*is* at Michelm*a*s, A*nn*o vt supr*a* (xxxij°, 1540).

Item, for Anthony Chobo, surgi*o*n ... ... x ti
Item, for docto*u*r Butte*s*, phesic*i*on, fee... ... ... xxv ti
Item, for docto*u*r Bentley, phesicon,[5] ... ... x ti
*(leaf 150) Item, for Thom*a*s Vicary, Surgion ... ... v ti
(leaf 151) Item, for docto*u*r Mighel de la so, phesicon ... ... } xvj ti xiij s iiij d
Item, for Nicholas Alcok, surgion ... ... 1 s

1 ? Edmund Harman.
2 (lf 109) Item, to docto*u*r L*u*ptons ser*u*aunt, in rewarde ... xiij s iiij d
3 (leaf 109) Item, to docto*u*r Lee ser*u*a*u*nt [? Dr. of Divinity] xij s iiij d
  (leaf 111) Item to Cornelis Hays, *th*at gave a shavi*n*geloth } vj s viij d
  wroght wit*h* gold ... ... }
4 (lf 136) Item for Basterd Falconbridge ... ... x li
5 (leaf 150) Item, for John Haywood, playo*u*r on y*e* virginall*es* 1 s. Also on other pages of the MS.

108 App. II. *Henry VIII's Payments to Vicary, &c.*

Item, for Iohn Aylyf, surgion ... ... v ti
(leaf 151, back) Item, for Thomas Bill, phesicon ... 1 s
Item, for Doctour Huic, phesicion ... ... 1 s
(leaf 152) Item, payd to Thomas Alsop, gentil- ⎫
man) poticary¹ to the kyngis maiestie, by the ⎪
kyngis warraunt, datid primo Septembris, ⎪
Anno xxxij° [1540], for the yerely payment ⎪
to him of xxvj ti xiij s iiij d, at iiij termes of ⎪
the yere, by even porcions, from the feast of ⎬ vj ti. xiij s. iiij d
Midsomer dicto Anno xxxij^do, during the ⎪
kyngis pleasur, the first part thereof to be ⎪
made to him at this terme of Michelmas, ⎪
vj ti. xiij s. iiij d for one quarter due to him ⎪
by vertue of the saide warraunt at this pre- ⎪
sent feast of saincte Michaell. ... ⎭

(leaf 161) Quarter wagis at Cristunmas, Anno vt supra (xxxij° : 1540).

Item, for Anthony Chobo, surgion, fee ... ... x ti
(leaf 161, back) Item, for Doctour Buttes, phesicion, fee xxv ti
Item, for Doctour Bentley, phesicon, fee² ... ... x ti
*(leaf 162) Item, for Thomas Vicary, surgion ... ... v ti
(leaf 163) Item, for Doctour Mighell de la Soo, ⎫ xvj ti. xiij s. iiij d
phesicion ... ... ⎭
Item, for Iohn de Sodo, poticary to yᵉ lady Mary vj ti. xiij s. iiij d
Item, for Nicholas Alcok, Surgion ... ... 1 s
Item, for Iohn Alyf, Surgion, wagis³ ... ... v ti
Item, for Thomas Bill, phesicion ... ... 1 s
(lf 163, back) Item, for Thomas Alsopp, gentleman ⎫ vj ti xiij s iiij d
potycary ... ... ⎭

(leaf 164, back) Rewardes geuen on Saterday, Newyeres day, at Hamptoncourte, Anno xxxij° (A.D. 1541).

Item, for Iohn Penn, Barbour, in rewarde ... ... xl s
Item, to Edmonde⁴, Barbour, in rewarde⁵ ... ... xl s
(leaf 165, back), Item, to Doctour Bentley, phesicon, ⎫ vj s viij d
seruaunt ... ... ⎭
Item to doctour Chambre seruaunt ... ... xiij s iiij d
(leaf 166) Item, to doctour Augustine seruaunt⁶ ... x s

---

¹ This 'gentleman poticary' is, we take it, in contrast with John Emmingway, the 'yoman poticary' who appears at pages 109, 113, 114, 117, 118, below.
² (leaf 161, back) Item, for Iohn Haywood, playour of yᵉ virginalles 1 s
³ (leaf 163) Item, for Rauff Stannowe, scholemaster to thenxmen v li
⁴ ? Edmund Harman.
⁵ (leaf 165) Item, to Anthony Tote, seruaunt, that brought the ⎫ vj s viij d
King a table [picture] of the storye of Kinge Alexander ⎭
(leaf 165, back) Item, to Bartlet, the kingis printer seruaunt vj s viij d
³ (leaf 166) Item, to doctour Le [? a divine] seruaunt, in rewarde xiij s iiij d

App. II. *Henry VIII's Payments to Vicary, &c.* 109

(leaf 180) Quarter wag*is* at o*ur* Lady day, a*n*no vt sup*ra* (1541).

Item, for Anthony Chabo, surgion	x li
Item, for docto*ur* Butt*es*, phesic*i*on, fee	xxv li
Item, for docto*ur* Bentley, phesic*i*on.	x li
*(leaf 180, back) Item, for Thom*as* Vicary, Surgion ...	C s
(leaf 181, back) Item, for docto*ur* docto*ur* [*so*] de la Soo, phesic*i*on	} xvj li. xiij s. iiij d
Item, for Io*h*n de Sodo, poticary to ye lady Mary	vj li. xiij s. iiij d
Item, for Nicholas Alcok, surgion	1 s
Item, for Io*h*n Aylif, Surgion	C s
(leaf 182) Item, for Thom*as* Bi*ll*, phesic*i*on, fee ...	1 s
Item, for Docto*ur* Huic, phesic*i*on	1 s
Item, for Thom*as* Alsop, gent*leman* poticary¹	vj li. xiij s. iiij d

(leaf 193, back) Quarter wag*is* at Midsom*er* a*n*no ut s*u*p*ra* (xxxiij°: 1541).

Item, for Anthony Chobo, Surgion, fee	v li
Item, for docto*ur* Butt*es*, phesic*i*on, fee	xxv li
Item, for docto*ur* Bentley, phesic*i*on²	x li
*(leaf 194) Item, for Thomas Vycary, Surgion ...	C s
(leaf 195) Item, for docto*ur* Migh*el*, phesic*i*on to the lady Mary	} xvj li. xiij s. iiij d
Item, for Io*h*n de Zodo, poticary to the lady Mary	vj li. xiij s. iiij d
Item, for Nicholas Alcoke, surgion	1 s
Item, for Io*h*n Aylof, Surgion	C s
Item, for Thom*as* Bill, phesic*i*on	1 s
(leaf 195, back) Item for docto*ur* Huic, phesic*i*on ...	1 s
Item, for Thom*as* Alsop, gentilman poticary	vj li. xiij s. iiij d
Item, for Iohn Em*m*yngway, yoman poticary [55s. 7½d.]³	} lv s. vij d. ob.

¹ (leaf 182, back) Item, for bastard Falconbridge x li. This entry is on other pages too.
² leaf 194 : Item, for Iohn Haywood, playo*ur* on the Virginall*es* C s
³ Item, for Thomas Sperin and his son, s*er*giant*es* of the beres [bearwards] lvij s q*uadrantæ* di*midium*. (Was this half-farthing a joke?)
In this MS, we notice that for the words 'rat-catcher and mole-catcher,' 'rattaker and molletaker' are used.
(leaf 6) in March 1538 (a°. 29).
Item, paide John Willis, the Kingis rattaker for his wagis after iiij by dey (from Sept. 8 to April 1) ... ... } lxx s. viij d
(leaf 151, back) Michaelmas, 1540 (a°. 32).
Item, for Iohn Wylle, rattaker ... ... lx s. xd
(leaf 170 : Feb. 1541) Item, for Iohn Whatson, molletaker ... ix iiij d
(leaf 182, back : Lady Day 1541) Item for Iohn Wylly, rattaker lx s x d
(leaf 193 : June 1541) Item, for Iohn Whatson, molletaker ... ... x s

110 App. II. *Henry VIII's Payments to Vicary, &c.*

1543-4. Further Quarterly Payments to Vicary, &c.

(From the Phillipps MS, No. 3852.)

The following payments of Henry to his Surgeons and Physicians, from Christmas 1543 to Michaelmas 1544, are taken from the late Sir Thomas Phillipps's MS, No. 3852, at Thirlestone House, Cheltenham, by his grandson Mr. T. Fitzroy Fenwick, who, we are glad to say, inherits his grandfather's care for MSS, and has been good enough to send us these entries :—

Receipts and Expenses of Hen. VIII, from Oct. 35th year, to Oct. 36th year, A.D. 1543-4.

Quarter wagis for *Cristm*as anno R*e*g*n*i Reg*is* Henrici tricesi*m*o quinto (A.D. 1543)

[Under this head, among other entries, occur the following]

Item, for Anthony Chabo, S*ur*geon	x li
Item, for Docto*ur* Butt*es*, Phisicion	xxv li
Item, for Doctour Benteley, phisicion	x li
*Item, for Thomas Vicary, S*u*rgeon	C s
Item, for Nich*o*las Alcok, S*ur*geon	L s
Item, for John Ayliff, S*ur*geon ...	C s
Item, for Thomas Bill, ffisicion ...	xii li x s
Item, for Docto*ur* huic, ffisicion ...	L s
Item, for Richard fferrys, S*ur*geon	C s

Quarter Wagis for *our* lady day, A*nn*o Re*gn*i Reg*is* Henr*ici* octaui tricesimo quinto (A.D. 1544)

Item, for Anthony Chabo, S*ur*geon	x li
Item, for Docto*ur* Butt*es*, phisicion	xxv li
Item, for Doctou*r* Benteley, phisicion	x li
*Item, for Thomas Vicary, Surgeon	C s
Item, for Nich*o*las Alcok, S*ur*geon	L s
Item, for John Ayliff, S*ur*geon ...	C s
Item, for Thomas Bille, phisicion	xii li x s
Item, for Docto*ur* Huyck, phisicion	L s
Item, for Richard fferrys, S*ur*geon	C s
Item, for Richard Asser, S*ur*geon	xlv s vi d

Quater Wagis for Midsomer, Anno R*e*g*n*i Reg*is* Henr*ici* ' viii,' xxxvi^{to} (A.D. 1544)

Item, to Anthony Chabo, S*ur*geon	x li
Item, to Docto*ur* Butt*es*, phisicion	xxv li
Item, to Docto*ur* Benteley, phisicion	x li

App. II. *First Payment of V.'s* 40-*Marks' Annuity.* 111

*Item, to Thomas Vicary, Surgeon	C s
Item, to Nicholas Alcok, Surgeon	L s
Item, to John Ayliff, Surgeon	C s
Item, to Thomas Bille, phisicion	xii ti x s
Item, to Doctour huyck, phisicion	L s
Item, to Richard fferrys, Surgeon	C s

Quarter Wagis at Mighelmas, Anno Regni Regis, Henrici octavi, xxxvi^{to} (A.D. 1544).

Item, for Anthony Chabo, Surgeon	x ti
Item, for Doctour Buttes, phesicion	xxv ti
Item, for Doctour Benteley, phesicion	x ti
*Item, for Thomas Vicary, Surgeon	C s
Item, for Nicholas Alcok, Surgeon	L s
Item, for John Ayliff, Surgeon	C s
Item, for Thomas Bille, phisicion	xii ti x s
Item, for Doctour Huick, phisicion	L s
Item, for Richard fferrys, Surgeon	C s
Item, for Cornelius Zefridus[1], doctour of phesik to the Lady Anne of Cleves	xi ti xiii s iiii d

---

Earliest[2] and Latest Payments of Vicary's Annuity of 40 Marks (£26 13s. 4d.), granted by Henry VIII on 29 April, 1530.

Tellers' Roll (Exchequer of Receipt), 27-28 Hen. VIII, No. 89.

[3]Easter, 28 Hen. VIII (A.D. 1536).

Thome Vycary, capitali Cirurgico Domini Regis, de Annui tate sua ad xxvj li. xiij s. iiijd. per annum, sibi debita a viij

---

[1] ? MS. Refridus.

[2] This is the last payment to Marcellus de la More in the Tellers' Rolls of the Exchequer:—
N°. 88. Easter, 27 Hen. VIII. (1535.)
 To Marcellus de la More, &c., by his own hands, by writ current (for the half year) ... ... £13 6s. 8d.
N°. 89.   Mich. 27 Hen. VIII. (1535.)
Nothing as to De la More in this and the following half-years down to Easter, 30 Hen. VIII. (1538.)
 [Can this one half-year's pay, Easter to Michs. 1535, be the *arrears* of th.s Annuity which Q. Mary orderd to be paid to Vicary on 20 Oct. 1553?—F.]
 In the Exchequer of Receipt, Auditors' Patent Books, vol. ii. ff. 198, 199, are entries of the payment of the Annuity of 40 Marks (£26 13s. 4d.) to Marcellus de la More, the King's Surgeon, granted him for life. The statements of payments made to him half-yearly run from Michaelmas, 19 Hen. VIII. (1527) to Easter, 25 Hen. VIII. (1534), when the Record stops. Most of the payments, including the last, are stated to have been made ' to his own hands.'

[3] Easter is not reckoned in these rolls according to the day on which the

112 App. II. *Payment of Vicary's* 40-*Marks' Annuity.*

die Septembr*is*, Anno xxvij^{mo} Reg*is* nunc Henr*ici* viij^{ui} [A.D. 1535], vsq*ue* fest*um* Pasche extunc p*r*oximo sequen*s*, acciden*s* xvj^{mo} die Apr*i*lis, Anno xxvij^{mo} [A.D. 1536], scil*i*cet, p*r*o CCxix dieb*us*, juxta Ratam p*r*edict*am*, Recep*tis* denar*iis* per m*a*n*us* p*r*op*r*ia*s*, per bre*ue* curren*s* xv li. xix s. iiij d.

Tellers' Roll, 28-29 Henry VIII, No. 90.

Michaelmas, 28 Hen. VIII. (1536.)

Thome Vycary, Capital*i* Cirurgico Do*m*i*n*i Regi*s*, de A*n*nuitate sua ad xl marc*as* pe*r* annu*m*, sibi debita pro fes*t*o Michae*l*is, Anno p*r*ed*i*cto, Recep*tis* dena*r*iis per man*us* p*r*oprias, pe*r* bre*ue* currens ... ... xiij li. vj s. viij d.

Easter, 29 Hen. VIII. (1537.)

————— [1] Vekery, surgia*n*ti [2] Dom*i*ni Regis, de feodo [3] suo ad xl m*a*rc*as* pe*r* annu*m*, sibi debito p*r*o med*i*etat*e* anni, finit*a* ad festum Pasche nu*n*c, pe*r* bre*u*e curr*ens*, Recep*tis* dena*r*ii*s* per manus Robe*r*ti Game ... ... xiij li. vj s. viij d.

Tellers' Roll, 29-30 Henry VIII, No. 91.

Mich. 29 Hen. VIII. (1537.)

Thomas Vicary 'surgia*n*t' to the King, &c., by the hands of John Swalowe ... ... xiij li. vj s. viij d.

Easter, 30 Hen. VIII. (1538.)

Thomas Vecary 'surgia*n*t,' &c., by the hands of Anthony Alyngton ... ... xiij li. vj s. viij d.

The payments doubtless run on regularly, half-year by half-year, but the rolls are very voluminous, and take a long time to go through. I therefore take further entries only from two of Edward VI, and the last ones of Elizabeth.

Tellers' Roll (Exchequer of Receipt), 5-6 Edw VI, No. 100.

Michaelmas, 5 Edw. VI. (A.D. 1551.)

Thome Vicarie, seruient*i* Chirurgor*um* Dom*i*ni Reg*is*, de feod*o* suo ad xxvj li. xiij s. iiij d. per annu*m*, sibi debit*o* ad festum Mich*a*elis anno v^{to} reg*ni* Regis Ed*w*ardi vj^{ti}, recep*tis* den*ariis* pe*r* manus prop*r*ias, pe*r* bre*ue* dorma*ns* ... ...
xiij li. vj s. viij d.

Easter, 6 Edw. VI. (A.D. 1552.)

A similar entry.

festival occurred, which varied so much as sometimes to cause *two* Easters to fall in *one* of the years of this reign ; but to avoid that inconvenience, Easter is here considered to be in the regnal year, *following* that in which the preceding Michaelmas occurred.—R. K.

[1] Blank.   [2] MS. surgiat'.   [3] MS. de feodo de feodo.

App. II. *Vicary's Annuities of* 40 *Marks &* £20. 113

Tellers' Rolls, 2 and 3 Elizabeth, No. 109.

Michaelmas, 2-3 Elizabeth. (1560.)

Thome Vicarie, de feodo suo ad xl^{ti} marc*as* per annu*m*, sibi [debito] pro di*mi*dio anni finito in festo Sa*n*cti Mich*ae*lis Arch*an*ge*l*i, Anno secu*n*do regine Elizabeth*e*, recept*is* denarijs p*er* manus pr*o*prias ... ... xiij li. vj s. viij d.

¹Thome Vicare predict*o*, de Annuitate sua ad xx li. p*er* annu*m*, sibi debit*a* pro di*mi*dio anni finito in festo Sa*n*cti Mich*ae*lis Arch*an*ge*l*i, Anno secu*n*do regine Elizabethe, recept*is* denarijs pe*r* manus pr*o*prias ... ... x li.

(m. 66) Easter, 3 Eliz. (1561.)
Similar entries to the above. The moneys were due at Lady Day.

There is no Tellers' Roll for 3-4 Elizabeth. (1561-2.)

Tellers' Roll, 4-5 Elizabeth, No. 110. (1562-3.)

The portion of this Roll relating to Michaelmas term 4-5 Eliz. (1562) has been searched, but I do not find anything as to Vicary. He no doubt died late in 1561, or early in 1562.—R. G. Kirk.

Vicary's Annuity of £20 for Wages and Medicines: its last Payments to Marcellus de la More; with its first and some later payments to Thomas Vicary, under a fresh Grant (not yet found) of Sept. 20, 1535.

(From the Wardrobe and Household Books, Exchequer, Queen's Remembrancer, Ancient Miscellanea, &c., in the Public Record Office.)

℣ 17-18 Hen. VIII, 30 Sept. an. 17, A.D. 1525, to 30 Sept. an. 18, A.D. 1526.

Account of Sir John Shirley, Cofferer of the Household.

(leaf 5 from end) Warant*um* Regis.

Marcello de La More, Capit*a*li Cirurgico Hospicij Do*mi*ni Regis, In Denar*iis* virtute Warrant*i* d*i*cti D*o*m*i*ni Regis, cuius dat*um* est apud Wyndeso*ur* ij^{do} die Ia*n*u*ar*ii anno Regni sui quinto [1514], durante beneplacit*o* soluendu*m*, p*r*o vad*iis* & Medicinis eidem Marcello, prout in eodem plenius continetur, infra tempus huius Compoti, xx ℔.

℣ 20-21 Hen. VIII, 30 Sept. 1528, to 30 Sept. 1529.

· Account of Sir Henry 'Guldeforde,' Comptroller of the Household. Marcellus de la More's Annuity is on the back of leaf 3 from end.

¹ This is Vicary's £20 annuity, or one of them, as to which see the entries following, on p. 114—122.

VICARY. I

114 App. II. *Vicary's Annuity of £20 (1535 Grant).*

$\frac{7}{5}$ 30 Sept., 22 Hen. VIII, A.D. 1530, to 30 Sept., 23 H. 8, A.D. 1531.
Marcellus de la More's Annuity is on leaf 3 from end.

$\frac{7}{6}$ 25-26 Hen. VIII, 30 Sept. 1533, to 30 Sept. 1534.
Marcellus de la More's Annuity is on the back of leaf 3 from end. This is the last payment found to Marcellus de la More in this set of Books. In $\frac{7}{8}$ 26-27 Hen. VIII, ? incomplete, there is no payment of annuities; nor is any in Book $\frac{7}{9}$.

The first payment found in these Books, to Thomas Vicary, is in Book

$\frac{7\frac{3}{6}}{16}$ 28-29 Hen. VIII, 30 Sept. 1536, to 30 Sept. 1537.
Account of Sir Wm. Paulet, Controller.
(back of leaf 4 from end) Warrantum Reg*is*.
Thome Vicars, Capital*i* cirurgico hospicij D*om*ini R*e*g*is*, in den*ariis* ei solut*is* v*er*tute w̄arranti d*ic*ti D*om*ini R*e*g*is*, cuius dat*um* est xx^{mo} die Septembris ap*u*d Byshopswaltham, anno Regni sui xxvij^{mo} [A.D. 1535], durant*e* vit*a* dicti Thome, soluend*um* pro vad*iis* et medicinis eidem Thome p*er* dictum Warrantum annuatim concessum, prout in eodem warranto plenius contin*e*tu*r*, infra tempus huius Computi, xx li.

The reader will see that this Annuity of £20 for Wages and Medicines, is not made under the original Grant of 29 April 1530 (p. 89), but under a fresh Grant of Sept. 20, 1535. As it is like Marcellus de la More's in being 'for Wages and Medicines,' we suppose that More must have disappeared after receiving his last £13 6*s*. 8*d*. at Easter 1535 (p. 111 above, note 2), and that Vicary got a fresh Grant from Henry on Sept. 20, 1535, to save him the trouble of proving More's death, or resolve not to come back to England, or otherwise act as Serjeant of the Surgeons. We presume that this Annuity was in substitution of the £5 a quarter which Vicary had as one of the Surgeons to the King during More's life (see Forewords); but it may have been an extra one. Readers must judge for themselves.

We go on with the Exchequer Q. R. Anc. Misc. extracts:

$\frac{74}{11}$ Controller's Account, A small Part of the grant to Thomas 'Vicars' of £20, is on back of leaf 4 from end.

App. II. *Vicary's Annuity of £20* (1535 *Grant*). 115

$\frac{7\frac{1}{2}}{1}$ Cofferer's Account, 30-31 Hen. VIII, 30 Sept. 1538, to 30 Sept. 1539. Vicary's £20 is on back of leaf 3 from end; and in $\frac{7\frac{1}{2}}{1}$, the Book of the Controller Sir Wm Kyngston for the same Period, on leaf 2 from end.

$\frac{7}{1}$ Exch. Q. R. Anc. Misc. Wardrobe and Household. Anno xxxj Regis Henrici Viij^{di} (A.D. 1539-40). Computus Edwardi Pekham, armigeri . . . ab ultimo die mensis Septembris, Anno dicti domini Regis xxxj^{mo} vsque vltimum diem mensis Septembris, Anno eiusdem domini Regis xxxij^{di} . . .

Thomas Vicary's Annuity of £20 for Wages and Medicines (under Warrant of Sept. 20, an. 27, A.D. 1535) is on the back of the 3rd leaf from the end. It is also in $\frac{7}{1}$, the Controlment book of Sir Wm. Kyngston, at the back of leaf 3 from the end:—

Warranta Regis.

Thome Vicarie, Capitali Chirurgico hospicij domini nostri Regis, in denariis ei solutis, virtute warranti dicti domini Regis, cuius Datum est xx° Die Septembris, apud Bysshopsse Waltham, anno Regni sui xxvij^{mo} [1535], durante vita Dicti Thome, per dictum warrantum annuatim concessum, prout in eodem warranto plenius continetur, infra tempus huius Computi, xx li /

1540-1. The like payments to Vicary of this £20 Annuity are in Sir E. Peckham's Account-book, *Exch. Q. Rem., Anc. Misc., Wardrobe and Household*, for 32-3 Hen. VIII (30 Sept. 1540-1), $\frac{7}{1}$ (at the back of the last leaf but 3), and in the Controlment Book for the same year, $\frac{7}{1}$, on the third leaf from the end.

Then for 1541-2 comes (Ex. Q. R. Anc. Miscell. Wardrobe and Household $\frac{7}{1}$) the

1541-2.

Account of Sir Edmund Peckham, Cofferer of the Household, from the year Sept. 30, an. 33 (A.D. 1541) to Sept. 30, an. 34 (A.D. 1542), back of leaf 5 from end.

Thome Vycarye, Capitalli Chyrurgico hospicii Domini Regis, in Denarijs ei solutis virtute Warranti Dicti Regis, cuius Datum est xx^{mo} Die Septembris apud Bisshops Waltham, Anno Regni sui xxvij^{mo} [A.D. 1535], durante vita dicti Thome, soluendis pro Vadiis & Medicinis eidem Thome per dictum Warrantum Annuatim concessum, prout in eodem Warranto plenius continetur, infra tempus huius Computi xx li /

116 App. II. *Vicary's Annuity of £20* (1535 *Grant*).

$\frac{7.6}{0}$. In the Book of Controlment of Sir Jn. Gage, Controller of the Household for the same year, Oct. 1, 1541, to Sept. 30, 1542 (an. 33-4), Vicary's Annuity of £20 is at the back of the last leaf but one of the MS. More than half the lower part of every leaf has perisht. For the next year, 1542-3, we have

Ex. Q. R. Anc. Misc. Wardrobe and Household $\frac{7}{7}^6$, 34-35 Hen. VIII. Account of Sir Edmund Pekham, Cofferer and Keeper of the Great Wardrobe of Henry VIII, for one year from Oct. 1, an. 34 [A.D. 1542], to Sept. 30, an. 35 [A.D. 1543], 4th leaf from end.

Warranta Do*mi*ni Regis.

Thome Vicarie, Capitali Chirurgico hospicij Do*mi*ni Regis, in denar*iis* ei solut*is*, virtute Waranti dic*ti* domini Regis, cuius Dat*um* est apud Busshoppe*s* Waltham, xx^{mo} die Septembris, anno Regni sui, xxvij^{mo}, durante vita dic*ti* Thome, soluend*is* p*ro* vad*iis* & medicinis, eidem Thome, xx ħ

In $\frac{7.6}{8}$, the Book of Controlment of Sir John Gage, Controller of the Household for the same Period, Oct. 1, an. 34 [1542] to Sept. 30, following [1543], the same payment is entered on leaf 4 from the end :—

[A.D. 1542-3] Warranta Do*mi*ni Regis.

Tome Vicarie, Capitali Chirurgico Hospicij do*mi*ni Regis, in denar*iis*. ei solut*is* virtute warrant*i* dicti D*omi*ni Reg*is*, cuius dat*um* est ap*ud* Bishoppe*s* waltham, xx° die Septembris, Anno Regni sui xxvij°, durante vit*a* dicti Thome, soluen*dis* pro vad*iis* et medecinis eidem Thome annuatim concessis, prout eode*m* warrant*o* plenius continetur, infra tempus huius Comp*uti*, xx ħ.

The next book (the Cofferer's) is of like kind, $\frac{7.6}{10}$, for the year 1545-6, Sir Edmund Peckham's Account; and in it, Vicary's annuity of £20 is on the back of leaf 4 from the end (not counting the Indentures fastend to the back of the last leaf). In the Controller Sir John Gage's book for the same year (Oct. 1, 1545, to 30 Sept. 1546), $\frac{7.6}{11}$, Vicary's payment is at the back of the 5th leaf from the end. And in the next and last book, $\frac{7.6}{14}$, of the Cofferer, Sir E. Peckham, from Oct. 1, 1546, 38 Hen. VIII, to March 31, of 1 Edw. VI, 1547, Vicary's half-year's payment is on the back of the 4th leaf from the end, partly on an erasure, 'viz. infra tempus hu*i*us Comp*uti*, x ħ.'

App. II. *Edward VI's Payments to Vicary, &c.* 117

Edward VI's Payments to his Physicians, Surgeons, Apothecaries, &c., from Midsummer to Christmas, 1547.

(Accounts of Sir W^m. Cavendish, Treasurer of the King's Chamber. *Misc. Books*, Augmentation Office, No. 439, leaf 26, back.)

Quarters wages for Midsomer, anno Regni Reg*is* Edwardi sexti Primo. [A.D. 1457.]

per Cade[1]	Item, to Docter Bentley, Phisic*i*on	... ...	x li exr.
per Knot	Item, to Doctor Huicke, Phisic*i*on	... ...	l s exr.
per Cade	{Item, to Cornelis zifridus, Docter of Phisike with the Lady Anne of Cleves[2]	... xj li xiij s iiij d	exr.
per Cade	Item, to Iohn de Sodo, Potycary	... vj li xiij s iiij d	exr
	Item, to Thomas Alsop, Potycary	... vj li xiij s iiij d	exr.
	Item, to Iohn Emyngwey, yoman potycary	lv s vij ob.	exr.
* per Knot	*Item, to Thomas Vycary, Surgeon	... ...	C s exr.
	Item, to Iohn Ailiff, Surgeon ...	... vij li xv s	exr.
[leaf 27]	Item, to Richard Ferres, S*u*rgeon	... ...	C s exr
per Knot	Item, to Nicho*l*as Alcoke, Surgeon	... ...	l s exr
per Cade	Item, to George Hollonde, Surgeon	... ...	l s exr.
	Item, to Thomas Gemynous, S*u*rgeon[3]	...	l s exr.

[leaf 43] Quarters Wages for Michaelmas, anno Re*gni* Re*gis*, E*dwardi* vj^e Primo. [A.D. 1547.]

per Cade	Item, to Doctor Bentley, phisic*i*on	... ...	x li exr.
per Knot	Item, to Doctor Huicke, phisic*i*on	... ...	l s exr.
per Cade	{Item, to Cornelis zifridus, Docter of Phisicke to the Lady Anne of Cleves	... xj li xiij s iiij d	exr.
per Cade	Item, to Iohn de Sodo, Potycarye	... vj li xiij s iiij d	exr.
per Knot	Item, to Iohn Emyngeway, yoman potycarye	lvs vijd ob.	exr.
per Cade	* {Item, to Thomas Vicary, S*u*rg*eo*n	... ...	C s exr.
	{Item, to Iohn Aylif, Surgeon	... ... vij li x s	exr
per Knot	Item, to Thomas Alsop, potycary	... vj li xiij s iiij d	exr.
[lf 43, bk]	Item, to Richard Ferres, Surgeon	... ...	C s exr.
Per Knot	{Item, to Nicho*l*as Alcoke, S*u*rgeon ...	... ...	l s exr.
	{Item, to George Hollande, S*u*rgeon ...	... ...	l s exr
	{Item, to Thomas Gemynous, S*u*rgoun[4]	...	l s exr.

[1] Cade and Knot were the men who took the fees for, or handed them to, the Officers. 'ex'' means 'examinatur,' when the account was checkt.
[2] I leave out here Nicholas Crasier, astronomer, C s, in all the entries.
[3] On leaf 27, back, are
per Cade Item, to Anthony Totto, Painter ... ... vj li v s exr.
Item, to Barthilmewe Penne, Painter ... ... vj li v s exr.
Item, to Misteris levyn Terling, Paintrix ... ... xli exr.
[4] Near the foot of the page is "Item, to Sir Thomas Paston, knight, for keping of the long gallery at Grenw*i*ch xvj li xiij s iiij d exr." On page 44, the

118 App. II. *Edward VI's Payments to Vicary, &c.*

[leaf 62] Yet Quarters Wages for Christem*a*s, anno Regni Regis Edwardi sexti Primo [A.D. 1547].

	Item, to Docter Benteley, Phisic*i*on	x ti ᵉˣʳ·
	Item, to Docter Huicke, Phisic*i*on	1 s ᵉˣʳ·
per Knot	Item, to Cornelis zifridus, Docter of Fisike wit*h* the Ladye Anne of Cleves	xj ti xiij s iiij d ᵉˣʳ·
	Item, to Io*h*n de Sodo, Potycarye ...	vj ti xiij s iiij d ᵉˣʳ·
per Cade	Item, to Thomas Alsop, Potycarye ...	vj ti iij s iiij d ᵉˣʳ·
Per Knot	Item, to Io*h*n Emyngwey, yoman Potycary	lv s vij d ᵉˣʳ·
*	Item, to Thom*a*s Vycary, Surgeon	C s ᵉˣʳ·
per Cade	Item, to Io*h*n Aylif, Surgeon	vij ti x s ᵉˣʳ·
	Item, to Richard Ferres, Surgeon	C s ᵉˣʳ·
per Knot	Item, to Nicho*l*as Alcocke, Surgeon	1 s ᵉˣʳ·
per Cade	Item, to George Hollande, Surgeon	1 s ᵉˣʳ·
	Item, to Thom*a*s Gemynous, Surgeon ...	1 s ᵉˣʳ·
per Knot	Item, to Henry Forest, Surgeon	x ti ᵉˣʳ·
	Item, to Henry Makereth, Surgeon	x ti ᵉˣʳ·

(On leaf 62, back, the painters Anthony Totto and Barthilmewe Penne, get their £6 5s. each, and 'Misteris Levyn Terling, paintrixe' her £10. And the MS. ends.)

Vicary's Annuity of £20 for the half-year, Sept. 1552, to March 1553; in Nov. 1552, and in Jan.—July 1554.

In 'The Boke of the Copies of the Certyficat made to the Kinges Ma*iesties* Counsell' from 19 Feb. 37 Hen. VIII [1546], to 2 and 3 Phil. and Mary [July 1555-6], given to the Record Office by Sir W. C. Trevelyan, Vicary's name occurs on p. 142, as entitled to his old wages of £20 a year. On p. 136 is the heading:—

Vlti*mo* Septembris, Anno E[dwar]di vj*ti* sexto (A.D. 1552).

The office of the Thresourer of y*e* Kinges m*a*iesties Chambre   The Declarac*i*on made the day and yere above written, by sir Willi*am* Cavendyshe knighte, Thresourer of y*e* kinges Maiest*ies* Chambre, To y*e* right reverende father in god, Thomas Bysshop of Norwyche, Sir Robert Bowys, and sir Walter Myldemay, knight*es*, and other y*e* king*es* m*a*iesties Comm*y*ssioners / of all the ordenary pay-

former payments are repeated to the painters, Anthony Totto and Barthilmewe Penne, & the paintrixe, Misteris Levyn Terlyn. On lf 44, bk, 'James Taillor, late son of the King*es* ol*d*e Waterman' gets 35 s. 5 d.

App. II. *Edward VI's Payments to Vicary, &c.* 119

mentes payable wit/iin his sayde office, for one hole yere ended at the Feaste of Ste Mychaell Tharchaungell, in the fyfte yere of ye Raigne of o*u*r saide soveraigne lorde [A.D. 1551]. The particularytes whereof more playnely hereafter is declared.

(page 142)
phesycons &
Astronomers

<u>mortuus</u>
{ Doctor Thomas Bylle, per *annum*, 1 li.
Doctor Huycke, x li. Cornelius Zefridus,[1]
xlvj li xiij s iiij d. Nich*o*las Crasyer, Astronomer, xx li }
Cxxvj li
xiij s
iiij d

Potycaryes
<u>mortuus</u>
{ Iohn de Zodoe, per a*nnum*, xxvj li xiij s iiij d.
Thomas Alsop, xxvj li xiij s iiij d. Iohn
Emyngwaye, xj li ij s vj d }
lxiiij li
ix s
ij d

Surgeons
* { Thomas Vicary, per *annum*, xx li. Iohn Aylyffe, xxx li. Richard Ferrers, lx li. Henry Forreste, xl li. George Hollande, x li. Thomas Gemynus x li }
Clxx li

There are earlier estimates of payments to Physicians, Apothecaries, Surgeons, grouped with other officers, on pages 95, 101, 109, 123; and on p. 150 (8 Nov. 6 Ed. VI, 1552) are the entries

To phisici*o*ns and Astronomers ... ... lxxvj li xiij s
To potecaryes ... ... xxxvij li xv s x d
To Surgeons ... ... Clxx li

for payments due for the year ending at Michaelmas, 6 Ed. VI, 1552.

On p. 153 is this heading (and on p. 159, Vicary's name):—

xx^mo^ die Novembris, Anno sexto R*e*gis Ed*w*ardi vj^ti^ (1552).

Here after is declared the names of all suche officers, men of Scyence, Artyficers, Craftismen, and other mynistres that arre payable wit/iin the saide office of Treasorer of the kinges ma*ie*sties most honorable Chambre, wit/i theire severall Feeze and wages, devidinge them in suche sorte as they, in theire severall romes doo serve or mynistre, wit/i the Bordewages, Ridinge Chardges, reparaci*o*ns, and other expences not certeyn), but as they happen; As also suche Anuytyes as are paid wit/iin the saide office, aswell to Inglysh men as to straungers, separatynge those *that* haue ye saide Anuyties Dueringe theire Lyves, from them *that* haue dueringe the king*e*s ma*ie*sties plesure, as by this declaraci*o*n hereafter followeinge shall appeare:

---

[1] 'Y^e^ lady Anne Cleves house' is written above his name.

120 App. II. *Edward VI's Payments to Vicary, &c.*

Officers & others mynisters	Phisicions & Astronomers	Doctore Huycke, phesicion, x li. Nycholas Crasyer, Artestronomer, xx li	xxx li
	Potecaryes	Thomas Alsop, potecarye, xxvj li xiij s iiij d. Iohn Emyngway, Potecarye, xj li ij s vj d	xxxvj li xv s x d
	Surgeons	*Thomas Vicarye, Surgeoun, xx li. Iohn Aylif, xxx li. Richarde Ferres, lx li. Henry Forreste, xl li. George Hollande, x li. Thomas Geminus, x li.[1]	CC li [170 £]

On p. 173 we find

| The Office of the Thresourer of the kinges Maiesties Chamber | A Brieff Abstract or an estymate what ys due within the Threasurers office of the chamber at the feaste of Midsomer, Anno vij^mo Regni Regis Edwardi sexti [28 Jan. to 6 July, 1553] |

| (p. 174) Ordynary paymentes payable quarterly and half-yearly | To phisicions and Artstronymers ... lxx li To potycaryes ... nihil quia solvuntur To Surgeons [sum right now] ... Clxx li[2] |

| (leaf 179) The Office of the Thresaurour of the Quenes Maiesties Chamber | Viij^mo Marcij, Anno primo Marie Regine [1554]. Brieffe Abstracte or estymate, what is due within the said offyce at the feaste of Thannuncyacion of our Blessed Lady the Virgen, next comynge [25 March 1554] |

| (lf 178) Ordinary paymentes paiable quarterly and half-yerelie | To phisicons and Artstronymers ... xxij li xv s To potycaries ... xxj li xiij s vj d ob. To Surgeons ... ... CClxvij li x s |

| (p. 202) Phesicion & Artstronomer | Doctor Huycke, phisicion, x li Nicholas Crasyer, Artstronymer, xx li | xxx li |

[1] On page 163, among the 'Annuyties of englishe men during plesure,' are Nycholas Backon, x li ; Nycholas Vdall, xiij li vj s viij d ; and among 'The lady Anne Cleves graces howsehold duringe plesure' is her doctor 'Cornelis Zifridus, xlvj li xiij s iiij d.'

[2] On p. 175, is a payment of £331 7s. 4d. to 'Sir gilbert Dethick, knight, Chester harrolde at Armes & rouge dragon pursyvaunt at armes, for their dyette and poste mony' (repeated on p. 179 and 190) ; and on p. 176, £160 'To the harroldes at armes, for their Dyettes in the progresse.'

App. II. *Q. Elizabeth's Payments of V.'s Annuity.* 121

potycaryes	{ Thomas Alsop, potycary, xxvj ti xiij s
	  iiij d. Johñ Emyngewaye, potycary, xj ti
	  ij s vj d
		} xxxvj ti
		  xv s
		  x d

Surgeons	* { Thomas Vycary, Surgeon, by yere, xx ti.
	  Johñ Aileffe, xxx ti. Richard Ferres, lx ti.
	  mortuus
	  Henry Forest, xl ti. George Hollande,
	  x ti. Thomas Gemynous, x ti¹
		} Clxx ti

Vicary's Annuity of £20. Its last payments in
1559-1561.

Book ⁷⁹⁄₁,² 1-3 Eliz., *Exchequer, Queen's Remembrancer, Ancient Miscellanea, Wardrobe and Household.* 30 Sept. 1559 (1 Eliz.) to 30 Sept. 1560 (2 Eliz.).

Computus Thome Weldon, Armigeri, Cofferarii, et Custodis Magne garderobe Hospicij Serenissime, invictissime principis, Domine nostre Elizabeth, Dei gracia, Anglie, Francie, et Hibernie Regine, Fidei Defensoris, &c., tum de omnibus et singulis Denariorum summis super expensis Hospicij predicti oneratis, quam de allocacionibus et solucionibus eorundem factis per vnum Annum Integrum, videlicet, ab vltimo Die Septembris Anno primo finiente, vsque primam Diem Octobris Anno iij° incipiente, prout in libro sequente plenius continetur.

In this Cofferer's Account, the Annuity of £20 "Thome Vicars" is on leaf 5 from the end. And in the Controller's Account for the same period, ⁷⁹⁄₂, Vicary's £20 is also on leaf 5 from its end.

The last payment of this 1535 annuity of £20 to Vicary is that of 1560-1.

In the Cofferer's (titleless) Account, ⁷⁹⁄₃, 1 Oct. 1560, to 30 Sept. 1561, Vicary's Annuity is on the back of leaf 7 from end; and in the Controller's Account (also titleless) for the same Period,—Oct. 1, 1560 to 30 Sept. 1561, 2-3 Eliz. ⁷⁹⁄₄, Ex. Q. Rem. Anc. Misc. Wardrobe and Household,—it is on the back of leaf 6 from end :—

Warranta Regine.
Thome Vicars, Capitali Chirurgico hospicii domine nostre Regine Elizabeth, in denariis ei solutis, virtute warranti Domini Regis Hen-

¹ Among the Annuities on p. 207 are Nicholas Backon x li, and Nicholas Vdall xiij li vij s vij d again ; and on p. 208, Doctor Cornelys has the m for *mortuus* over his name, though the sum xlvi li xiij s iiij d follows it.
² '79 upon 1' this seeming fraction is cald.

rici viij sancte memorie defuncti, Cuius datum est apud Bysshops Waltham, xx° die Septembris, Anno Regni dicti Domini Regis xxvij°, durante vita dicti Thome solvendis, pro vad*iis* et medicinis eiusdem Thome, per predictum warrantum Annuatim Concessis, prout in eodem warranto plenius Continetur, infra tempus huius Computi, xx ti //

The Book for 3 and 4 Eliz., Oct. 1, 1561 to Sept. 30, 1562 is unluckily missing; tho' in it we should hardly find the wonted *mortuus* when a payee died after the Account was made up, as Vicary must have died late in 1561, or early in 1562.

In the Account for 4 and 5 Eliz., Oct. 1, 1562, to 30 Sept. 1563, Vicary's name is of course not among the Annuitants on the back of leaf 6 from end, and on leaf 5 from end. His Will was proved on April 7, 1562.

III.

EXTRACTS FROM THE CITY OF LONDON REPERTORIES, JOURNALS, &c. AT THE GUILDHALL.

1. *Those relating to the Foundation of Bartholomew's, and to Vicary, and to his Governorship of the Hospital.*

The Act 37 Hen. VIII, ch. 28, past on Feb. 4, 1536, gave the King all the small Monasteries, &c. whose land was not worth above £200 a year. After this, the larger Monasteries, &c. were gradually surrenderd to him more or less voluntarily. The Act 31 Hen. VIII, ch. 13 (of the Parliament held, 28 April to 28 June, 1539), vested in the King the lands of all Monasteries, &c. theretofore[1] or thereafter dissolvd. The Priory and Hospital of St. Bartholomew's, &c. were surrenderd to Henry VIII on Oct. 25, 1539.[2] Foreknowing this, the City of London saw that it would be left without any houses for its poor, well or ill, and accordingly askt the King to give them some.

1539, Feb. 11. The City Petition to Henry VIII for the Hospitals, &c.[3]

(Repert. 10, lf. 79, bk.) *Martis, xj februarii, anno 30 H. 8.* (A.D. 1539).

Forman. [*Present*] M*ayor* [William Forman, haberdasher], Re-
[Mayor] *corder,* Waren) [Ralph, mercer], Gresham, Denham, Paget, Bowyer [draper], Laxton) [grocer], Tolos,[4] Sadler, Aleyn), Wylford ....

---

[1] 645 abbeys, 152 colleges, and 129 hospitals.—Toone.

[2] Dugdale does not say expressly when the Hospital was surrenderd to Henry; but as it was originally 'given to the neighbouring priory, and was in many things subject to it,' tho' it had a distinct estate (*Monast. Angl.* vol. vi, pt. 2, p. 626, col. 1), we assume that it past to the King with the surrender of the Priory by Robert Fuller on Oct. 25, 1539, 31 Hen. VIII.—*Monast. Angl.* vi. II. 291, col. 2.

[3] This first Petition to Henry VIII is (we find) printed from the City's Journal 14, leaf 129, as the Appendix No. I to the "Memoranda .. relating to *The Royal Hospitals,*" 1863, p. 1—4, and in the Charity Commission Report, No. 32, 1840, Part VI, p. 344.

[4] John Tholouse, sheriff in 1543.

124 App. III. 1. *City Petition for Bartholomew's, &c.*

London)
for y^e
Freres of
London

Item, that a suplicacion shalbe made, yn the name of the mayer & cominalty of london), to the kinges highnesse, for the iiij howses of fryers, that ys to say, Augustynes, blakke Freres, Grey Freres, & whyte Freres, & also for the iij hospitalles, that ys to say, saynt bartylmew yn smythfeld, saynt Mary hospytall without bysshoppesgate, & saynt Thomas spytell yn Suthwerk.

1539. (Repert. 10, lf. 81, bk.) Sabbati, 23 februarii, 30 H 8.

Forman
[Mayor]

[*Present*] Mayor, Recorder, Waren), Gresham, Denham, Dormer, Cotes, Dauncy, Bowyer, Laxton), Hamcottes, Tolos, Aleyn), Wylford ....

London

Item, the booke devysed for the iiij freres, whyte, blakke, grey & Augustynes, & also iij hospitalles—saynt Mary without bysshoppes gate, seynt Thomas yn Suthwerk, & seynt bartylmew spytell,—was Redde ; & agreed that my lorde mayer, master Waren), master Gressham, master Recorder, master Dormer, & master Rauf Aleyn), shall knowe[1] whyther the seyd booke shalbe exhybytted vnto the kynges highnesse by the Right honourable lorde privye seale / by my lorde Mayer / or by some other of the Cytye.

This Book or Petition sent to the King, is enterd in Journal 14, leaf 129, between an entry of 4 March, 1539, and another of 6 March, 1539, so that we may perhaps date the presentation of it, 5 March, 1539. It is printed in the *Memoranda* relating to *The Royal Hospitals* 1836, and its reprint of 1863, Appendix, p. 1, where its date is given as 1538, without any note of 'old style.'

1539. (Repert. 10, lf. 96, bk.) Jouis, xxiiij die Aprilis, anno 30 [*i. e.* 31][2] H. 8. [A.D. 1539].

Forman

[*Present*] Mayor, Waren, Gresham, Denham, Dormer [mercer], Paget, Cotes, [John, salter], Kytson), Bowyer, Dauncy, Laxton) [grocer], Heberthorn) [merchant-tailor], Bowes [goldsmith], Tolos, Sadler, Alen), Wylford ....

Freres

Item, that the kynges highnesse, & lorde privy seale, & other of the kinges most honourable counsayll, be moved for the iiij places of Freres.

For 5 years Henry did not move: see below. (The next 3 entries refer to Vicary, and not to Barts.)

[1] know or learn whether. The MS. is awkward. Dr. Reginald Sharpe kindly read it for us.
[2] The 30th year of Hen. VIII ends on 21 April 1539. Leaf 97 of the Repertory is rightly dated 26 April 'a° 31 H 8', that is, 1539.

App. III. 1. *Vicary's demand for a Felon's body.* 125

1540. Vicary and other Surgeons demand a Felon's dead body for Dissection.[1]

(Rep. 10, lf. 186) Adhuc Martis. 14. Decembris, Anno 32 H 8.
(A.D. 1540)

Roche [Mayor] Felons, & other that suffer deth by the lawes, hare not to be buryed by the Shreves of London

Item, yt ys Agreyd, Att the request & petycion of the right worshipfull Master Laxton & Master Bowes, nowe Shreves of this Citye of London, made vnto this Court for & concernyng the buryall of suche Felons As nowe be, & herafter shalbe, comyttyd or Atteynted of Felony, Murdre or treson within this Citye of London, or the Shere of Middlesex, that the bodyes of all suche persones, & namely[2] of them that shalbe nowe next putt in execucion of dethe att Tyburn, in the sayd Countye of Middlesex, shall eyther be buryed by the inhabitauntes of the Tounshipe of Padyngton, Or els the same ded bodyes to be suffred to hange there styll, &c./

For the delyuerye of A ded bodye by the Shreves to [Thos. VICARY &] the Surgeons, &c.

Item, Master Laxton & Master Bowes, Shreves of this Citye, prayed the Advyse of this howse for & concernyng the Delyuerye ouer of one of the dedde bodyes of the Felons of late condempned to dethe within this Citye, And requyred of the seyd Master Shreves by Master Vycary & other the Surgeons of this Citye for Annotamye, Accordyng to the fourme of An Acte of parlyament therof lately made / And Agreyd that the same Acte be first seen / & then Master Shreves to worke ther after, &c/.

---

24 March, 1542. Vicary (as Warden of the Surgeons) before the Common Council.

(Repertory 10, lf. 239) Martis 24 / 3 / Anno 33° H 8 / (A.D. 1542)   Dormer, *Mayor.*

[*Present*] Mayor [Sir Michael Dormer, mercer], Waren [mercer], Gresham, Denham, Cotes, Bowyer [draper], Dauntsey, Laxton, Bowes, Hamcotes [fishmonger], Tolos,[3] Sadler, Wylford, Lewen, & Judde [skinner] / . . . .

(lf. 240, bk.) Surgeons

Item, yt ys Agreyd that the Wardeyns of the Surgeons be warnyd to be here the next Court day, Aswell for the Stey of theyr sute in the Escheker Ageynst John Margetson, Bruer, As Also for & concernynge the certificat of the peryll & Jeopardye of Richard Pygott,

---

[1] Under the Statute, p. 205, below.   [2] Specially.
[3] John Tholouse, sheriff in 1543.

126 App. III. 1. *Vicary advises the Lord Mayor.*

Vyntener, to be made to my lorde Chaunceler; whyche Pygott was lately hurte & woundyd by one Thomas Eton), yoman, nowe beynge in warde within thys Cytye for the same.

1542. (Repertory 10, lf. 241) Jouis / 26 / 3 / Anno 33° H 8.
Dormer. [26 March, 1542]

[*Present*]   Mayor [Sir Michael Dormer], Recorder, Waren), Gresham, Forman), Cotes, Bowyer, Daunsye, Laxton), Bowes, Hamcotes, Tolos,[1] Sadler, Wylford, Lewen, Judde /
[Vicary]   Att thys Courte came Master Vycars, seriaunt of the
Pygrtt &   Surgeons, & declaryd to thys Courte, that As towchyng
Eton)   the certificat to be made by my lorde Mayer vnto my lorde Chaunceler, for the hurte done vnto one Rychard Pygott, Vintener, by one Thomas Eton), yoman) / that he wolde not advyse my seyd lorde mayer to make eny suche certificat as yett / for he doth sum-what doute of the Recouerye of the seyd Pygott; And that he wyll so declare & report hym) self vnto master Bryan), master vnto the seyd Eton) /

For 5 years after the above City Petition or Petitions of 1539 (p. 124), nothing was done by Henry in answer to them. Then he issued Letters Patent of 23 June 1544, creating a new Bartholomew's Hospital, a Corporation of a Master (a priest) and 4 Chaplains, to whom he gave the site, buildings, and church of the old Hospital of St. Bartholomew's the Less, and all its jewels, goods, and chattels, but without any other endowment. (The englishing of these Letters Patent of 23 June 1544 is printed as Appendix II to the *Royal Hospitals*, (1836, and) 1863, p. 4—7. The Patent itself is in the Patent Rolls of 36 Hen. VIII, part 2, membrane 41.

The City of course wanted its Hospitals endowd, in part at least. On Nov. 23, 1545, Parliament met, and by the Act 37 Hen. VIII, ch. 4, confirmd all Surrenders of Monasteries, &c. made to the King, set aside all fraudulent and other grants, leases, &c. of Monastery lands, and empowerd his Commissioners to enter and seize such lands. In Dec. 1545, the City appointed a Poor-Relief Committee. In 1546 they agreed to endow the Hospitals jointly with the King. In 1547, they got the work well under way; and in 1548 appointed their first Surgeon-Governor of Barts, Thomas Vicary, who soon became Resident Governor, and (practically) Chief Surgeon.

App. III. 1. *A City Poor-Relief Committee.* 127

1545. Appointment of a Hospital-Committee, or Governors, for the Relief of the Poor: 10 Dec. A.D. 1545.[1]

(Journal 15, leaf 213.) Bowes *Maiore.*

Common Council of 10 Dec. 1545.

Com*une* Consilium tent*um* decimo Die Decembr*is* Anno regni Reg*is* Henr*ici* viij^{ui} xxxvij^{mo}, coram Martino Bowes, Mil*ite*, Maiore Ciui*tatis* London*ie*, Roberto Broke armige*ro*, Record*atore* eiusdem Ciui*tatis*, Rad*ulpho* Waren, Mil*ite*, Ricardo Gresham, Mil*ite*, Joh*anne* Cotes, Will*elmo* Laxton, militib*us*, Henr*ico* Hoberthorn, Joh*anne* Tolos, Joh*anne* Gresham mil*ite*, Joh*anne* Wylford, Rolando Hyll, Mil*ite*, Thoma Lewyn, Andrea Judd, Ricardo Dobbes, Ricardo Jerves, Thoma White, Roberto Chertesey, Will*elmo* Lok; & Georgio Barne & Rad*ulpho* Aleyn vicecomites[2] / ac maiore p*ar*te Comm*un*iari*orum* de commun*i* consilio Ciui*tatis* pred*ic*te exist*entis* &c./ ....

[leaf 213, back]
Provysyon
for the
Releif of
the poore

Item, Thom*as* Barthelett, Stacyone*r*, John Wyseman, Skynne*r*, Humfrey Pakyngton mercer, Thom*as* Bacon, Salter, John Royce, m*er*cer, Will*ia*m Garrett, haberdash*er*, Stevyn Kyrton mercha*un*nt*tailor*, And Augustyn Hynde, ar' this day no*m*inatted by the said hole Common*e* counsell here assemblyd, to ioyne w*i*t*h* my lorde Maire and suche iiij of his worshipfull brethern, thalderm*en*, as his lordship*e* and his said brethern, thaldermen shall therunto name & apoynt, for the inuentyng & devysyng of som*m* good, charitable, & godly wayes & meanes, wherby the very pore, indigent, syk*e* & wek*e* p*er*sons of this Cittie, not able to lyve of themselff*es*[3] may charitably be ayded, comforted, & releyvyd, by the deuocyon and charitable Almes of the good & well disposed Citizens & inh*a*bitaunt*es* of the same Cittie, in suche wyse that they or eny of theym shalnot haue eny iuste cause or nede hereafte*r* to begge or aske eny Almes openly, either in churches or elleswhere within the said Cittie, as they now vse to do ///

---

[1] They continued to act till Vicary's appointment on Sept. 29, 1548, and then some retired. See p. 132 below.
[2] sheriffs.
[3] Though these words are general, yet the next entry below shows that the present provision was meant mainly for St. Bartholomew's, or the House of the Poor in West Smithfield.

128   App. III. 1.   *The City agrees to endow Barts.*

1546. Acceptance of Henry VIII's Offer of the Hospitals and 500 Marks a year, on the City finding another yearly 500 Marks (13 April 1546).

(Journal 15, leaf 244.)   Bowes [Mayor].

Common Council of 13 April 1546.

Comm*u*ne Consilium tent*um* xiij° die Ap*ri*l*is*, Anno R*eg*ni Regis Henrici viij*ui* xxxvij° [A.D. 1546], coram Martino Bowes, Mil*ite*, Maiore Ciui*tatis* Londoni*e*, Rad*u*lpho Warcñ [*rest blank*]

(leaf 245.)   Bowes Ma*iore*

London for the poore

A*s* Henry VIII gave the City some P*o*orhouses,

and endowd them with 500 marks a year,

on condition that the City gave 500 marks more,

We enact

that the City shall covenant to pay this fresh yearly 500 marks.

¹Item, forasmoche as it hath pleased the Kyng*es* high̅nes, of late, of his most vertuous & godly disposicio*u*n, not only frely to gyve & gra*u*nte to this Cittie certeyne convenyent plac*es* for the Receyte, comforte & lodgyng of the pore people of the said Cittie / but also to in̅dowe the same plac*es* towar*des* the mayntena*u*nce & Releif of the said poore people with lond*es* & tene̅ment*es* to the clere yerely value of D. m*er*k*es*, v*ppon)* condicio*u*n that the Citizens of the said Cittie wylbe bounden) yerely forcu*er* to gyve other D. m*er*k*es* to the said vse & intent / It ys therfore enacted, clerely assentyd & agreyd, by the said com*en* Counseⱶⱶ, & by thauctoryte of the same, That the said Citizens & their' Successo*ur*s, by their' Wrytyng sufficient in lawe, vnder their' comm*en)* seale, shalbe bounden) for the yerely payment of the said som) of D m*er*k*es* to the vse aforesaid accordyngly, &c /

The long Deed of Covenant made (in pursuance of the Resolution above) between Henry VIII and the Mayor, Commonalty and Citizens of London, respecting the Hospitals, and bearing date the 27 Dec. 38 Hen. VIII, A.D. 1546, is printed in the *Royal Hospitals* (1836), Appendix IV, p. 8—21 (1863, App. IV, p. 8—19), and is abstracted in the Charity Commission Report, No. 32, 1840, Pt. VI.

1546.   Ma*rtis*, quinto die Octobr*is*, A*n*no xxxviij° H. 8.

(Repertory 11, lf. 310, ink, bk., 287 pencil, bk.)

Bowes [Mayor]

[*Present*] Mayor, Recor*der*, Roche, Forman), Cotes, Laxton), Wylford', Judde, Dubbys, Hyⱶⱶ, Barne, Chertsey, Lok, Hynde, Turke ; Ac Jervys, vnus vice*comes* /

ᴸ This is (we find) printed also in the *Royal Hospitals*, ed. 1863, Appendix III, p. 8.

App. III. 1. *The Preparation of Barts Hospital.* 129

[leaf 311 or 288]   Item, this day my lorde Mayer, for the very good love
The newe   that he baryth to this Cytie, Att the hartye desyer of
Condytes &   the hole court here, dyd Agree & graunted / to take
Hospytall   payne wyth such othere of my Maistres the Aldermen)
for the pore.   & Comeners As beyn) Alredy Apoyntyd, & with
   Maister Sturgeon), haberdasher, both to Fynyssh the
   Newe Condytes, & also Aboute the creccion & con-
The completion of   sumacion of the newe hospytall in Smythfeld for the
Barts.   pore, Aswell after the tyme of his Maryalte, As he
   hath hytherto done.

---

5 Oct. 1546.   The City not in complete possession of Bartholomew's.

(Repertory 11, lf. 310 ink, bk., or 287 pencil, bk.) Martis, quinto die Octobris, Anno xxxviij° H. 8 / (A.D. 1546).

(lf. 311, or 288   Item, the letters of the ryght honourable lorde privye
pencil)   Seale & other, dyrectyd to this Court, in the Fauour of
Paladye   Rychard Paladye for the Stuardshipe of lytle seynt
   Barthilmewes in Smythfeld were red : And therupon)
   Aunswere made hym), that when) the Cytie shalbe perfytly
   in possessyon) of the seyd howse, they wyll make hym) a
   further Aunswer therin) /

Then come the very long second Letters Patent of Henry VIII, 13 Jan. 1547, containing the endowd Grant and Establishment of Bartholomew's and the other Hospitals, turning the churches and parishes of St. Nicholas and St. Ewin's into the new parish of the church of *Christ* within Newgate, &c., printed as Appendix V in the *Royal Hospitals*, 1836, p. 22—49 ; 1863, p. 20—45, and abstracted in the Charity Commission Report, No. 32, 1840, Pt. VI.

---

26 April, 1547.   Henry VIII's Letters Patent for Bartholomew's brought into the City Court.

(Rep. 11, lf. 345, bk.) Martis, xxvj*to* die Aprilis, Anno primo Edwardi vj*ti* [A.D. 1547].

(leaf 346, ink;   Item, sir Martyn) Bowes, Knyght, brought in this
322 pencil)   day in-to the Court here, the lettres patentes of our
Hoberthorn)   late soueraygne lorde, kynge Henry the viij^th, of the
Mayor.   foundacion & newe ereccion of the hospytall in
The kynges   Smythfelde, & of Crystchurche wythin) Newgate ;
letters pa-   whiche lettres were forwyth Delyuered ouer to the
tentes of   sauffe Custody of Master Chamberleyn) / And Agreyd
Thospytall in   that the seyd Master Bowes shalbe truely recom-
Smythfeld   pensed, wyth thankes, of & for All suche money As
VICARY.   K

130 App. III. 1. *Henry VIII's Endowment-Deed.*

he hath dysbursed Aboute the pryses of the seyd lettres & othere the affayers of this Cytie; And Further, that there shalbe An especiall Court here holden vpon Fryday comme sevyn nyght, for the herynge & perusynge of the seyd letters patentes.

6 May, 1547. The Hospital-Indenture of Henry VIII and the City, brought in. Its provision as to the Beadles' pay varied.

(Repertory 11, lf. 349, bk., ink; 325, bk., pencil) Veneris, vj^{to} die Maij, Anno primo Edwardi vj^{ti} [A.D. 1547].

Hob[er]thorn [Mayor]	[*Present*] Mayor, Recorder, Waren, Laxton, Bowes, Tolos, Wylford, Judde, Dobbys, Barne, White, Hynde, Lyon; ac Jervys, vicecomes (Sheriff). . . .
London: the late grey Fryers & lytle seynt Bartholomewes	Item, this day the indenture made bytwene our late soueraygne lorde, kyng Henry the viijth & the Mayer & Cominalty & Cytezeins of this Cytie, of & for the howse of the late grey Fryers & the hospytall of lytle seynt Bartholomewes, was red; And Agreyd that sir Martyn Bowes, knyght, & thother Aldermen & Cominers hertofore Apoynted to travayll therin, shall take the paynes to abridge¹ both the yerely revenues & profyttes of the seyd howse & hospytall, & also the yerelye charges apoynted to be borne out
In[tratur]	of the same, & to make reporte therof to this Court with As convenyent spede as they can in wryting.
Bedylles for the pore.	Item, yt is orderyd & Agreyd, that euery of the viij bedylles that be apoynted to Attende vpon the house of the pore, & the syke & impotent people therof,
Camerarius.	shall yerely haue of the Chamber of this Cytie, in lieu, stede, & recompence of there v markes whiche they are apoynted to haue yerely by the Indenture concernyng the fundacion of the seyd howse for the
Beadles to have 4 Nobles a year, a Livery Gown, and standing for 1 Car.	pore / iiij nobles in redy money, one lyuerye gowne, & one Carre rome² to be occupyed with-in the seyd Cytie & the lybertyes therof by their deputyes or assignes / wyth as moche lybertye as eny other person or persones doth enioye the lyke rome, duryng the tyme that they shall contynue in their seyd romes & offyces.

Next is the Grant by the Common Council, on 29 Sept. 1547, of one half of a Fifteenth on the Citizens and Inhabitants of the

---

¹ Make a list or short statement, abridgement, of them.
² Room for the standing of a Car or Cart.

City towards the Maintenance of the Poor in St. Bartholomew's, with power to raise this tax or levy by distress. This is printed from the City's *Journal* 15, leaf 325 back, in Appendix VI to *The Royal Hospitals* (1836), p. 49—50; 1863, p. 45—6.

3 Nov., 1547. (Journal 15, leaf 317.) Huberthorne, Ma*iore*. Tercio die Nouembris, A*nn*o pr*i*mo Edwardi vj^{ti} (A.D. 1547) [entry of a Bond; then on the back].

[leaf 317, back]

The dispo*si*c*i*on & bestowinge of seynt Nich*ol*as churche and seynt Ewyns commytted to the lorde Mayere & other /

the Governors of Bartholomew's,

who may manage, sell, or let these Churches

and their sites,

which Henry VIII

in 1547

gave the City,

for the Hospital poor.

It*e*m, Att this c*om*en counsell yt ys ordeynyd, enactyd, Assentyd and Agreed by the Auctorytye of the same c*om*en counsell, that the lord Mayer and Aldermen) of this citye that now are, or the more of theym, wit*h* the Advice & consent of suche Alldermen) and com*en*ers of the seyd citye as are hertofore, that ys to sey, at and by the c*om*en counsell here holden) the xth daye of December, An*n*o 37 Hen*ri*ci .8. [A.D. 1545] Assygned and Appoyntyd to be of counseyll wit*h* the pore w*i*t*h*in the hospitall of the pore lately foundyd and establysshyd in west smythffeld in the s*u*burb*es* of the seyd citye by o*u*r late most redowtyd soue*re*ygn) lorde, Kinge Henrye the viijth, and S*u*rveyo*u*rs of the revenues of the same hospytall, shall fullye and hoolye haue the orderynge, bestowinge, sellinge, dymysyng, or otherweyse by their good and sage wysdomes and discrecc*i*ons, bothe of the late p*a*rishe churches of seynt Nicolas in the shambles, and of seynt Ewyns wit*h*in Newgate of the same cytye, And allso of the Sight*es* or Soyles wheruppon) the same ij churches Are nowe sett and buyldyd; w*hi*ch ij churches, wit*h* All the londes and Ten*n*ementes to theym and either of theym belongynge, wit*h* all their Appurten*a*unces, o*u*r seyd late soue*r*eigne lorde Kynge Henrye the viijth, by his most gracyouse le*tt*res patentes berynge date the [thirteenth] daye of [January] in the [thirtyeighth] yere of hys most noble reigne, [A.D. 1547] Amonge diuerse and menye other londes, tene*mentes* and possessio*u*ns, gave and gr*a*untyd to the Mayer, Co*min*altye and Citezens of the seyd cytye and to theyr successours, for the charytable Ayed and Relyff of the pore wit*h*-in the seyd hospitall for the tyme beinge, and for the Maynteyn*a*unce of dyue*r*se other godlye vses and intent*es* wit*h*in his highnes seyd le*tt*res patentes menc*i*oned and expressyd /

132 App. III. 1. *Barts Governors of* 1545 *continued.*

[*Continuance to* 1548, *and Future Election, of the Hospital Governors of Dec.* 10, 1545.]

Counseyllers & Surveyours of the pore in the hospitall of The Pore // and of the revenues of the same.

The old ones shall continue for 1 year.

After that, 2 Aldermen and 4 Commoners shall be chosen yearly,

to act with the Lord Mayor as Aiders of the Hospital-poor, and Surveyors of Revenue;

all working gratis.

And yt ys allso enactyd and Agreed by the seyd Auctorytye, that the lorde Mayere of this cytye for the tyme beynge, and those Alldermen and commyners that were Assygned and Appoyntyd at and by the seyd comen counseyH holden the seyd x^th daye of December in the xxxvij^th yere of the reigne of o*u*r seyd late sou*e*reygn) lorde Kynge Henrye the viij^th [A.D. 1545][1] to be Ayders and of counseyH for the pore wit*h*-in the seyd hospitaH, & S*u*rveyours of the revenues of the same / shall, for and by the space of one hole yere now next ensuynge, stond, remayne and contynew in the same their rome and office /
And that frome thensforthe there shall yerlye be newlye electt and chosen by the co*m*en counseiH of the seyd cytye, ij Alderme[n] and iiij Comme*n*ers of the same cytye / w*h*ich, wit*h* the lorde Mayer of the seyd cytye for the tyme beynge, shaH Allweyes duelye, iustlye, and dylygently, vse, execute and excercyse the seyd rome and office of Ayders & counseyllers of and for the pore wit*h*iñ the seyd HospitaH for the tyme beynge, and be Surveyo*u*rs of the Revenues of the same for ever, all Franklye & frelye wit*h*owt anye mane*r* of thinge or thinges claymynge or demaundyng for eny their labours or paynes by theym, or anye of theym, at anye tyme herafte*r* to be takyn or susteynyd by reason of the excercysyng, vsyng & excecu*t*ion of the same theyr seyd office and rome /

---

15 Nov. 1547. City Committee on the Bill in Parliament for St. Bartholomew's Property.

(Repert. 11, lf. 389, ink ; 365, pencil. 15 Nov. 1 Edw. VI)

London) pro terris pauperum

It*e*m, it is agreyd that M*ais*te*r* Crayforde, M*ais*ter Atkyns, & M*ais*ter Goodyng, shall pe*r*use advysydly the draught of A certein) Boke devysed to passe by Acte of p*a*rlyament, bytwen) the kyng*es* maiestie & this Cytie for the assurau*n*ce of suche Land*es* as were geven) by o*ur* late soue*r*aygn) Lorde, kyng Henrye the viij^th to the hospytaH of the poore /

[1] p. 127, above.

App. III. 1. *New Governors of the new Barts.* 133

1548. First Appointment of Vicary as a Governor of St. Bartholomew's, 29 Sept. 1548.

[*Journal* 15, *leaf* 383, *back.*]

Common Council of 29 Sept. 1548.

Commune Concilium tentum
Die Sabbati, Videlicet, xxix° die Septembris, Ac in festo Sancti Michaelis Archiepiscopi / Anno regni regis Edwardi, dei gracia, sexti, &c., Secundo / Coram / Johanne Gressham milite, Maiore Ciuitatis Londonie / Roberto Brooke armigero, Recordatore / Willelmo Laxton), Martino Bowes / Militibus / Henrico Amcotes, Johanne Wylford / Andrea Judd, Georgio Barnes, Roulando Hill milite / Ricardo Dobbes, Willelmo Lock / Augustino Hynde / Ricardo Turke / Thoma Whyte / Roberto Chartesey / Johanne Lyon) / Johanne Lambard / Willelmo Garrad aldermanis, Ac Willelmo Lock / et Johanne Ayliff, tunc Vicecomitibus[1] / Ac Maiore parte Communiorum Communis concilij Ciuitatis predictæ existentis // . . . . .

[*leaf* 384] J. Gresham, Maior.

Gubernatores domus pauperium in West Smithefelde, et possessionum eiusdem.

In order that the Rules made by the City Managers of the House of the Poor in West Smithfield (St. Bartholomew's), may be duly kept,

The Common Council order

1. that 4 Aldermen and 8 head Commoners of the City shall rule and manage the said House

Item, to thentent that suche good and necessarye ordres, rules, And constitucions as hytherto (with gret Industrye, studye, and paynes) haue beyne devysyd, made, and sett furthe, by suche wurshippfful Aldermen) and commoners of this Cytye as haue hadd the surveye, rule, and gouernaunce of the house of the poore in westsmithefeld, in the suburbes of the seyd Cytye, (for the obseruacion or mayntenaunce and contynuaunce of good and godlye rule, order, and lyvinge within the soyd house, and for the gouernaunce and preseruacion of the same house, and of the landes and Tenementtes, renttes, revenues, goodes, and catalles, therunto belonginge,) maye allways from hensfurthe be dulye, iustlye, and fyrmelye obseruyd and kept, and putt in due execucion / with-out the which, all lawes and ordenaunces, be they neuer so good, ar butt baryn, ded, and vayne / yt ys therfore ordeynyd, enactyd, & establyshyd by the lorde Mayer, Aldermen), and commens of this present comen Counsayll Assemblyd, And by the Authorytye of the same, that four Aldermen) of this cytye for the tyme beinge, and viij of the hed Cominers of the same, shall Alweis From Hensfurthe for ever haue the Surveye, rule, order, and gouernaunce of the seyd house,

[1] Sheriffs.

134   App. III. 1. *Vicary a Governor of Barts.*

and its property;

2. that the Lord Mayor and Aldermen shall appoint them;

3. that they shall act as Governors for 2 years;

half of them retiring at every Michaelmas,

but electing 2 fresh Aldermen and 4 fresh Commoners, to take the places of the retiring Governors.

4. That for next year the 6 old Governors who've drawn up the Hospital Rules,

[* leaf 384, back]

shall continue in office;

and that 6 new Governors,— THOMAS VICARS (or VICARY), one of them—shall join the 6 old ones.

And of all the londes and Tenementtes / rentes, reuenues, goodes and cattalls nowe belonging, or that herafter shall belonge, or in enye wise Apperteyn) to the same / And that the Lord Mayer and Aldermen) of the seyd Cytye for the tyme beinge, shall Alweyes have Full powre and Authorytye to nominat, elect, and Appoynt the seyd foure Aldermen) and viij Cominers from) tyme to tyme, when and as often as to theym shall Seame mete and expedient / And that all the foure Aldermen) and viij cominers so elect, nominatyd, and apoynted, shall Allweys stonde, remayne, and contynue in the seyd rome and office by the space of ij hole yeres ; and by all the same tyme shall diligentlye indeuoyr theym selfes, and euerye of theym, as they maye conuenyentlye Attende to the due execucion and excersyse of the seyd rome and Office / And that the seyd lorde Mayer And Alldermen) for the tyme beinge, shall yerly alweyes at the Feast of seynt Michaell tharchangell, or within xiiij dayes next before the same Fest, remove and clerlye dyscharge from the seyd rome and office, suche ij of the seid foure Aldermen), and suche iiij of the seyd viij cominers as then shall have stondyn and contynued in the same office or rome by the space of ij hole yeres. And in their stedes and places, Then newlye to elect, nominate, and Appoynte other ij Aldermen) and iiij comminers to be associate with ye other ij Aldermen) & iiij comyners, which then shall remayne and stond still in the seyd office for one other hole yere then next ensuyne, for that th[ere] they have then excersisyd and executyd the seyd rome and office but by the space of one yere / And For the parfyte establyshementt and confirmacion of this presentt Acte, the nominacion, election), and Appoyntment of master Austyn Hynde and Master William Garrard, Aldermen) ; William Rawlins and Thomas Lodge, grocers ; Thomas Berthelet, stacioner ; and Thomas Bacon), Salter, whoe *hertofore (with other) haue taken) gret and manyfold paynes and labour in the devisinge and makinge of the seyd ordres And constitucions, and in executinge and diligent excersyse of the seid rome and office, yet to remayne and contynue one hole yere longer in the seyd office ; and the nominacion, eleccion, and newe Appoyntment of master Willsforde and master Dobbes, Aldermen) ; and Thomas Vicars, barbour Surgeon ; William Chester, Draper ; William Clarke, skinner ; & Stephen) Cobb, Haberdasher, to Joyne and be Associate with the seyd master Hynd

App. III. 1. *Barts 500 Marks a year to be paid.* 135

and other Aforenamyd, by all the tyme Afore rehersyd, made by the seyd Lord Mayer and Aldermen) at this present, ys lovinglye ratyfyed, Approvyd, and confyrmyd by this hole court of commen Counsayle, And by the Authorytye of the same.[1]

On Dec. 20, 1548, the Common Council past an Act, ordering the payment of 500 Marks a year to St. Bartholomew's or 'the House of the Poore in Westsmythfeld,' and assessing the 59 City Companies to the same, in the several sums set after their names at the end of the Act; the Barber-Surgeons being down for £5 6s. 8d. This Act is printed as Appendix VII in the *Royal Hospitals* (1836), p. 51-6; ed. 1862, p. 46-51. It was enforced by the Precept of 22 Dec. 1548 printed below.

Precept of 22 Dec. 1548, to each City Company, bidding it comply with the Act of Common Council, 20 Dec. 1548, assessing each Company with its Proportion of the Bartholomew's Hospital 500 Marks a year.[2]

Amcotes Maior (Journal 15, leaf 401[3]).

By the Maire.

A precept directyd to the Craftes of this citie of London), for payment of their Sessment vnto the poore //
[of St. Bartholomews].

For-Asmuche As yt was lovyngly grauntyd, Enactyd[4] & Aggreyd by Aucthoritie of A Comon) Councelt holden) at the Guildhall of the sayd Citie the .xxᵗⁱ. daye of this present moneth of December, that your Company shold yerely gyve & paye towardes the Sustentacion, Releif & comfort of the poore people within the house of the poore lately fownded in West Smythfeld in the Suburbes of the same Citie, for the tyme beyng .Nˡⁱ. of good & lawfull money of England, to be payd yerely at the iiijᵒʳ vsuall termes of the yere (that ys to saye) At the Feastes of the Birth of our lorde god, Thannunciacion of our Lady, The natyvytie of

---

[1] 'This act of Common Council is referred to in Mr. Firth's *Memoranda* as existing in *Liber Legum*: as will be seen from the reference, it has been found in the Journal (15, leaf 384), and is now printed at length.'—Note to the first print of this Act in the 'Supplement to the *Memoranda relating to the Royal Hospitals*' (1867), p. 1. This first print has a few mistakes, which we have set right by the MS. Journal 15.
[2] This Act is printed in the *Royal Hospitals* (2nd ed. 1862), p. 46-51.
[3] At the top of the leaf is a Precept to the Wardmote Inquests to make a return of all the aged, impotent, and lame folk in their respective Wards who live by begging.
[4] Every final *d* has a curl to it, as others have generally in the City MS Books we have used.

136 App. III. 1. *Power to vary the Barts Regulations.*

Saynt John Baptist, & St. Michell tharchaungell, by even) porcions, The First payment therof to begyn) at the said feaste of the birthe of our lorde god next Commyng / We therfore straytly charge & commaund you / that ye, Immediately vpon) the Recepte hereof, Cause suche Taxacion & order to be taken) emonges your sayd Company, that ye fayle not to make redy payment of .O.ⁱⁱ, parcell of the said .Nⁱⁱ. now payable at the said feaste of the birth of our lord god next commyng, to the gouernours of the said house of the poore, or to their Sufficient Deputie / And so from) hensforth quarterly, vntyll other order shalbe taken) for the dischargyng therof, Accordyng to the te[rmes ?] of the said Acte / As ye will Answer at your perill / Dated at the [*MS. torn*¹] aforesayd, the .xxijᵗⁱ. daye of December in the .ijᵈᵉ. yere of the R[eigne] of our Soueraign) Lorde, Kynge Edward the vjᵗᵉ. [A.D. 1548].

Blackwell [Town Clerk].

---

1 Aug. 1549 (3 Edw. VI.). Lord Southampton directs a License to be drawn up for the City to vary Henry VIII's Regulations for Bartholomew's (Journal 16, leaf 26, back).²

<small>As the City say that the Royal Regulations for Barts give too large fees to superfluous Officers,</small>

After our hartye Commendacions / Havinge given vs to vnderstande, by the Maior and comynaltie of the Citie of London, that the foundation of the hospitall of St. Bartylmewes in weste Smythfelde, cannot in all pointes so be obserued, as was mente by the foundation thereof, by cause moost of the Revenewe to the same Assigned, is consumed in feez and wageez to stipendarye preestes and other superfluous officers / the whiche abuses can in no wise be reformed, but onely by aucthoritie or dispensation from the kinges Maiestie /

<small>We tell you to draw a License from K. Edw. VI authorising the City to vary those Regulations,</small>

Theis shalbe therfore to will and requyre yow (after full knowledge had of the foundation and state thereof) to drawe a booke of Lysaunce from his Maiestie, to the Maior and Auldremēn of the same Cytie, aucthorisinge them by the same, to transpose, alter, and chaunge the said nomber of preestes, and all other offices & thinges whiche shalbe thought by them not necessarye for the mynisterie of the said hospitall, vnto some other kynde of mynysters or vses, as to them shall be thought more

---

¹ 'Guildehall' one would expect, but the two letters shown look like vn.
² This is also in Letter-Book R, leaf 26.

App. III. 1. *The Barts Orders and Surgeon.* 137

meter and convenient for the better sustentation and for the better help of the Poor.
comforte of the diseased and impotent persons within
the said hospitall; and that the same be sent hether to
vs, warraunted withe your handes, forseing alwaies that
the kinges Maiestie susteine no losse by the same, and
also that thei contynewe charged, aswell withe the
nombre of the poore, as the fyve hundreth markes But the City's
yerely, the whiche thei be now bounde to dispende, for yearly 500 Marks is still to be paid.
the sustentacion of the said poore people / and So byd
you fare well: from westminster the first of august,
Anno 1549.

E. Somersett.[1] Your Louinge frende
R. Riche, Councell. Thomas Southampton.

In 1552 the City's Order for Regulations of St. Bartholomew's were printed, and are reprinted (with sidenotes) at the end of this Appendix. In 1557, a revision of these, 'The Order / Of the / Hospitalls of K. Henry / the viij th and K. Ed- / ward the vi th, / viz. / St. Bartholomew's. / Christ's. / Bridewell. / St. Thomas's. / By the Maior, Cominaltie, and Ci- / tizens of London, Governours of / the Possessions, Revenues and / Goods of the sayd Hospitalls. / 1557. /' were printed; and are reprinted as Appendix XIII to the *Royal Hospitals* (1836), p. 83—107; ed. 1862, p. 77—100. They are also in the Charity Commission Report 32, Part VI, 1840.

1557. The Bartholomew's Surgeon & Orders.

(Repertory 13, No. 2, lf. 506) Adhuc Jovis, tercio decimo die maij, Annis tercio & quarto &c. [Philippi & Marie, A.D. 1557].

Surgeons. Item, it was agryed that the Wardens of the Surgeons[2]
(The Wardens not to interfere with the Surgeon of Bartholomew's.) shalbe warnyd to be heare the nexte Courte day to shewe cause why they go aboute to interrupte the Surgeon of the howse of the pore[3] to practyse those thinges that he dothe lawfully meddle withall.

---

[1] The Protector.
[2] We suppose that these were the representatives of the Fellowship of Surgeons, not more than 12 in number. See the Statutes, and Supplement to them, below. But see the Order of 24 March 1542, above, when 'the Wardeyns of the Surgeons' had to appear, and Vicary does so. Yet he cannot have interrupted the Surgeon of Barts. He was Resident Governor, and must have always been practically Chief Surgeon of the Hospital; and in Jan. 1552 was made Governor for life.
[3] We assume that this 'howse of the pore' means Barts, and not St. Thomas's or Bridewell. In the Minute (Sept. 2. 1589) of the sealing of a Lease of a House and Shops in Ship Alley, Little Wood Street, belonging to

138 App. III. 1. *Vicary's Governorship of Barts.*

1557. (Repertory 13, No. 2, lf. 545) Adhuc Martis 28 Septembris, Annis 4 & 5 [Philippi et Marie, A.D. 1557]. Offley Maiore.

| Ordenaunces concernynge the gouernaunce of the Cytyes Hospytalles. | Item, it was agryed that all the seuerall artycles and ordynaunces hereafter mencionyde and expressyd, and openly red to the Corte here this day,[1] concernynge the Gouernaunce and orderynge from hensefurthe of the howse of the pore in weste Smythefeld and the hospytalles of this Cyty, lately devysyd by Sir martyn Bowes and Sir Rowland Hyll, knightis, and dyuers other of my Masters, thaldermen, and the Comyners of this Cyty (beynge governors and surveyors at this present of the sayde howses, and of all the landes and other Revenues of the same, what so euer), shulbe here enteryd of Recorde, and Frome hensfurthe be put in due execucion from tyme to tyme, accordynge to the true meanynge and purporte of the same. |

1558. Vicary's Governorship of St. Bartholomew's Hospital, under the Corporation of the City of London.

A.D. 1558.

(City of London Records, Repertory XIV. leaf 72, back.)

Mercurii, 28 Septembris Anno 5° & 6° &c. [Philip and Mary, 1558].

| Curtes, Mayor. gouernours of the hospitalles &c / | Item, this day the names of my Masters thaldremen & Commoners of this Cytie newelye nominated, electe, & chosen by the gouernours of the hospitalles & howses of the pore of this Cytie & of Bridewell, to ioyne & travaille with certen of the olde gouernours of the same hospitalles remayning in the seid offyce for the yere insuinge, presentyd here by Richard Grafton, grocer, |

Barts, the Lessors are described as 'the Maior, Commynaltye and Cyttizens of the Cyttye of London, Masters and governours of the howse of the pore, commonly called lyttle Saint Bartholomewes hospitall, in west smythefelde neare London, Curryor.' Repertory 22, leaf 91. There are entries in the Repertories of many (? all) other Leases of Barts property : see, for instance, in 1606, Rep. 27, leaf 194 (190, pencil).

On May 5, 1614, Barts is still the House of the Poor. See the entry 'Hospitall Leases' in Repertory 31, no. 2, leaf 303 ; 'this day, seaven Indentures of leases made by the Maior and Cominaltie and Citizens of London, Governours of the house of the poore commonly called St. Bartholomewes Hospitall, neere West Smithfeild, London, of the foundacion of King Henry the Eight . . . were here sealed with the vsuall seale for sealing of hospitall leases.'

[1] These are doubtless the revised Ordinances or *Order* of 1557, printed as No. XIII in the Appendix to the *Memoranda on the Royal Hospitals*, p. 77-100, ed. 1862. Also in ed. 1836, and the *Charity Com. Report* 32, Part VI, 1840.

App. III. 1. *Vicary, a Governor of Barts.* 139

one of the seid old gouernours, werre here red, ratefyed & allowyd : which names, togither with the names
of the seid houses & hospytalles whervnto they arre
seuerally allottyd & appoynted, herafter insue, videlicet :—

Sir Marten) Bowes, knight, Comptroller generall /
Sir Rowland Hyll, knight, Surveyour generall /

Sᵗ Bartholomews Hospytall

Sir John Lyon), knight	Mʳ Wallys
Mʳ John Whyte, Aldreman)	Mʳ Bushe
Mʳ Alderman) Malorye	Mʳ Dane
*Mʳ Vycars	Mʳ Ramsey
Mʳ Style	Mʳ Fletcher
Mʳ Atkinson)	Mʳ Ambrose Nicholas

The Governors of Christes Hospytall, Sᵗ Thomas Hospytall, &
Brydewell, follow on leaf 73.

Sᵗ Thomas Hospytall

Sir William Chester, knight	Mʳ Thomas Pyerson)
Mʳ Draper, Alderman	Mʳ Wythers
Mʳ Altham, Alderman	Mʳ Hayward
Mʳ Sayer	Mʳ Bonde
Mʳ Cater	Mʳ Onslowe
Mʳ Dychefeld	Mʳ John Olyff

1559. Repertory XIV. (leaf 216) Leigh, Maiore.

Jouis, 28 / Septembris, Anno primo domine Elizabethe Regine, &c.
[A.D. 1559]

[*Present*] Recorder, Bowes, Hill, White, Lyon), Garrard, Curtes,
Huet, Lodge, Harper, Johannes White, Altham, Malory, Draper,
Martyn), Foulkes, Rowe, Avenon), Cowper, Baskerfeld, Alyn); ac
Halse & Champyon), Vicecomites [or Sheriffs] . . . .

Item, the nominacion and elleccion of my maistres the Aldermen)
and worshipfull commoners of this Cytie appoynted by the right
worshipfull Sir Martyn) Bowes and Sir Roland Hill, knyghtes, and
other their assocyates, gouernors at this present of all *The gouernors of*
the Cyties hospytalles, to stonde and be gouernors of the *the houses of the*
pore & of the Cy-
sayd hospytalles for the yere now next insuynge, here *ties hospytalles.*
presentyd this day by the sayd Sir Martyn) Bowes and his com-
panyons, was ratefyed and confyrmed in euery poynt by the hole
Court ; The tenour wherof heereafter insueth in thes wordes : " Yt
may please your Lordship to be aduertysed that, the xxvijᵗʰ day of
September, anno 1559 / we, the gouernors of Thospytalles of this Cytie
of London), assembled together at Christes Hospytall accordynge to

140 App. III. 1. *Vicary, a Governor of Barts.*

our accustomed maner, haue nominatyd, and appoynted and elected, certeyne Aldermen) [*leaf* 216, *back* 1] And Citizens to serue in the sayd hospitalles for the yere ensuynge, most humbly beseechinge your Lordshipe and bretherene to ratefye and confyrme the same nominacion and eleccion

Sir Martyn) Bowes, knyght, Comptroller generall.
Sir Roland Hill, Surueygher generall./

The names of those that continued one yere, & must remayne another./ ...	The names of those that nowe are electyd to serue for the yere insuynge ./ ...
St Bartholomewes Hospytall./	St Bartholomewes Hospytall /
Mr John) White, Alderman)	Sir William Garrard, knyght
Mr Malorye, Alderman)	Mr Beswyke, draper
* Mr Vikers	Mr Fowler, grocer
Mr Busshe	Mr Lambertt, grocer
Mr Dane	
Mr Ramsey	
Mr Ambrose Nycholas	
Mr Atkynson)./	
St Thomas Hospytall	St Thomas Hospytall
Sir William Chester, knyght	Mr James Bacon)
Mr Draper, Alderman)	Mr Medcalf, goldsmyth
Mr Altham, Alderman)	Mr Spryngham, mercer
Mr Wethers	Mr Thomas Blanke, Junior
Mr Dychefeld	
Mr Anslowe	
Mr Oleffe	
Mr Thomas Pyreson)	

A.D. 1560.

(Rep. 14, leaf 391) Martis / 15 / Octobris anno secundo domine Elizabethe Regine, &c. [A.D. 1560].

Chester, Maior de nouo, nuper clericus, Hill, White, Garrard, Offley, Leigh, Harper, Jo. White, Malorye, Champyon), Martyn), Avenon); Baskerfeld, Alyn), Chamberlyn); ac Draper & Rowe, Vicecomites [= Sheriffs] . . . .

[*leaf* 391, *back*] Item, this day Mr Alderman Bowes brought in here the
The gouer- names of all the gouernors of the cyties hospytalles here-
nors of the under namyd, that are appoynted and newly electyd
Cyties accordynge to the forme of the actes & ordenaunces of
hospytalles this cytie in that behalf prouyded and made, to serue,

---

1 At top is 'Leigh Maiore. Adhuc Jouis 28 Septembris, Anno primo domine Elizabethe Regine, &c.'

App. III. 1. *Vicary, a Governor of Barts.* 141

and take paynes and travayle for the gouernaunce of the same hospytalles for the yere insuynge : the names of all the w*h*ich gouernors hereafter insue in these wordes./ .

Gouernours elected the xiiij*th* of October, an*n*o 1560, for the gouernement of Chrystes, S*t* Bartholomeus, Brydewell, and S*t* Thomas Hospytalles. /

    S*i*r Martine Bowes, knyght, Comptroller gen*e*rall.
    S*i*r Roland Hill, knyght, S*u*rueyour generall./ . . .

S*t* Thomas Hospytall   [*leaf* 392]   S*t* Bartholomeus Hospytall./

M*r* Sayer	S*i*r Will*i*am Garrard, k*n*ig*ht*
M*r* Oleffe	M*r* Jo*hn* White, Alderman)
M*r* James Bacon	M*r* Malorye, Alderma*n*)
M*r* Spryngham	*M*r* Vikers
M*r* Tho*mas* Blanke	M*r* Rich*ard* Lamberts
M*r* Thomas Pierson	M*r* Beswyke
M*r* Medcalf. /	M*r* Fouler
Gouernors newly electyd [*l*/ 392]	M*r* Ramsey
M*r* Lodge, Alderman)	M*r* Ambrose Nicho*l*as
M*r* Champyon), Al*d*er*man*	M*r* Atkynson
M*r* Allyn), Alderman)	M*r* Skot*t*
Master Mynors	Gouernors newly electyd [*in marg.*]
M*r* Chaire. /	M*r* Brystowe

        Gouernours gen*e*rall
           M*r* Wethers
           M*r* Foulk*es*. /

On 24 April, 3 Eliz. 1561, 'A Precept for the Poore yn the Hospitalls' was issued by the Lord Mayor, appointing Committees to sit with the Governors of the Hospitals, to collect weekly Alms for the poor in the several City-Wards, and not allow foreign (or strange) beggars or other poor to beg in the parishes. Among the Governors told off to act with the Committees for the Wards of Farringdon Without, Aldersgate and Cripplegate, was the Resident Governor of Bartholomew's, "M*r* Vycars," our Thomas Vicary. This Precept, with its names of Committees and Governors, is printed from the Guildhall-Records *Journal* 17, lf 310, in *The Royal Hospitals* (1836) as Appendix XIV, p. 107—111 ; ed. 1862, p. 100—104. See an extract in our Forewords.

142 App. III. 1. *Vicary, a Governor of Barts.*

A.D. 1561. (Repertory XIV, leaf 534.)

Jouis, 25 Septembris, anno 3° Domino Elizabethe Regine, &c.

[A.D. 1561] Chester M*aiore*.

[*Present :*] Recorder, Bowes, Hill, White, Leigh, Harper, John White, Champion, Avenon, Cowper, Baskerfeld, Alyn, Chamberlin, Gilbert ; ac Draper et Rowe, Vicecomites [= Sheriffs]. . . . .

[*leaf 534, back*] **Gouernours of the Cityes Hospitalles**   Item, the names, aswell of the gouerners of y*e* Cities hospitalles *that* weare lately newly elected, as also of those that haue allredy seruid by y*e* space of one hole yeres past, presentid vnto this Courte here this Day by M*aster* Alderman Bowes & other of my masters the Aldermen, gouernors of the said houses, as hereafter ensueth, in Christes hospitall. The gouernors *that* haue remayned one yere & do contineve still /
Sir Thomas Offley, knighte, M*r* Martyn, Alderman Basford, M*r* Peirson, M*r* Mabbe, M*r* Kynge, M*r* Whithornes.

[*leaf 535*] S*t* Bartholomews
  Sir William Garrard, Knight
  *M*r* Vikers
  M*r* Ambrose Nicholas
  M*r* Bristowe /
  M*r* Atkinson
  M*r* Scott /
S*t* Thomas Hospitall /
  M*r* Champion, Alderman /
  M*r* Alen, Alderman /
  M*r* Sawyer /
  M*r* Chare /
  M*r* Spryngham
  M*r* James Bacōn /
  M*r* Mynoures
  M*r* Thomas Peyrsoñ /
Brydewell /
  M*r* Harding
  M*r* Boxe
  M*r* Harrys
  M*r* Pers /
The names of them *that* are newly elected /
  M*r* Chamberlyn, Alderman
  M*r* Vincent Randall /
  M*r* Thomas Garden), Goldesmyth
  John Keale, Goldesmith
  Richard Johnsoñ /
  Jeames Mastoñ /
  William Albeney
  John Jaksoñ /

S*t* Bartholomevs
  Sir Thomas Leigh, knight
  M*r* Bankes, Alderman
  Robert Soole
  Thomas Lave [Lawe]
  John Lute
  Robert Hulsoñ /
S*t* Thomas Hospitall
  Sir William Huett, knyght
  M*r* Lorymer /
  M*r* Golston /
  M*r* John Baker /
  M*r* Thomas Huett /
  Rychard Violett /
  Rychard Morrys /

Brydewell
  M*r* Hayward }
  M*r*. Gilbertt } Aldermen
  M*r* Thomas Bonde
  M*r* Roger Bamsted
  M*r* Thomas Bannyster
  M*r* Nicholas Wheller
  M*r* Kyteley
  M*r* Edward Dove
  Richard Taylor
  William Gybbons /

[*As we do not know for which Hospitals the new Governors were severally elected, we print the entries as they stand in the MS.*]

App. III. 1. *Governors of Barts in* 1562-3. 143

The entry above, of Sept. 25, 1561, is the last in which Vicary's name appears. It is of course not in the next, of Sept. 24, 1562, as his Will was proved by his Widow on April 7, 1562 (see p. 194 below). Yet we add the 1562 List.

A.D.1562. (Repertory XV, leaf 124, back.)

Harper Maiore   Jouis, 24 Septembris, anno iiij^{to} Domine Elizabethe Regine [A.D. 1562].

[*Present:*] Recorder, Bowes, Garrard, Offley, Leigh, Huett, Lodge, Champiōn, Cowper, Chamberlin, Banckes, Jakman; ac Avenon & Baskerfeld, Vicecomites [= Sheriffs].

[*leaf* 125, *back*]   Item, this day Sir Martin Bowes, knight, Controller
The gouernors   generall of all y^e Cities hospitalles, Christ, the house
of the Cities   of y^e poore [St. Bartholomew's], Bridwell, & St
hospitalles   Thomas in Suthwerke, brought in the names of all
                  the gouernours & Surveiors of ye same houses, newly
                  elect & chosen for the yere insueyng, according to
                  thordere heretofore taken [*leaf* 126] for the same, whose
                  names hereafter do destinctly and seuerally insue, &c.

Gouernors elected and chosen for ye gouerment of Christes, St Barthelmewes, Bridwell, & St Thomas Hospitall, the xxj of September, 1562 . . . . . .

St Barthelmews.	St Barthelmews	St Barthelmews
Sir Wm Garrard, knight	Mr Banckes, Alderman	Mr Jakman, Alderman
Sir Tho Leighe, knight		
Mr Ambrose Nicolas	Mr Bristowe	Mr Witton Sowene
Mr Lute	Mr Atkinson	Mr Jeames Hawes
Mr Loo		
Mr Soole		
Mr Scot		
Mr Howland		

St Thomas Hospitall	St Thomas Hospitall	St Thomas Hospitall
Sir William Hewet, knight	Mr Allen, Alderman	Sir Wm Chester, knight
Mr Champion, Alderman	Mr Sayer	
Mr Jeames Bacon	Mr Springhām	Mr Offlee
Mr Lorymer	Mr Chare	Mr Boxe
Mr Colston	Mr Tho Hewett	Mr Francis Barnham
Mr Baker		Mr Nicholas Love
Mr Richard Violett		Mr Welles
Mr Richard Morris		
Mr Mynors		
Mr Tho Peirson		

144   App. III. 1. *Governors of Barts in* 1563-6.

For the next year, Sept. 1563-4, the old Governors continue: Rep. XV, leaf 281, back:

governours of the hospitalles   Item, this daye the governours of aH the citiez hospitalles here being present, dyd gently agree to stande stiH in the same their offices one other yere more; and yt was orderyd that the audytours appoyntid to take thaccomptes of the sayd offyceres shaH take the same wyth convenyent spede.

On Tuesday, Sept. 25, 1565 (*Repert.* XV, leaf 472, back), two of the Bartholomew's men (Mr Rychard Foukes, Clothworker, and John Jaxson, founder, two of the gouernours of the house of the pore) brought in the list of the Governors of the several Hospitals elected at the Meeting at Christ's Hospital on Sept. 21, 1565, and besought the 'Court to ratefye and alow the same'; which, 'after good & mature consideracion thereof,' the Court did. But the Bartholomew's men were only the old ones [leaf 473]:

of such as haue Contynewed but one yere & yet remayne	Such as haue Contynewed ij yeres That remayne
S^t. Bartholomewes	
Mr Aldreman Martyn	Sir William Garret, presydent
Mr Aldreman Chamberlyn	Mr Ambros Nicholas, Tresurar
Mr Thomas Banester	Mr William Wytton
Mr Edward Bryght	
Mr Rychard Barnes	[There is no Barts entry in the 3rd column, headed " Such as ar new elected."]
Mr Thomas Gore	
Mr Richard Yonge	
Mr William Cockes	
Mr John Hardyng	

The general Officers are given on the back of leaf 472:—

Sir Martyn Bowes, knyght, Comptroller
Sir Thomas White, knyght, Surveyor } generall

Mr Lawrence Wether
Mr Rychard Folkes } gouernors & Audytours generall
Mr Robert Hardynge

The names of such as haue Contynued but one yere, & yet remayne

Mr Alderman Lambert
Mr Wylliam Leonarde, mercer
Mr Henry Sutton, Goldsmith
Mr Christofer Edwardes, haberdasher

App. III. 1.   *Later Governors of Barts.*   145

Many Lists (probably all the early ones) of the Barts' Governors are given in the Repertories. In turning over some of their leaves we came on a few. For those of 1582 and 1583, see Repertory 20, leaf 235, back, lf. 458, bk. ; for those of 1584, -85, -86, -87, Rep. 21, lf. 87, bk., lf. 213, lf. 335, lf. 470 ; for 1589, -90, Repert. 22, lf. 100, 212 ; for 1593-4, -94-5, -95-6, Rep. 23, lf. 100 (or 103), bk., 295, bk., 441.

In 1598, the *Repertory* 24, leaf 283, shows that the Governors of Bartholomew's were Sir John Harte, knight, president, Sir John Spence, knight, 5 Aldermen, M^r. Thomas Smith, Thre*surer*, Will*iam* Massham, Esq*uire*, 4 other men, 4 mercers, 4 grocers, 1 draper, 3 goldsmithes, 3 Skynners, 6 merchant*tailors*, 2 haberd*ashers*, 1 vintener, 1 Clothworker, 2 diers, 1 letherseller, and 1 Cooper. There were 4 Auditors. (Why the number of Governors was so increast, we don't know.)

In 1599 (*Rep.* 24, leaf 460, back), the Barts' Governors were Sir Stephen Soame, knight, Maior, Sir John Harte, knight, president, Sir John Spencer, knight, 2 Aldermen, M^r. Thomas Smithe, Thre*surer*, William Masham, Esq*uire*, 5 men entitled to be cald 'Master,' 3 me*r*cers, 6 Grocers, 1 Draper, 4 Goldsmithes, 5 me*r*chaunt-tayleres, 2 Skynne*rs*, 2 habe*r*dashers, 1 Vintene*r*, 2 Clothworkers, 2 diers, 1 letherseller, 1 Cowper ; and there were 4 Auditors, as before.

For the Governors for 1600-1, see Repertory 25, lf. 154 ; for 1605-6, 1606-7, Rep. 27, lf. 85 (81, pencil), lf. 274 (271, pencil) ; for 1611, Rep. 30, lf. 182 ; for 1613, Rep. 31, no. 1, lf. 166 ; for 1614, Rep. 31, no. 2, lf. 417, bk. ; &c. &c.

1614, April 19.   Bartholomew's not a House for bringing-up Children.

(Repert. 31, no. 2, lf. 297)
St. Bartholomewes Hospitall /
It is charged with the keep of 3 Foundlings, 2 left in the Hospital Cloisters, 1 in Watling Str.
VICARY.

Item, this day, Thomas Juxon, Thre*sorer*, and others of the Governo*u*res of St Bartholomewes Hospitall, London, exhibited their humble petic*i*on to this Court, Intymating thereby that the said Hospitall hath bin chardged w*i*th the keeping of three Children,—two of them being left by pe*r*sons vnknowne, in the Cloyster of the said Hospitall, and the other being left in Watlingstreete in the p*a*rishe of St John Evangelist*es*,

L

146   App. III. 1.   *Barts not a Children's School.*

sent to the Hospital by the Lord Mayor to be cured.
Mayn't the 3 Children be sent to Christ's Hospital?

and after sent by the Lord Maior to the said hospitall to be cured of her infirmitie,—and desiryng that the said Children may forthwith be received into Christes hospitall, to be kept there; alleadging that it is contrarye to the foundacion of the hospitall of St Bartholomewes to keepe or bring vpp any Children:

Committee of 3 Barts Governors, and 2 of Christ's, appointed to report on the case.

Wherevpon it is ordered by this Court, that Sir Thomas Lowe, Sir William Craven, Sir Thomas Hayes, knightes and Aldermen, Master Alderman Leman, Master Alderman Stile, or any three of them, shall forthwith meete and consider of the said peticion, and certifie to this Court in writing vnder their handes, whether they shall finde it against the foundacion of the said Hospitall to keepe Children, and of their opynions therein: And John Savage to warne & attend them./[1]

We don't find any further entry relating to this matter, but have no doubt that the Committee's decision was in favour of Barts, and that the children were shifted to Christ's Hospital, especially as Sir Thomas Lowe was then President of Barts, and Sir Thomas Hayes and Master Nicholas Stile, Alderman, were Governors: see Repertory 31, No. 1, lf. 166. Sir W^m. Craven was President, and Alderman John Leman was a Governor, of Christ's Hospital.—*ib.*

1624, Sept. 3.   A Petitioner for the Hospitallership [or Chaplaincy] of Barts.

(Repert. 38, leaf 229)

Peticio

Roberts recommended for the reversion of the post of Hospitaller of Barts, &c.

Item, this daie the humble peticion of William Robertes preferred vnto this Court, to haue a revercion of the Hospitlers place of Saint Bartholomewes, and to haue a Clark or Vicar Choralls place in Christ church, is by this Court referred to the favourable consideracion of the President and Governours of the said Hospitall of Saint Bartholemewes./

---

[1] Dugdale, *Monast. Angl.*, vol. vi. pt. 2, p. 627, ed. Ellis, says that 'the foundation was for a Master, brethren, and sisters, and for the entertainment of poor diseased persons till they got well; of distressed women big with child, till they were delivered, and able to go abroad; and for *the maintenance (till the age of seven years) of all such children whose mothers die in the House.*'

App. III. 2. *Smithfield Encroachment and Lease.* 147

2. Supplementary Extracts from the Guildhall Records as to Bartholomew's.

1512, March 16. The Master to pay rents for his Encroachments on Smith-Field.[1]

Rogerus Acheley, Maior

(Letter-Book M[2], lf 189, bk)
Seint Bartholomeus Spitel

The Master to pay 6s. 8d. rent for the site of his pale,

and 13s. 4d. for that of his gatehouse and porch.

Sextodecimo Die Marcij, Anno regni Regis Henr*ici* octaui *ter*cio : Maior, Tate, Aylemer, Kebul̄l̄, Copynger, Monox, Butler, Exmew, Reste, Basford̃, Brugis, Mil-bo*ur*ne ; Fenrother, Holdernesse, vic*ecomites*./ Atte this Court of aldermeɲ, it is agreid̃ that the Master of seint Bar*tholomeus* Spitel̄l̄, for the grounde that his pale stand̃itl̄l̄ vpoɲ), shal̄l̄ paye yerely to the Chamberlayɲ) vj s viij d̃ ; And for his gate hous and porche newely bildid̃ upoɲ) the Co*m*en grounde, xiij s iiij d̃, to the vse of the Co*m*i*n*altie of this Citie. &c.

1515. Lease for life, to the Master of Barts of the Common Ground of the City.

(Rep. 2, leaf 209 (208 pencil), bk.) Martis xvj° Die Januarij [6. Henry VIII, A.D. 1515].

Monoux [Mayor]

M*agister* Hospit*alis* sa*n*cti

[*Present*] M*ay*or, Capell, Haddoɲ), Aylem*er*, Jenyns, Boteler, Rest, Exmewe, Myrfyɲ), Milburɲ), Sheltoɲ), Fenrother, Alder*n*es, Baldry, Bayly, [ac] Yerford̃, Mundy, Vice*comites* [Sheriffs].
At this Court yt ys agreed̃, that the M*aste*r of thospital̄l̄ of seynt Bartilmewe in Westsmythfeld̃, londoɲ), shal̄l̄ haue a lees of the Co*m*en ground̃ of this Citie[3]

---

[1] There are of course many entries in the Guildhall Records as to the early Barts Hospital and Priory. We give only 2 or 3, mainly to show how Smithfield was gradually encroacht on, and dockt of its old extent in Chaucer's days.

[2] On lf. 246, bk, of this book, Sept. 6, 1515, £400 is to be paid to the two 'Surveyours of the lazare houses, Called Seynt Gyles in the Feld lookes, & Kyngelond, of the Foundacioɲ of this Citie . . . Towardes the Reparacio*n*s of the seyd ij houses, that ys to sey, lookes & kyngeslond.'
On lf. 247 is an entry, that on Sept. 21, 1515, the Cardynall of Englond sends a message that the King has orderd a *Te Deum* to be sung at St. Paul's that day at evensong, because 'the Quenes gr*a*ce, beyng of late Conceyved w*i*t*h* Childe, ys nowe (thanked be our Lord !) quykened of the same, to the grete Ioye & Comforte of all*e* the kyng*e*s treu & lovyng Subgett*es* ; wheruppon the seid Maier & his Brethern, in Scarlet, went to the seid Church, & their taryed tyll Te Deum was sunge, meane betwen*e* evensong & Compleyn was Fynysshed.'

[3] We take this to mean the open part of West Smithfield.

148 III. 2. *Encroachment on Smithfield. Privileges.*

Bartho*lom*ei Ex assens*u* *domini* Mayor*is*.
for the te*r*me of ce*r*teyn) yeres, as more playnly apperyth the xxx day of March in the tyme of Mayralte of M*aster* Accheley [Nov. 1511-12], Prouided' alweyes, that yf yt happen) the seyd' Mest*er* to decesse w*it*hin the seyd te*r*me, that then) the seyd' te*r*me & lees to be vtte*r*ly voyde; And the seyd' lees to begyn) at Mighel*mas* last passyd'. And' as touchyng' tharre*r*e of the seyd' Rent, the seyd' Mast*er* p*r*omytteth to stand' & obey to such order & direcc*i*on as this Court shall award' in that behalf.

1515. Encroachment. A Forge built on West Smithfield, granted to Barts for a year.

(Repertory 2, lf. 210 ink, 209 pencil.)

Smythes Forge next to Hertyshorn)
Ite*m*, yt ys agreed' that the Smyth which nowe settyth vp a Forge next to the Hertyshorn) in Westsmythfeld', vppon) a pa*r*cell of the Co*m*en ground', That the Chamb*er*ley*n* shall viewe the seyd Co*m*en ground', & sett a Rent the*r*uppon) by his discrec*i*on), & to make to the M*aster* of thospitall of Seynt Bartolo*mewes*, own*er* of the hous wherunto the seyd' Forge adioyneth, a graunt therof, oonly for this yere.

1541. Privileges of Bartholomew's as to Arrest in the Hospital.

(Repert. 10, lf. 216) Martis, 19 Julij, A*n*no 33°, H. 8 (A.D. 1541).

Roche Mayor.
[*Present*] Ma*y*or, Waren), Gresha*m*, Forma*n*), Denha*m*, Dorm*er*, Pagett, Cotes, Bowyer, Dawnsy, Hob*er*thorn*e*, Tolos, Aleyn), Wylforde, Lewen), J. Gresha*m*, Judde; Ac Laxton) & Bowes, Vic*ec*omites (Sheriffs) . . . .

Seynt Bartho*lo*- *mewes* hospytall Claims freedom from arrest, and that Widow Bromeby shall be set free.
Ite*m*, the most gracyous le*tt*res patent*es* of kyng Edward the seconde, made & graunted to seynt Bartholomew*es* hospytall in Smythfeld, were Alowyd, for & conc*er*nyng suche pryvyledg*es* As they do clayme therby att the p*res*ent tyme / whiche ys, that none of the officers or Mynistres of this Citye shulde do or execute eny arrest within the p*re*cyncte of the seyd hospitall / And that the arrest made by Vnderhyll, one of my lorde Mayers s*er*iaunt*es* vpon) Alyce Brom*e*, wydowe, dwellyng' w*it*hin the seyd hospytall, shalbe dyscharged, &c.

App. III. 2. *Barts Privileges, from Arrest, &c.* 149

1541. Barts Privileges as to Arrests allowd by the City.

(Letter-Book Q, lf 34, bk) Roche, Maior.

Martis, xix° die Julij, Anno xxxiij° Henrici viij, in Repertorio.

Saint Bartholomewes hospytaH

Its claim for freedom of its indwellers from arrest, is allowd,

and Underhill's arrest of Widow Browne is discharged.

Item, the most gracyous lettres patentes of kinge Edwarde the seconde, made & graunted to saint Barthilmewes hospytall in Smythfeld, were allowed, for & concernyng suche pryvyleges as thei doo clayme therby at this present tyme / which ys, that none of the offycers or mynysters of this Cytie shulde doo or execute any arrest within the precincte of the said hospytaH / And that the arrest made by VnderhiH, one of my lorde Mayers seruauntes, vpon Alyce Browne, wydowe, Dwelling within the sayd hospytaH, shalbe dyscharged &c/.

1542. A Governor of the Hospitals surrenders his post.

(Rep. 10, lf. 269) Adhuc Martis, primo Augusti, Anno 34, Henrici viij¹. (A.D. 1542). Dormer, Mayor.

Master Gallard

[surrenders his post as Governor of the Spital Houses]

Item, Att this Court, Master Gallard, paynterstayner, being¹ one of the vysytours & gouernours of the SpyteH howses nere adioynyng vnto thys Cytye, hath thankefully, by the mouthe of master Hayes, Comptroller of the Chambre of this Cytye, surrendred hys sed Rowme & offyce into the handes & dysposycion of this Court / most hertely desyring the same to Apoynt some other hable man) for the due execucion of the same rowme.

On Jan. 23, 1543 (Rep. 10, lf. 303), 'John Nyk,—who lately had the gouernaunce & ordering of the poore people being in the lazar house Att Kyngysland, & of the ymplementes of the same house / And also the Colleccion of the charitable Almes of the people gevyn) vnto the seyd poore men,'—was reinstated in his office, which he had lost by absence for a time. On 22 May, 1543 (Rep. 10, lf. 334, bk.), 'Mr Rychard Holte, Cytezen) & merchaunttayller' is appointed 'one of the gouernours & Vysytours of the lazarhouses.'

---

15 Nov. 1547. The Vicar of St. Bartholomew's Hospital.

(Repert. 11, lf. 387, ink ; 363 pencil, bk. 15 Nov. 1 Edw. VI)

seynt Bartholomewes

Item, yt was agreyd that the vycar of saint Bartholomewes hospytaH, resortyng to maister Judde & thother Aldermen & commyssyoners for the poore, shalbe assured of aH suche thynges as he, doyng his duetie, ought to haue.

150 App. III. 2. *Claims against Barts.*

12 Jan. 1548. Dr. Howell's Claim on the Bartholomew's Governors for his Stipend.[1]

Jo. Gresham, Mayor. (Rep. 11, lf. 380, pencil; 404, ink.)
Howell   Item the lettres of sir Edward North, knyght, and other the lerned counselers of the Court of the Augmentacion of the Revenues of the kynges Crown, dyrected to this Court in the favour of doctour Howell, Physycion, for the stypend or Fee by hym demaunded, were delyuered, by the order of the Court, to Maister Judde, Aldernan, to the intent that he, & thother Aldermen hauyng the gouernaunce of the hospytall of the pore, shuld make an aunswere thervnto /.

---

24 Jan. 1548. Bridge-money for Bartholomew's. Dr. Howell's claim.

(Rep. 11, lf. 408, ink ; 334, pencil, bk.) Mart*is*, xxiiij° die Januar*ii*, A*n*no primo Edwardi vj° / [A.D. 1548].

the howse of   Item, it is agreyd that the Brydge-maisters for the
y*e* poore   tyme beyng, shall from hensforth paye vnto my maisters thaldermen nowe havyng, & that hereafter shall have, the gouernaunce of the howse of the pore,
In*t*ratur   all suche money as they heretofore were wont yerely to paye to the late maister there / And further that the seid Brydge-maisters shall, w*i*th convenyent spede,
A Cage /   cause A good stronge Cage to be new made att the further ende of the seid Brydge for the due punysshement of Vagabundes therin /.
Howell   Item, this day, M*aiste*r Bowes, M*aiste*r Judde, M*aiste*r Hyll, M*aiste*r Barne, M*aiste*r Jervys, M*aiste*r Hynde, & M*aiste*r Garrard, Aldermen, are assygned by the Court here, to repayre to M*aiste*r Chaunceler of the Augmentacion of the Revenues of the kynges Crown, for the aunsweryng of his lettre wryten in the favour of Howell y*e* physycion /.

---

20 Sept. 1548. Mr. Losse's Claim against Bartholomew's.

(Rep. 11, lf. 492, ink ; 470 pencil, bk. 20 Sept. A° ij° E(dw.) 6 /.)
J. Gresham, Mayor.
Maister   Item, M*aiste*r Losses byll exhybyted to my lorde
Losse   Mayere & my maisters the Aldermen, for certein yerely Fees that he claymeth out of the late Hospytall of saint Barthylmewe & the late grey Fryers, red /

---

[1] He perhaps did work at the Hospital before Vicary's appointment.

App. III. 2. *Barts Governors' Gateway.* 151

Yt was agreyd that he ¹shuld repayre to Maister Judde & thother Aldermen, gouernours of the house of the poore,² for his aunswer therin, accordyng to the ordre here lately taken for the same /.

1552. **Bart's Governors to have a Gateway thro' the City Wall into the 'House of Work'.**

A.D. 1552. Common Council, 1 Aug., 6 Edward VI.
(Journal 16, leaf 201, ink ; or 195, pencil, back.)

It was also this day (for dyuerse good & necessarie consyderacions & causes especially moving this honourable Courte of comen counsaill) ordeined, enacted, graunted & agreyd by thassent & aucthoritie of the same, that it shalbe lefull to & for the lorde Maire & Aldermen of the said Cytie that nowe are, & to their Successours, Maires & Aldermen of the same Cytie for the tyme being, at their free will & pleasure to pull & breake downe, & cause to be pulled downe, asmuche of the Cyties wall stonding on the Backsyde of Crystes churche in the warde of Faringdon within, as to their sad dyscrecions & wysedomes shall seame meate and convenyent for the making of a gate or dore thurrough the same wall, for the Apte, commodyous, & meate passage of the gouernours of the house of the poore of the fundacion of kinge Henrie the viij^th in west Smythfeld, nere vnto London, & other Cytezens of the sayd Cytie, to & from the same house, vnto & from the house of the said Cytie, ordeined by the same Cytie for the relief & socour of the poore, called the house of woorke / eny Lawe, Acte, ordenaunce or decree heretofore made & ordeined to the contrarie, in eny wyse notwithstanding/.

(lf. 203, ink, or 197, pencil)

Enacted that

the Lord Mayor and Aldermen

may break thro' the City Wall at the back of Christchurch,

and make a gateway for the

Governors of Bartholomew's, &c.,

to pass from the Hospital to the

House of Work for the poor.

1552. Bartholomew's to have a second Rent from all City houses turnd into Alleys.

(Journal 16, leaf 127.) Primo die Octobris, Anno Regni Regis Edwardi vj^ti v^to
[A.D. 1552].

By reason of turnyng, converting & transporting of capytall mesuages & houses into alleis, wherby great nombre of beggers, vagabundes, idell & suspecte persons are incresed within this Cytie / & the comen welthe therby muche impaired, & evell rule muche-

As the turning of Houses with their grounds into Allies has increast the number of vagabonds, &c.,

¹ leaf 493, ink.   ² 'of the poore' struck thro'.

152 App. III. 2. *Barts Rent from Houses made Alleys.*

inhaunsed & growen), & the abylytie & suertie of the said Cytie muche decresed & mynysshed / For reformac*i*on wherof, be yt ordeined & decreed, by aucthorytie of this c*o*men counsaitt, that where any pryncypatt Meases[1] or howses shatt, at any tyme herafter be converted & turned into any alley or alleis / that eue*r*ie seue*r*att inha*b*itaunt wit*h*in eue*r*ie Rowme & place wit*h*in suche Alley or Alleis for the tyme being for eue*r*, shatt yerely yelde & paye to the house of the poore in Westsmythefeld of London) yerely the hole value by the yere of eue*r*ie suche Rowme & Rowmes as eue*r*ie of theim dothe or shatt dwett in [in] any suche Alley or Alleis made or to be made, so longe as the same shalbe vsed as an Alley / the same to be paid quarterly by evin) porc*i*ons, afte*r* suche rate as thei paye in yerely rent for the s*a*me to the landelordes therof[2] / or as the same shalbe estemed to be yerely worthe by iiij indyfferent men) of the said Cytie, to be aucthorysed by the Maire of London) & A futt court of Aldermen for the tyme being / & that it shalbe laufutt for any offy*c*er of the said Cytie, by the co*m*maundement of the Maire of the same for the tyme being, to dystreine for the same / or elles the goue*r*nou*r*s of the said hospitatt for the tyme being, to haue an acc*i*on of dett[e] for the same / wherin no wager of lawe[3] shatt lie ne be admytted / And be yt further ordred & decreyd by aucthorytie aforesaid, that where, at eny tyme wit*h*in .x. yeres now last past, ij dwelling howses or moo, haue bein) converted into one dwelling house, and where at any tyme herafter ij dwelling houses or moo shalbe converted into one, that in eue*r*ie suche case, thinha*b*itaunte or i*n*ha*b*itauntes of the same shatt from hensforthe doo bere & paye att suche & asmuche scott, lott, [*leaf* 127, *back*] charge, payment & other thing*e*s as was accustumed to be done, paid & borne for eue*r*ie of the same when) thei were vsed / as seue*r*att dwelling howses / & as shuld, or of right ought to be done, borne & paid for the same, in case thei were stitt vsed as seue*r*att dwelling houses //

Side-notes: We order, that in every such House-converted Alley, every tenant shall pay to St. Bartholomews the whole yearly value of his Room or Rooms quarterly, according to the actual rent, or that fixt by the City's valuers, such value to be recoverd by distraint by any City Officer, or the Governors of St. Bartholomew's. And when any 2 houses have been or shall be converted into 1, the inhabitants of it shall pay the old scot, lot, &c., for the old 2 houses.

---

[1] messuage, holding: see *Catholicon*, page 1 (= 50, of Introduction) and p. 232; and 'a mese of landes & tenementes' in *Fifty earliest English Wills*, E. E. T. Soc. (ed. F. J. F.), p. 126.

[2] MS. of therof. This makes a double rent payable, 1 to Bart's, besides the other to the landlord.

[3] See the bit from Jacob's *Law Dict.* on page 98, *Statutes*, note, below.

App. III. 2. *Barts Gifts. Incontinent Folk.* 155

A.D. 1557. Gifts to St. Bartholomew's.

(Repert. 13, No. 2, lf 552)   Martis decimo nono Octobris, annis quarto
& quinto [Philippi et Marie, 1557]. Offley maiore.

London et Bennet.

(3 houses given to Bart's)

At this Courte, a deade of a graunt of Annuyte of v ti vj s viij d to be made by the Mayer and Comynalty and Citizens of this Cytye, to one Constance Bennet, gentleman, for Terme of his lyfe, for and in consideracion of three howses set & beynge here within the Cyty, by hymme, the same Constance, already frely gevyn) to the same Cytye, to the vse of the pore within the howse of the pore in weste Smythefelde, was read ; & agryed that the same shalbe sealyd and delyueryd over accordingely.

Another gift of lands, by Katheryn Hall, 'to the vse of the pore within the hospytalles of the saide Cyty' is on the back of leaf 552 ; and at the top of leaf 553 (date as above), is a Lease of (seemingly) Hospital lands in Oxfordshire :—

Smythe. Item, it was agryed that the Counterpane of¹ the lease here red this day, grauntyd by the Chamberlyne and iiij of the Gouernors of the howse of the poore in weste Smythefelde, to one Alyce Smythe, wydowe, of the Cytyes messuage or Tenemente, and certayne Landes lyeng and beyng in Heathe[1] in the County of Oxford, nowe in the holdinge and occupacion of the said Alyce, or her assignes, for Terme of certayne yeres, shalbe sealyd and delyueryd over accordingely.

In Repertory 22, leaf 107, is the following Minute of the Common Council Meeting on Oct. 14, 1589 :—

Legacies to thospitall.

(by Rd. Walter, girdler.)

Item, yt is orderyd that Warninge shalbe presently gyven to the Threosorer and governors of the fower severall hospitalls of this Cyttye, to haue a speciall Care to looke for suche legacyes as haue lately byn gyven and bequeathed to the same hospitalls by the last will and testament of Richard Walter, girdler, deceassed.

A.D. 1562. Incontinent Folk at Bartholomew's.[2]

(Repertory 15, lf. 59, bk.) Adhuc martis, 7 Aprilis anno iiij^to Domine Elizabethæ Regine [A.D. 1562].

Item, it was orderid that master Fulkes, and othere the gouernours of the house of the poore, shall cause the

---
[1] Near Bicester, in Ploughley Hundred.
[2] This extract, the Cesspool one of 1567, &c., are given as specimens, to show how complete was the control of the City over Barts and the other Hospitals.

154 App. III. 2. *Barts Cesspool. Misdoers turnd out.*

Incontinent
Lyuers

man & yᵉ ij women *that* they haue nowe remaynyng in their kepinge for vicious & inco*n*tinent liuyng, of whome they informed the Court here this day, to be inquired of by the wardmote inquest where they did offend : And that they then shalbe pu*n*nished according to the lawe /

1564. Enclosure of the City's Garden near the House of the Poor.

(Rep. 15, lf. 327, bk. ; 324, inner margin) Adhuc 23ᵈⁱᵒ Marcij, a*n*no 6 D*om*ine Eliza*bet*h*e* Regine, &c.

Me*m*orandum, that the xxvᵗʰ daye of Marche, in the vjᵗʰ yere of the regne of o*u*r sayde sovereign) Lady &c., M*aiste*r Laurence Wyther, Saulter, M*aiste*r Richarde Fulkes, Clothworker, & the Chambe*r*lein were appoyntyd by my lorde mayre and my M*aiste*rs thaldermen, to talke w*it*h M*aiste*r Haddo*n*, one of the m*aiste*rs of the Request, for & concer*n*yng*e* the cytyes gardeyne at the house of the poore adioynynge to his lodginge there / & to move him quyetly to suffer the cytyseins & governo*u*rs of the seyd house to inclose, vse, & enioye the same, beinge their owne p*r*opre grounde, to their most benyfytt and com*m*odyte, & to make reporte to my seyd lorde mayre & aldermen w*it*h convenyent spede, of his answere therein /

1567. The Cesspool at Bartholomew's to be clenzd.

(Rep. 16, leaf 261, back¹) 29 July, 1567 (An. 9 Eliz.).

the Sewer
at the house
of yᵉ pore

Item, it was this day ordered by the Court here, that m*aste*r Chambe*r*lyn, & m*aste*r Bright, Iremonge*r*, and ij or iij more of the goue*r*nors of the house of the pore, & Thomas Wheler, drape*r*, shall viewe the comen Sewer or vawt at the seid house, & consider how and by what meanes the same Sewer may be clensed, And make an estymate as neight² as they can, what the Charges of the doyng thereof will amount vnto, &c. /

[*The Hospital Surveyors to reform Hospital Abuses.*]

Su*r*veyors
of the Cities
Hospitall*es*

to turn out misdoers ;

Item, this day, m*aste*r Garrard, m*aste*r Offley, m*aste*r Chester, m*aste*r White, m*aste*r Rowe, m*aste*r. Becher, and m*aste*r Hardyng, Aldermen, were assigned to pe*r*vse, viewe, & vnderstand pe*r*fetly the estat*es* & condycions of all the Cyties hospytall*es*, and to reforme & avoyd all such misdowers as they shall fynd in eny of theym, either in the resceipte and maynten-

---

¹ ink no. ; 252, pencil.   ² nigh, near.

App. III. 2. *Barts Extracts.* 3. *Barber Bits.* 155

to see the right number of poor and children kept,

and to reform things amiss.

*au*nce of eny more or greater n*u*mber of pou*e*rtie or Children in eny of theym then ther ought to be, or otherwise by eny wayes or meanes ; And to take such order for the[1] custodye of the money & Tresure of eue*r*y of the seid houses, And for the reformac*i*on of all such thing*e*s as they shall fynd amys, as to theym shall seme Convenyent.

1573. Lease to Dr. Freer, of Dr. Caius's old House.

(Repert. 18, lf. 106, bk.) 16 Nov. 1573.

Cam*erarius* Doctou*r* Freres leas*e* /

Item, at this Courte it was ordered that the gouerno*u*rs of S*t* Bartillmewes hospitall shall make a lease for xxj*tie* yeres of the house wherein Docto*ur* Keies lately dwelled, to M*r* Docto*u*r Freer, for xx*li* fyne, & the rent of v*li* by yere, notw*i*t*h*standinge any Acte heretofore made for not lettinge of any of thospitall land*es*, other then to fremen &c. /

1574. A Dishonest Hospital-Collector discharged.

(Repert. 18, lf. 264) Cur*i*a Sp*e*cialis. Sabba*ti*, vndecimo die Septembr*is*, Anno xvj*to* Elizabethe Regine. [A.D. 1574]

Edwardes /

Item, Wylli*a*m Edward*es*, Skynner, one of the gather[er]s of the money due to the hospytall, for that yt was substancyally provyd vnto this Co*u*rte, that he had verye lewdly & dysceytfully vsed and behauyd him selfe in thexecutio*n* of his sayd offyce, was therefore by this Co*u*rte cleyrlye dyschardged of and from the same offyce and the executio*n* thereof for ever /

---

3. *Extracts from the Guildhall Records relating to Barbers, to foreign and unlicenst Surgeons; and to the Plague in London.*

1496, Jan. Two Aldermen to examine the Statement of the Barbers and Barber-Surgeons.

(Rep. 1, lf. 12.) [*Present*] M[ayor[2]] Colet, Brou*n*, White, Mathue, Remyngto*n*, Isak, Broke, Pemb*er*ton, Purchas, Welbek, Shaa, Woode

It is agreed at the same co*u*rt that M*ai*ster Purchas and M*ai*ster Woode shall haue the exam*i*nac*i*on of the bill of barbo*u*rs and barbo*u*rs surgions, and to speake with the said barbo*u*rs, and to make report ageyn to the court.

[1] MS the the. [2] Henry Colet.

156 App. III. 3. *Unlicenst Medicals reported.*

1513. The Archbp. of Canterbury's Barber to be a Freeman of the City without fee.

(Rep. 2, lf. 158) x° die Maij, Anno v° H. viij' (1513).

Int*ratur*. [*Present*] M*ayor* [Sir W*m* Browne], Reco*rder*, Capell, Kneseworth, Aylem*er*, Acheley, Monoux, Boteler, Exmewe, Reest, Basford, Bruge*s*, Shelton, Dawes, Ambo Vic*ecomites*.[1]

The bisshope of Canterbury barbo*ur*. At this Court, At the instance of the right Reue*r*end Fader in god / the Archebusshoppe of Canter*bury* And Cha*u*nceler' of Englond, it is graunted to Thom*as* Hill, his se*r*ua*u*nt & barbo*ur*', that the seid Tho*mas* shall be fre man) of this Citie w*ith*out ony thyng paieng there*for* to the Chaumber*lein*.[2]

1514. The Surgeons not to be assest with the Barbers.

(Rep. 2, lf. 173, bk.) xvj° die Marcij (5 Hen. VIII, 1514).

Int*ratur*. Memora*n*dum, that the Surgeons from hensforth be Surgeons not ioyned w*ith* the barbours when) charges shall be sessed up-on) them).

1524. Three unlicenst Doctors reported, and one Surgeon stopt.

(Rep. 4, lf. 201) Jouis, xxj° die Sept*embris* (an. Hen. VIII, 16 ; A.D. 1524).

[*Present*] M*ayor* [Sir Thos. Baldire], Recor*der* / Butler / Milburn) / Brown) / Lambert, Askue, Pecok, [& Michael] Englyshe, Vic*ecomes*.

Drs. Bentley and Yakesley, the Examiners in Medicine and Surgery, complain of 3 incompetent Practitioners, At this Courte, camme Doctour Bentley & doctour Yakesley, docto*urs* of phisik, & examyners Admytted to hable or disable suche as practise phisik & Surgery in London); And by weye of Compleynt, ce*r*tyfied to this Courte, that the thre p*er*sones vndernamed, dayly practise phisik, [*leaf* 201, *back*] havyng no maner speculacion) & cu*n*nyng' that to doo / Wherfore Wylliam Nycholson) ys commaundyd to warne those iij p*er*sones to be here on) Tewesday next cu*m*myng', At which day the seyd ij doctours haue promysed to be here & c'

Smith, ⎧ Roger Smyth, Appotecary
Roys, ⎨ Roys, at the Grey Friers
Wescott. ⎩ Wescott, in) Seynt Swith[i]nnes lane

[1] The Sheriffs were John Dawes, John Bruges, Roger Basford.
[2] The next entry relating to the Surgeons is on "xij° die Maij" (1513) :—
Also it is co*m*maunded that warnyng be geuen) to the Wardeyns of surgeons of this Citie to appere at the next Co*u*rt of Aldermen, that is to sey, the tuysday after trinite sonday next co*m*myng./ [No further entry.]

III. 3. *Misbehaving Surgeons. Spittle-Houses.* 157

(Rep. 4, lf. 201, bk.) Martis, 27 die Septembris (1524).

[*Present*] Mayor / Recorder / Prior Ecclesie Christi / Boteler / Exmewe / Milburn / Mundy / Aleyn / Seymer, Partryche / Rudstone / Skevyngton / Dodmer / Broun.

(lf. 202)

Roys, the incompetent Surgeon,

is forbidden to practise physic.

Item, At thys Court camme the seyd Roys, Surgeon, dwellyng' At the Grey Fryers; to whome was Redde the Acte of Comen Counsell made the xxviij day of Aprill last passed[1] concernyng' Phisik', & Also the Certificate of the Doctours of phisik / Wherfore, Accordyng' to the same Acte, At the Request of the seyd Doctours, Iniunccion ys geuen to the seyd Roys, that he shall no more occupie Phisik, vppon payn lymytted in the same Acte, & c'

1536. City Control of the Spittle-Houses.

27 March 1536 (27 Hen. VIII), Repertory 9, leaf 117, back.

T. Barnwell appointed a Visitor,

at the old pay.

Item, that Thomas Barnwell, gentleman, shalbe one ot the visitors of the spyttelhowses, or lazar cotes, about thys Citye, yn as large & ample maner as Henry Clydero, late Cytezyn & of London, deceassyd, occupyed the same, with all the profittes & commodytyes therto belongyng / as longe as the sayd Barnwell well & truely behave theym selffes [? himself] yn the same.

1536. Richard Smith, a Doctor-Surgeon, expeld the City for a misdemeanor.

In 1536,[2] one Richard Smyth, a Doctor of Physic and a Surgeon, had committed some misdemeanor,—wrongfully troubled an Alderman, &c.—which the Wardmote Quest reported to the Common Council. The following entries are in Repertory 9 :—

(lf. 169) Jouis, xxvij die Aprilis, anno 27 (28) H. viij (A.D. 1536).

Smyth Item, the verdyt of the Wardemote enquest concernyng Rychard Smyth, yn the Warde of Colmanstrete of London, was Redde ; & agreed that a quest shalbe chargyd, & to trye the sayd matters by & by, without dylay.

(lf. 177) Jouis, xviij die Maij, anno 28 H. viij. (A.D. 1536).

Smyth Item, Richard Smyth, doctor of physyk & Surgeon, hath faithfully promysed to obserue the Judgement gyven ageynst hym the xxvij day of Aprill last past / & bycawse

---

[1] There is no entry of this Act in the Minutes of the Council held on April 28, 1524, on leaf 150 of this Repertory 4.

[2] The heading 'anno 27 H viij' must be a mistake for '28 H viij.'

158   III. 3. *Troublesome Smith. Foreign Surgeons.*

           the day ys past, he promyseth to departe owt of this
           Cytye afore the xx day of may, thys present moneth /
           & that no copye be delyuered concernyng the seyd matter

      (lf. 189) Martis, xxix die Augusti, anno 28 H. 8 (A.D. 1536).

Smyth   Item, the lorde pryvye seale made request to thys courte
           for Doctor Smyth, who of late was banysshed the Cytye
           for his mysdemeanure.

      (Rep. 9, lf. 230) Jouis, xiiij die Decembris, anno 28 H. 8. (1536).

Smyth   Item, a peticion of Rychard Smyth, Surgeon, was
His Petition Redde, touchyng the restitucion of hym self to hys
refused, because howse, & c¹; & bycause he hath wrongefully trowbled
he troubled Alder-
man Denham. Maister Denham,[1] Alderman, & others of the Warde-
           mote enquest of the warde of Colmanstrete, & yet
           continueth yn hys wylful mynde, wherfore hys seyd
           Request ys refused.

      (lf. 234) Jouis, xj die Januarij, anno 28 H. 8 (A.D. 1537).

Smyth   At thys courte a peticion by Richard Smyth, surgeon,
           requyryng to be restored to theyre favours, & to comme
           ageyn ynto thys Cytye.

      (lf. 236) Jouis, xviij die Januarii, anno 28 H. 8 (A.D. 1537).

Smyth   Item, at thys courte, a byll of Compleynt was exhybytted
           to thys Courte by Richard Smyth, Surgeon, to have lycens
           to comme to hys howse to see hys goodes

A.D. 1539. Leave for a Dutch Surgeon to practise jn London.

      (Repert. 10, lf. 163[2]) Jouis, xix? die februarii, anno 31, H. 8 (A.D. 1539).

Mastryk   Item, that vnder the scale of office of London, Henry
[Maestricht] of Mastryk, Surgeon, shall have lycens to exercyse
           hys connyng withyn the libertyes of London, without
           empeschement of eny person, as farre furthe as the
           Authorytie of thys Courte may graunt the same.

1538. A French Surgeon who has done great Cures.

      (Repert. 10, lf. 64) Jouis, vij Novembris, anno 30, H. 8 (A.D. 1538).

Forman   Item, the matter of John Lesture of fraunce, who hath
[Mayor]   occupyed phisyk & surgery; & dyuerse honest persones

---

[1] William Denham, Sheriff in 1534-5. He was not present at this Court of Dec. 14, with the Mayor (Sir Ralph Warren), the Recorder, Aleyn, Mundy, Ascue, Champeneys, Hollyes, Forman, Dormer, Cotes, Monnoux, Dauncy, Gresham, Bowes, and the Sheriffs R. Paget and Wm. Bowyer, who constituted it.

[2] The first 163. After lf. 175, the next leaf is numberd 156, and the numbering starts again from it.

App. III. 3. *Norman Physician. Lytster's Cures.* 159

Lesture	declaryd̾ the honesty & great Cures doon̄) by hym; Wherfore it ys agreed̾, that the Wardeyns of Surgeons shalbe here vpon̄) tuysday next com̄myng, & there shewe theyre gryeff; & the matter to be orderd̾ by thys *Courte*.

There is no further entry on this subject on Tuesday, Nov. 14 (lf. 65), or on any leaf near the Minutes of that day, unless the John Lytster of 1542 is the Jn. Lesture of 1538. The next Surgeons' entry is on leaf 66, back, (die) 'Jouis, xxj die Novembris' (A.D. 1538).

Surgeons Smythe	Item, at thys *courte* cam̄me doctor Yaxley & the Wardeyns of the Surgeons with̄ complaynt vpon̄) Smyth, mere surgeon. And it ys agreed that they shaH put agaynst the seyd̾ Smyth yn wrytyng: And than̄) the seyd̾ Smyth to answerr to the same yn wrytyng.

[No further entry on this.]

1539.   Malyard, a Norman Physician, allowd to practise for a year.

(Rep. 10, lf. 10) Jouis, xxix die Novembris, anno 29 H. 8 (1539).

Malyard̾	Item, Roger Barker, & Robert Nycolles, brewer, the swordeberer & Brygg-seriant, shewyd̾ how they were holpen̄) of theyre diseasses by one John̄) Malyard̾, straunger, phisicion, a Norman̄) borne: To whome, by thys *courte*, licens ys gyven̄) for a yere, to occupye his facultye, so that they¹ shaH com̄me to euery person̄) that woH.

1542.   Wonderful Cures done by John Lytster.

(Rep. 10, lf. 237) Adhuc Jouis 12/3/ Anno 33° H 8 (March 12, 1542).

Lytster Twenty-two City traders, and a gentleman,	Item, Att thys Court can̄) Thomas Trappys & Wylliam Calton̄), goldsmythes / John̄) Wendon̄) & Wylliam Mathewe, grocers, Rouland̾ Goodman̄) & Wylliam Berde, Fysshmongers, Wylliam Machyn̄),² vpholder, Robert Huntley, Skynner, John̄) Kynge & Rauf Marshall, Taloughchaundelers / George Hynde, plomer, William Lambe, gentleman, Rafe Hamersley, Clothworker, Robert Herdye, John̄) Clerke, Wylliam Smyth̄, John̄) Chundeler & John̄) Trett, drapers, Wylliam

---

¹ ? he, or he and his assistants.
² ? A relative of Henry Machyn of the *Diary* 1550-63 (Camden Soc. 1848), who was an undertaker and furnisher of funerals, and belongd to the Merchant-Taylors.

160 App. III. 3. *Surgeon Ferres.* 1547 *Plague.*

declare that Jn. Lytster has done notable cures to them;

and they ask that he be allowd to practise, tho' he isn't licenst by the Bp. of London.

The Court back him.

Preyst, pulter, Thomas Hawes, founder, Edward Burlasye, mercer, Walter Porter, & Thomas Neveson), haberdasshers, And declaryd vnto the same Court, that one John) Lytster, that practyseth physhyk' within thys Cytye, had, by the helpe of God, done dyuerse soueraH notable Cures vnto theym) / desyrynge the Ayde of the seyd Court for hys contynuance for the exercyse therof / forasmoche as he ys interrupted therin by the physycions of thys Cytye / bycause he ys not Admytted so to do by the Bysshope of London), accordyng' to the lawe / And Agreyd that there shalbe asmoche done theryn As thys Court may do for hys furtheraunce.

1544. Hen. VIII's Application for a City Post for his Surgeon, Richard Ferres.

(Rep. 11, lf. 75 ink, 73 pencil) Martis xxvij^{mo} die Maij, Anno xxxvj^{ti} Henrici viij. (1544).

Waren)
[Mayor]
(lf. 76, or 74)
Henry VIII asks that his Surgeon Rd. Ferres be made a Common Appraiser of the City.

Answer: there is no such post.

[*Present*] Mayor, Gresham, Dormer, Cotes, Laxton), Amcottes, Hoberthorn), Wylford, Sadler, Lewen), Judd, HyH, Jervys, Ac Tolos & Dobbys, vicecomites. Item, A lettre dyrected to sir Wylliam Bowyer, late lorde Mayer [part of 1543-4], & Master Recorder, from) the Kynges grace, in the Favour of Rychard Ferres,[1] one of his graces Surgeons, for to be one of the Comen preysers in this Cytye, was red / And Agreyd that An) Aunswer' shalbe made vnto hym), that there ys no suche Offyce, & that Sute Afore tyme hathe bene made for the same to the Comen) CounsayH by other persones to haue suche Offyce / but they in no wyse wolde assent to the same /

---

15 Nov. 1547. Crosses to be set on Plague-stricken Houses.[2] Gutters to be flusht.

Rep. 11, lf. 387, ink ; 363, pencil (by which Alchin's Index goes) Martis, xv^{to} die Nouembris, anno primo Edwardi vj^{ti} (A.D. 1547).

Gresham
(Mayor)

[*Present*] Mayor, Recorder / Waren / Roche / Laxton) / Bowes / Hoberthorn) / Amcottes / Tolos / Wylford / Judde / Dobbys / HyH / Barne / Lok / Hynde / Goodeve / Lyon) / Garrard / ac Whyte & Chertsey, vicecomites /

[1] See him in the cut of Holbein's picture.
[2] See the Section of the printed *Remembrancia* Calendar on this.

App. III. 3. *Plague of* 1547. *Sir John Aylyf.* 161

Crosses to be sett vpon mens dores for the declaracion of the plage

Item, for asmoche as my Lorde Mayer reported that my Lorde Chauncelar declared vnto hym that my Lorde protectours graces pleasure ys, & other of the Lordes of the Counseyll, that certein open tokens and Sygnes shulde be made & sett furth in all suche places of the Cytie as haue of late bein vysyted with the plage / Yt is therfore agreyd that preceptes shall fourthwith be made furth to euerye of my maisters thaldermen, that thei shall cause euerye howseholder of their seuerall wardes / which, syth the fest of All sayntes last past, hath bein vysyted with the plage, or that, ouerthyssyde the Fest of the puryfycacion of our Ladie nowe next commyng, shall happen to be vysyted with the seid dysceas, shall cause to be fyxed vpon the vttermost post of their Strete dore A certein Crosse of saynt Anthonye devysed for that purpose, there to remain xl dayes after the settyng vp' therof;

In every Ward, Wells and Pumps to be drawn thrice weekly, and 12 bucketsful of Water pourd down the Street-Gutters.

& also to cause all the welles & pumpes within their seid wardes to be drawen iij tymes euerye weke, that ys to saye, Mondaye, Wednesdaye, & fryday / And to cast down into the canelles att euerye suche drawyng, xij buckettes full of water att the least, to clense the stretes wythall /[1]

1548. (Rep. 11, lf. 458 ink, 466 pencil) Martis, xxviij° Augusti, Anno ij° E. /6/ (A.D. 1548).

J. Gresham Mayor. Adiournement of yᵉ courtes.

Item, this day, by reason of the vyolence of the plage att this tyme, & for dyuerse other reasonable consyderacions movyng the court, yt ys ordered & agreyd by the same, that this their sayd court shall cease & be adiourned vntyll thys day fourtnyght /.

1550. Sir John Aylyf leaves the Barber-Surgeons' Company.

(Rep. 12, No. 1, leaf 251 ink, 249 pencil) Jovis, xvij° Julij, Anno predicto [4 Ed. VI, 1550].

Hill, Maiore. [*Present*] Mayor, Laxton / Hoberthourne / Judde / Dobbis, Barne / Whyte / Locke / Hynde / Lamberd, Woodrof¹ / Kyrton / Ofley / Wythers, ac Turke vice-comes /

Barbour Surgeons

At this Courte, the Wardeins & Assistauntes of the felowship of the Barboursugeons, gave their assentes, accordyng to the order of the seid Courte, for the Removyng of Sir John Aylyf, knyght & Alderman, from their seid Felowshyp /

---

[1] See the extract as to Street-Cleaning, Nov. 1535, in the Street Series below, p. 170.

162 App. III. 3. *Barbers' Freemen. Dutch Surgeon.*

1550. A young Tailor-barber, Jn. Gardener, to decide whether he'll be a real Barber or not. He says No.

(Rep. 12, No. 2, leaf 278 ink, 276 pencil, back,) Jovis, 23 Octobris, Anno 4ᵗᵒ Edwardi vjᵗⁱ [A.D. 1550].

Hill, M*ayor*.
Barbours

Item, the yonge man vpon whom the barbours did here complayne, for that he, being free of the *mer*chaunt-tailors, occupyeth barbarye / hath day vntill twysdaye next, to make a p*r*ecyse Au*n*swer, whither he wilbe translatyd vnto theymͻ, or els leve barbary or not /

(Rep. 12, No. 2, leaf 281 ink, 279 pencil) Martis, 4ᵗᵒ Nouemb*ris*, An*n*o 4° Edwardi 6 [A.D. 1550].

Judde,
M*ayor*.
Gardene*r*

Item, Johnͻ Gardene*r*, *mer*chaunttayller / who heretofore hath occupied their barbours occupac*ion*, wherof the wardeins of the barbours dydᵗ here latelye complayne, hath this day here declared that he is contentydᵗ clerely to leve the saide occupac*ion*, & no more to medle wit*h* barbary /

1550. A Barber can't be admitted Freeman by redemption, save for £20.

(Rep. 12, No. 1, lf. 177, bk.) Mart*is*, xxvj Novemb*ris*, A*n*no iij° pred*i*cti Re*g*is (Edw. VI, 1550).

[Sir Rowland]
Hill, M*ayor*.
A Barbour

The le*tt*res of certein) of the Kynge*s* most hono*u*rable counseH in the favo*u*r of a barbo*u*r to be admyttyd into the libe*r*ties of this Citie by redempc*i*on were red, & redelyue*r*yd to my Lorde Maye*r* ; the contente*s* therof concernyng his admyssionͻ otherwise then for xx ɫi / clerely denyed /¹

1562. *A Dutch Surgeon allowd to set up his Bills on Posts.*

(Rep. 15, lf. 156) Jovis, 3° Decembris, an*n*o vᵗᵒ Do*m*i*n*e Elizabethe Regine, &c. (A.D. 1562).

Lodge, M*ayor*.
[*Present*]

Lyon, Huett, Chester, Harpe*r*, John White, Malory, Halse, Drape*r*, Rowe, Avenon, Baskerfeld, Bankes, Gilberd ; ac Alyn et Chambe*r*lyn, Vic*e*comites.

¹ Under May 6, 1606, is the following entry in Repertory 27, lf. 195 (191, pencil) :—

Kingman
Came*r*a*r*ius.

This daye, at the request of Elizabeth Stowe, widow, and according to a former graunt of this Court, It is ordered that Phillip Kingman shaɫbe made free of this Cittye by Redempc*i*on in the Companie of barborsurgeons,² paying to Ma*ster* Chambe*r*len, to the Cittyes vse, vjˢ˙ viijᵈ˙

² 'Pewterers' was first written, then crost out, and 'barborsurgeons' written over the line by another hand.

App. III. 3. *Barber-Surgeons' Cess.* 1563 *Plague.* 163

Van Dura*n*   At this Courte, Peter van Duran, a straunge*r* borne, who p*r*ofessethe y*ᵉ* knoledge & science of surgery, was licensed by the same Courte to sett vp bylles vpon postes,[1] in suche p*a*rtes of this Cytye as to him shall seame good, to geve the people knolege of his said science. And he agreid & graunted to the said Courte, to deale very honestly w*i*th all theym *th*at he shall take vpon hym to cure, for their charges conce*r*ninge y*ᵉ* same.

1563. The Barber-Surgeons having paid 2 cesses for 1, are let off another.

(Rep. 15, lf. 211, bk.) Adhuc Jovis .11. [Marcij] anno v*ᵗᵒ* D*o*m*i*ne Elizabethe Regine & c (1563).

Barbours & surgeons   Item, forasmoche as, vpon due examinac*i*on made, it did appeare vnto the Courte here this day, that y*ᵉ* felowship of the barbors & surgeons of this Cyty did, by a certeyne oversight & errou*r*, disburse & prest as moche redy money at y*ᵉ* lone made by the Cytezens of this Cyty in Octobre last past towarde*s* the provision of wheate & Rye, as they ought & have byn accustomed to be charged w*i*thall at ij severall lones, was this day graunted & agreid* by the Courte here, that at this p*r*esent last lone for the like p*r*ovision, shalbe spared & clerely discharged.

1563. London Plague Regulations.[2] Blue Crosses to be set on infected Houses; Gutters to be flusht; Bedding burnt.

(Repertory 15, lf. 259, bk.) Adhuc sabbati, 3º die Julij, an*n*o v*ᵗᵒ* dom*i*ne Elizabethe Regine, &c. [A.D. 1563].

Lodge M*aiore*

Came*r*ar*ius*

Blewe }
Crosses }

Item, it was ordered that there shalbe CC blew hedles Crosses made wit*h* all convenient spede by the chamberlyn, to the Intente that one of them may be sett vp vpon the vttermoste parte of the dore post at every

---

[1] Mr. Sidney Young believes that the Barber-Surgeons' Minute-Book has an Order about pulling down these Bills of Van Duran's.

[2] Mr. Baddeley of the Guildhall Library Committee, Churchwarden of St. Giles's without Cripplegate, says that there are over 4000 entries of deaths in his Church-Register for this year. Almost all are of poor folk, and enterd as dying of the plague or fever. The few richer ones wouldn't acnowledge to the plague, and are enterd as dying of dropsy, &c. See Mr. Baddeley's forthcoming book on his Church. This Plague gave rise to Wm. Bullein's Dialogue of the Feuer Pestilence, 1564, now editing for the E. E. Text Soc. by Mr. A. H. Bullen and his cousin Mr. Mark Bullen, from the edition of 1578.

164 App. III. 3. *London Plague-Regulations of* 1563.

mansion[1] howse of this Cyty that hathe of late, or shalbe visited this Sommer season with the plage;[2] And that every of my maisters the aldermen, having a competente number of the same Crosses, shall cause them to be sett vp as aforesaid by the constables or bedylles of their said wardes, as occasion shall require.

(Rep. 15, lf. 260, bk.) Adhuc martis, 6° Julij, anno v^{to} Domine Elizabethe Regine, &c. [A.D. 1563].

Camerarius. blew crosses. Item, it was ordered that the Chamberlyn shall cause CC hedles blew crosses more to be made with sped, at the Cytyes charges, to be vsed according to the order here taken the last Courte day for the same.

(Rep. 15, lf. 263, bk.) Adhuc Jovis, 8° Julij, anno v^{to} Domine Elizabethe Regine, &c. [A.D. 1563].

Lodge, Maiore.

[Blue Crosses for Finsbury.]

The donge hill at fynnesbury, & the plage.

Item, Laurence Nasshe, bayly of fynnesbury, had thie day, blew crosses delivered vnto him by the Courts here, to be sett vpp there at fynnesbury, vpon the vttermost Postes of the Dores of suche howses there as are visited with the plage; & he was also commaunded to cause the filthie donghill lyinge in the high way nere vnto fynnesburye Courte, to be removed & caried away; & not to suffer any suche donge or fylthe, from hensfurthe, there to be leyde.

(Rep. 15, lf. 281) adhuc .26. Augusti. anno. 5to Elizabethe Regine. &c. [A.D. 1563].

Lodge, Maiore

Adiournacio curie Maioris et Aldermannorum ad tempus &c. [15 Sept. 1563].

Item, yt was this day orderyd & agreyd by the courte here, that the same courte,—in consideracion of the greate plague that yt hath pleasyd almyghty god sharpely to vysyt & towche this citie with-all, at this presente, and of the absence of a greate number of my maysteres thaldermen from the sayd cytye, for theschuynge of the greate Daunger & perill of the sayd plague yet fyersly reygnynge /—shall stey & cease vntyll the xv.th. daye of September next comm-

---

[1] dwelling.
[2] See p. 56 of Bullein's *Dialogue on the Feuer Pestilence* (1564), ed. 1578, E. E. T. Soc. 1888 :—

'Good wife, the daiely ianglyng and rynging of the belles, the commyng in of the minister to euery house in ministryng the communion, in readyng the Homelie of Death, the diggyng vp of graues, the sparring in of windowes, & the blasyng forth of the blewe crosse, doe make my harte tremble & quake. Alas, what shall I doe to saue my life?' And compare what follows this in Bullein, with the Aldermen's going into the country to avoid the Plague of 1563 : **extract of 26 Aug. on this present page.**

The citeezens feare.

App. III. 3. London Plague-Regulations of 1563. 165

ynge, except yt be for somme greate & vrgent cause, which shaH necessarely requyre expedycion.[1]

(Rep. 15, lf. 281, bk.) Mercurij 29. Septembris. anno. 5to Elizabethe Regine. &c. [A.D. 1563].

Lodge, Mayor.

[*Present*] Lyon, Huet, Harper, Avenon, Baskerfilde, Alyn, Chamberlein; ac Bankes et Heywarde, Vicecomites [= Sheriffs]

Camerarius. The orderinge of the beddynge & clothes of the infectyd with the plague./
Yt was this daye orderyd by the courte here, that ij honest poore men shalbe appoynted by my Lord mayer, to burne & bury suche strawe, clothes, & beddynge as they shaH fynde in the fieldes nere adioynynge to the citye or with-in the same cytie, wheruppon eny person vysited with the plague hath lyen) or dyed. And that they shalbe recompensyd by the Chamberlein for their paynes therin.

(Repertory 15, lf. 287, 2 Dec., A.D. 1563.)

a proclamacion for the stey & lettynge of houses
Item, yt was agreyd that the proclamacion devysed for the steyinge of thowneres of thinfectyd mansyon howses within this cyty, from the lettynge of the same for a tyme, & here redde this daye, shaH tomorrow be openly proclaymyd thurrough the citye.

1564. (Rep. 15, lf. 301) adhuc Jovis. 20. Januarij, anno. 6. domine Elizabethe Regine.

White, Mayor.

Item, yt was orderyd that preceptes shaH furthwith be made to euery one of my Masters thaldermen, to caH aH thinhabitauntes of theyr severaH Wardes withoute delaye before them, & to gyve streyght charge and commaundement, with aH dylygence to ayre, clense &

---

[1] On September 28, of this Plague year, 1563, there was a City Gift of £60 to the Poor of London (Repertory 15, leaf 281, back) :

Adhuc Martis .28. Septembris. a° 5. Elizabethe Regine, &c. [A.D. 1563].

Camerarius the poore London /
Item, forasmuche as thinhabitauntes of this citie beinge of eny wealth, are not well hable to releve & succour the poverty of the same city in many places therof / yt is therfore orderid & agreyd by the courte here this day, that the Chamberlein, at the citiez charges, shaH disburse .lx li towardes the relyef of the sayd poore, at the order & appoyntment of my lorde mayre./

In *Repert*. 14, lf. 465 (27 March, 3 Eliz. 1560), the Treasurer of Barts is orderd to pay £100 to St. Thomas's, because it had then 140 poor there, 40 over its after-prescribed number. And on leaf 512, back (24 July, 3 Eliz. 1560), the title-deeds—'Evidences, wrytynges and munementes'—of all the City Hospitals, 'and the house of the pore in Smythfeld,' were orderd to be deposited in the Guildhall. These entries are printed (we find) in the *Supplement to the Memoranda* (1867), p. 43-4.

166 App. III. 3. *Plague-Regulations. Pest-House.*

preceptes and *proclamacion* for ayringe & *purginge* of howsez & other thinges./.

purge all theyre howsez, beddynge & apparrell, for the daunger of thinfecc*i*on of the sycknes of the plague, forseinge neu*er*theles, & takynge care, that they or eny of them doe neyther hange or beate oute, or cause to be beaten out or hanged, eny mane*r* of beddynge or apparrell that hath beyn) or come nere to the daunger of infecc*i*on of the sayd sycknes / & that a p*r*oclamac*i*on of lyke substaunce & effect shall furthw*ith* be drawen, & openly p*r*oclamyd to morowe, for the generall admonyshement & warnynge of all p*er*sons w*i*t*h*in y*e* seid cyty to doe y*e* lyke /

See, in Journal 18, leaf 184, the Precept of the Lord Mayor & Aldermen dated February 12, 1564 (6 Elizabeth), forbidding the setting forth or playing of 'eny maner of enterlude or stage playe, at eny tyme hereafter, w*i*thout the specyall lycence of the said Lord mayor Fyrst hadd & obteyned for the same, vppon payne of imprisonement of their bodies, at the discret*i*on of the said Lord *Mayor* & Aldermen).' The plays were not to be acted in any 'mansione house, yarde, gardyn, orchard, or other whatsouer place' in London or its liberties; and this, because 'the greate and frequent confluenc*es*, congugac*i*ons and assembles of greate nombers and multytud of people pressed together in smale Rowmes [was] very daungerous' for spreading the plague. See also the Precept there following, dated Feb. 14, 1564, for the inspection and watching of infected houses, and the supply of food to their inmates.

1611, Sept. 10. Compensation to the Surgeon of the Pest-house.

(Rep. 30, lf. 170 bk.)
Kinge, Chirurgeon at the Pesthowse. Camer*arius*.

Wm. King says he has so diligently attended Plague patients at the Pesthouse, that his own friends won't *use* or employ him: he asks for a Pension.

Item, this day, William Kinge, the Chirurgeon belonginge to the Pesthowse, p*r*esented a petic*i*on to this Court, shewinge his great Care & diligence in Curinge of such p*er*sons *a*s haue beene sent thither; and that, by reason of his attendance & imployment there, his fryndes & former acquaintance do vtterly refuse to vse him in his profession; and therfore desireth some reasonable allowance & yearly penc*i*on from this Citty, for his better maintenance, and the more to encourage him to continue his former care and indeavo*u*r in helpinge such p*er*sons as come to the Pesthowse: It is thervpon ordered by this Court, for the Considerac*i*ons aforesaid, that the said William Kinge shall haue

App. III. 3. *Plague of* 1625. *Doctors & Surgeons.* 167

yerely paid him, out of the Chamber of London, the some of iij^li for a stipend, so long as he shall Continue his place of a Chirurgeon at the Pesthouse w*i*th that Care & diligence as heretofore he hath vsed. The same to be paid him quarterly; the first payment to begynne at Michelmas next. And this shalbe M*aste*r Chamberlens warrant for the payment therof. *[The City give him £3 a year, as long as he works at the Pesthouse, to be paid quarterly.]*

1625, June 28. Physicians appointed by the City to attend Plague-stricken Folk.[1]

(Repertory 39, 1f. 255) Martis, Vicesimo Octavo die Junij, 1625, Anno *Regni Regis* Caroli, Anglie &c, primo /

Item, it is thought fitt, and so ordered, by this Court, that S*i*r John Leman, S*i*r Edward Barkham, S*i*r Martyn Lumleye, m*aste*r Alder*m*an Johnson, m*aste*r Alder*m*an Hamsleye, m*aste*r Alder*m*an Cambell, master Alder*m*an Ducie, and m*aste*r Alder*m*an Moulson, or any foure or more of them, shall meete this afternoone att the Guildhall, and conferr and treate w*i*th S*i*r Will*ia*m Paddie, knight, and others, Doctors of Phisicke, for and about one or two skillfull & sufficient Phisitions to bee interteyned and ymployed by this Cittie for the cure of those visited w*i*th the Plague; And to consider what recompence is fitt to be made vnto them for their advise and paines in that behalfe; And to certifie this Courte in writing vnder theire hands of theire doeinges and opinions therein; And John Olliffe to warne and attend the said Comittees. *[(1f. 255, bk.) Committee of 8 appointed to confer with Doctors about the City employing 1 or 2 Physicians to cure folk ill of the Plague; to fix the Physician's pay, and report to the Court.]*

1625, July 4. A Spanish Doctor, and English Surgeons, for the Plague.

(Rep. 39, 1f. 279 bk.) *Spanish Docter:* Item, this daie, the right hono*ura*ble the Lord Maior informed this Court, that hee hath agreed w*i*th the Spanish Doctor Pone [? MS.] one hundred

---

[1] James I died on March 27, 1625. On May 11, Charles I was married by proxy to the Princess Henrietta Maria of France in Nôtre Dame. An English fleet brought her to Dover, where Charles I met her, took her to Canterbury, thence to Gravesend, and by a grand procession up the Thames to London, which, says Oldmixon (*Hist. Engl.,* Jas. I and II 1730, p. 75, col. 1), "was in Mourning and Lamentation; the most dreadful Pestilence that ever had been known in Europe then raging there, above 40,000 dying this year of the Plague .... fatal predictions were not wanting on the Queen's Entry in such a calamitous Conjuncture, as if she had brought in her Retinue all the Scourges that were to make the Kingdom desolate (Larrey, p. 16)."

168 App. III. 4. *Punishment of Bawds and Scolds.*

<small>100 markes per Annum.

Heath Surgion 50ᵘ per Annum to cure yᵉ poore of the Plague:

Smith: surgin for yᵉ Pesthouse:</small>

markes per Annum, and master Heath, Surgion, to bee with him in some convenient [*blank*] within the Cittie, for Fiftie pounds per Annum, to doe theire best endevours for the curinge all the poore infected with the plauge, for nothing; and of the better sort infected, for some reasonable recompence; And also hath agreed with one master Smith, a Churgion, for xxxˡⁱ per Annum to abide att yᵉ Pesthouse for the cure of those sent thether visited. Wherevpon this Court, haueinge formerly referred this busines to his Lordshippe, doth now ratifie and Confirme that his Lordshippes doeinge./

---

4. *Street-Scenes: Punishment of Culprits, Public Rejoicings, Scavenging, Archery Meeting.*

A.D. 1523.   A Proclamacion for Bawdes & Scoldes.[1]

<small>(Letter Book N, leaf 233)

As Roger Gill and Jn. Inman and his Wife,

have practist lechery and bawdry,

Gill being bawd to his Wife,

and Inman and his wife being bawds or panders for Priests and other folk,

the 3 shall be taken to their prison, thence to Newgate, and thence (with pipes, pots and</small>

Munby, Maior.[2]

Forasmoche as Roger Gyll, Sadler, Iohn Inneman & [*blank*] his wif, that here stonde, been laufully Convict before my lord the Maire of this Citie, & his Brethern thaldermen of the same, by solempne processe after the Custome of this Citie, of that / that they be persons not dredyng god, ne shame of thys worlde, But contynually vsyng the Abhomynable Custome, mayntenaunce & Conceillyng of the foule and detestable synne of lechery & bawdry / That is to seye, the said Roger Gill, for beynge Bawde to his wif / And the said Iohn Inneman and his wif, for that / that they be Comon Bawdes for prestes[3] & Mennys wiffes, wedded Men and Syngle women / Yt is therfor adiuged by my saide Lorde Maier and his brethern, that the said iij persones soo atteynt, accordyng to the Lawes & Customes of this Citie in that behalff vsed, & owt of tyme of mynde contynued, shalbe conveied to the prison[4] that they cam froo / And from thens to Newgate / And from Newgate they to be conveied with Mynstralcy, Basyns and pannes Rongen afore theym, thorugh Chepe,

---

[1] This is given for its 'Mynstralsy, Basyns and Pannes rongen afore' the Culprits, as a sample of the London street-sights that would come under Vicary's eyes. And the Vagabond extracts below are added for the like reason.

[2] He was elected in Nov. 1522.

[3] Of the long list of men taken in adultery from 2 Henry IV, onwards, in Letter Book I, leaf 288, almost all are Chaplains.

[4] ? MS. persone.

App. III. 4. *Festivities on Francis I's Capture.* 169

and soo to the pillory in Cornehill / And then) the said iij persones to be sett in the said Pillory by A certeyn space / And then) and ther' the said Causes to be proclamed / And so from thense to be conveyed too Algate, and then) to be voided owt of this Citie / And god saue the kyng /¹ · {*pans rung before them,) to the Pillory in Cornhill (to be pelted), then to Aldgate, and there turnd out of the City.*}

1525, March 11. Bonfires, Music, and Festivities, for the taking of the French King, Francis I, at the Siege of Pavia, on Feb. 24, 1525.²

By the Maire³ (Journal 12, lf. 329).

Bayly Mayor. Intratur. — We charge and commaunde you,⁴ on the behalf of our' soueraigne lorde the kyng, that anon), vpon) the sight herof, ye do prepare, and cause to be made,

We bid you have Fires made at 7 p. m. in your Ward, — within your said warde this present Saterday, at vij of the Clok in the Evennyng, certayn) Fires, after the maner of Midsomer fyers,⁵ or better, by your discrecion) ;

and let the young Children be well drest, and sit round the Fires, with Music, while the Householders drink joyously together. — and that the yong Childerne of the same your warde, be goodly garnysshed, and so to sitt vpon) the stalles aboute the said Fiers, after the maner of a Somer game, with mynstralsy accordingly / and the housholders, with their seruauntes attendyng vpon theym), be neybourly drynkyng to-gethers at the said Fiers In Ioyous maner ;

---

¹ See another entry of like kind against Richard Wyer of Bread-Street Ward on May 25, 1529 (21 Hen. VIII), in Journal 13, lf. 141, bk. He was 'a Comyn brynger & Conveyer of certeyn sengle Women to merchauntstrangers places within the said Citie . . to vse & occupie the fowle & detestable synne of lechery & Bawdry, to the high displeasure of almyghty god, & to the perelous example of other good & well disposed persons, & Contrary to the Auncyent liberties & Custumes of the said Citie.'

² *Newes of the siege of Pauia, & the taking of the French king prisoner.* On thursdaie the ninth of March [1524-5], at seauen of the clocke in the morning, there came a gentleman in post from the ladie Margaret, gouernesse of Flanders, which brought letters, contening how that the foure and twentith of Februarie, the siege of Pauia (where the French king had lien long) was raised by force of battell, and the French king himselfe taken prisoner . . . .

(Bonfires and Triumph in London.) Bounfires and great triumph was made in London for the taking of the French king, on saturdaie the eleuenth of March ; and on the morow after, being sundaie, the twelfe of March, *(Henry VIII at St. Paul's.)* the king came to Paules, and there heard a solemn masse ; and after the same was ended, the queere sang *Te Deum*, and the minstrels plaid on euerie side. [An account of the Siege of Pavia follows, from Guicciardini's History.]—1587. Holinshed's *Chron.* iii. 884, col. 1.

³ Sir Wm. Bailey, Nov. 1524-5. ⁴ The Alderman of each Ward.
⁵ See Stow's *Survey of London*. p. 39, col. 1, ed. Thoms, 1842.

170 App. III. 4. *Festivities on Francis I's Capture.*

<small>Have the Watch well drest and armd,</small>
<small>with Lights before them.</small>
<small>You be in scarlet and on horseback at the Guildhall at 7 p.m.</small>

and that you cause aH the Constables within your' said warde to be in harneys and other goodly appareH, and to be furnyssed with his Watche accordingly, with Cressett ligʜt borne before them, and to kepe the watche oonly in their said warde / And that you your self, beyng apparelled in Scarlett,[1] and oñ horsback, be redy at the Guihald at the said houre of vij at the furthest, then) and there to attende vpon vs / Nott fayling herof, as you wiH aunswer at your pereH / Youeñ at the said Guihald this present·Saterday the xj[th] day of Marche, &c.

Halle says in his *Chronicle*, p. 633, ed. 1809:

'Saterdaie the xi daie of Marche, in the citee of London, for these tydynges [the defeat of the French, and the taking of <small>A triumph for the taking of the Frenche kyng.</small> their king, Francis I, by the Emperor's and the Duke of Bourbon's forces, at the Siege of Pavia], wer made greate fiers and triumph; and the Maior and Aldermen road about the citee with Trumpettes, and much wyne was laied in diuerse places of the citee, that euery man might drynke; and on Tower hill the Ambassadours of Rome, of Flaunders, and Venice, had a greate banket made in a goodly tent, whiche pleased theim well; and as thei returned homewarde, all the stretes were full of harnessed men and Cressettes, attendyng on the Constables, whiche they praised moche.'

Lord Berners, writing from Calais on Wednesday, March 8, 1525, says he has just heard the news of the capture of Francis I, and prays to God that it may be true. (Brewer's *Calendar*, vol. IV. Pt. 1, p. 514, No. 1167.) The Emperor Charles thankt God, but forbade any public rejoicings. (*Cal.* iv. Introduction, p. xl.) Henry VIII was told of it on March 9, and was in high spirits. (*Cal.* iv. Introd. p. lxx.)

Nov. 1535. House and Street-Cleaning in London.[2]

(Rep. 9, lf. 134 bk.)

Raker   Item, that the Raker yn euery Warde, that ys to say, wekely, euery Munday, Wedyns day, & Saturday, shaH

---

[1] The Alderman's state colour.
[2] We have not lookt for entries about cleansing the Thames, but having come on the two following, we give them in a note. The pulling-up of the weirs was doubtless to get a good scour for the river as well as to clear it for navigation.
   Oct. 9, 1606. At the Court of Common Council held this day, (Repertory 27, leaf 284 (281, pencil),
'Item, it is ordered that the Waterbaylif shall presently pull vp all the weirs,

App. III. 4. *Street-Cleaning at Furnivall's Inn.* 171

Intratur	have a horne, & blowe at euery mannes doore, that they may have warnyng to lay owt theyre offaⅡ of theyre howses ynto the opon) streates euery day afore v of the clokke afore nyghte, vpon) payn) & peryⅡ that
Camerarius	shaⅡ faⅡ therevpon) / & that Master Chamberlayn) shaⅡ
Hornes	provyde hornes for the sayd seueraⅡ Rakers at the
Intratur	costes of thys Cytye.

1536. Complaint of the non-Scavenging of Furnivall's Inn and Ely Place.

(Rep. 9, lf. 183, bk.) Jouis, xiij° die Julij, anno 28 H. viij (A.D. 1536).

Ely Furnyvalles Inne.	Item, forbycawse compleynt was made by one of þᵉ scavagers of yᵉ Warde of Faryngdon), for kepyng of the stretes there vnclene, & yᵉ gentlemen of Furnyvalles
The Gentlemen of the Inn and the Bp.'s tenants won't clean their bits of Holborn, or pay for having them done.	Inne & tenauntes of yᵉ Bysshope of Ely woⅡ not amende, nor pay theyre Duetye for the clensyng thereof afore the Bysshoppe of Elyes rentes & afore Furnyvalles ynne; & by thys courte it ys ordered that Master the Chamberlayn & Towneclerk shaⅡ go to my lorde of Ely & the company of Furnyvalles Inne, & to knowe theyre myndes yn that byhalf; & therof to make reporte therof to thys courte.

1536. Streets to be kept clean, and Wells drawn.

(Letter-Book P, lf. 98) Aleyn, Maior, secundo tempore.

xxj° die Augustij, Anno regni Regis Henrici viij^{ui}, xxviij°.

streetes to be kepte clene	Item, that my masters the Aldermen shall Resorte ynto their wardes, to see and cause the stretes and lanes withīn their sayde wardes be clensed of almaner of Fylthe; And that the[y] cause the welles to be Drawen accordyngly /.

stoppes, and hatches in the river of Thamys betwene Colne Ditche and London bridge, And that he take care—as he will answere it in this Court, if anye default in him shalbe found,—That none of them be hereafter suffered to continue againe : & Master Chamberlen to paye the charge therof.'
The Colne runs into the Thames at Staines in Middlesex. On Nov. 14, 1609, and 23 April, 1610 (in consequence of a charge from Jas. I's 'owne mouth'), the Common Council appointed Committees to guard against the Plague, to see to new buildings and their inmates, &c. 'And alsoe for taking care of apprehending of all sortes of Rogues, vagabondes, and idle persons, to be punished and delt with according to the lawes and Statutes of this Realme, Or otherwise for sending such of them as shalbe found within the Cyttie, to Bridewell, there to be sett on worke, for clensing the ryver of Themes /'
On May 3, 1611, order was made for the continuance of the Committees and their work : Rep. 30, leaf 112, back.

172 App. III. 4. *The grand March-Past, May* 1539.

1539. Muster and March of London Citizens before Henry VIII at Westminster.

On May 8, 1539, was a grand Muster of the Citizens of London before Henry VIII at Westminster. Armd and in gala array, they marcht from Aldgate in 3 battalions, and the function is described enthusiastically by some predecessor of Ben Jonson in the office of City Chronologer or Chronicler, afterwards held by Thomas Middleton and Francis Quarles (*Remembrancia*, 305, 306). His description takes up 7 pages of the Letter Book P, leaves 202-5.[1] Had not our Appendix been so full of other details, we should certainly have printed (or reprinted) this picture of martial City life; but as matters stand, we must content ourselves with an extract showing where the 'Surgeons' (the small 'Fellowship of Surgeons') were, for the Barber-Surgeons do not appear. We at first supposed that the Chronicler had naturally sunk the less dignified 'Barbers' on this magnificent occasion, but we now think that the Barber-Surgeons, as a poor Company, could not, or would not, go to the cost of the gay white sarcenet coats which the other Citizen-soldiers bought for this grand march-past.

The cause of this manifestation of loyal feeling was, that the King, having been informd by trusty friends '*that* the cancarde & venemous serpent, Pawle, Bysshop of Rome [Pope Paul III], by that Archetraytou*r* Reignolde Poole, enemye to Gode*s* worde & his owne natra*ll* countrey, had moved, excyted & styrred dyuerse greate Prynces & Potentates of Crystendome, not alonely to envade this Realme of Englan*d* w*i*th morta*ll* warre, but also by fyer & sworde to extermyn & vtterly to destroy the hole nac*i*on & gene*r*ac*i*on of y* same.' Henry had accordingly gone to the coast, built blockhouses, got his navy ready, orderd musters of all able men, reports of armour, &c. all over the country and in London. These musters had been made in London, and all the ablest men pickt out. The King promist to see the Londoners march past him at Westminster. So they bought silk coats, silk helmet scarves, brooches, feathers, chains, gilded their armour and poleaxes, and at 6 a.m. on the eventful 8th of May, musterd in the fields in the East of London, which 'were a*ll* cou*e*red w*i*th men in bryght harnes w*i*th glystering wepons.' They formd 3 Battalions. Vicary was, we fear, not let into the Second,

---

[1] A copy of it on parchment, A.D. 1826, is in the Guildhall: see the Library MS. Catalog. All the opening and ending passages of this Account were printed by Grafton in *Hall's Chronicle*, p. 828-830, ed. 1809, without acknowledgment.

App. III. 4. *Citizens march past Henry VIII.* 173

with the 'upper ten,' the Surgeons. In the second Battalion or 'Battayle' marcht first the light Ordnaunce, and Gunners, with a Standard, under an Alderman captain. Then the Archers; 3. the Pikes; 4. the Billmen, five and five in a rank, with their Captains in front; 5. the Constables and Whifflers; 6. five Drums (dromslettes) and Fifes ('all apparelled in whyte Satten puffed out with crymsen sarcenet,) which made a warrelyke noyse'; 7. '.v. talle persones .. in whyte Sarcenet ruffyd & pouncyd very gorgeously,' with five Banners, which 'waving & Strayned with the wynde... made a goodly Showe;' 8. the Swordbearer in white damask on a good horse, freshly 'trapped,' his scabbard 'sett full of oryent perle'; 9. the Lord Mayor, Sir Wm. Forman, in gilt armour, and over it a coat of black velvet with a rich cross embroiderd on it; a massive gold chain round his neck, and on his head a black velvet cap with a rich jewel in it; his horse had crimson velvet trappings embroiderd with gold, and he was attended by four footmen in white satin hose puft with white sarcenet; 10. his two Pages in crimson velvet and cloth of gold, on prancing coursers trapt with bells and buttons of goldsmith's work; 11. sixteen halberdiers in white satin hose and doublets puft with crimson sarcenet, white leather coats slasht, white caps and feathers, and gilt halberds; 12. the Recorder in fair armour and a coat of black velvet, bearing 'a two-hande sworde on his sholder,' a chain round his neck, and four halberdiers in attendance; 13. five ranks of Constables in silk, Attornies' Clerks, and Guildhall Law-Officers, all in white silk with gold chains and brooches; 14. (no Barbers, tongs or razors allowd), unarmd :—

Than folowed all the surgeons of the Cytie, without harnes, in whyte cotes, with their bendes of whyte & Grene bawdryke-wyse, & their splatters ouer the bende (which ys their accustomed cognysaunce[1]) in verye good ordre & apparell /.

15. the two Sheriffs, Wm. Wilkinson and Nicolas Gibson, in coats of black velvet, followd by halberdiers, billmen, five Captains, &c. Then came the third Battalion with the great ordnance in its rear.

In this ordre the fyrst battle entred in at Algate before ix of the clock, the same day being thursday / And so passed thorough the Cytie in good ordre after A warlyke facion tyll thei camme to Westminster, where the Kinge & all the nobylytie stode & beheld the mustre, before whom, as well the great Gonnes as the hande gonnes

[1] This Badge was given them by Henry VIII. See Dethick's Arms for the Barber-Surgeons in South's *Craft of Surgery*, opposite the title-page, and the blazons on pages 352, 353; 358, 359. *Splatter* is a short spatula.

174 App. III. 4. *Citizens' March-past. Vagabonds.*

of euerye battayll, shott very terrybly[1] / and so all thre battailles, in the ordre before rehersed, one after and other, passed thorough the great Sanctuarye at Westminster, & so abowte the Parke at saint Jamys, into A great feld abowt the same place, where the Kinge, standing in his Gate-house at Westminster, myght bothe see theim that camme forward, and also theim that were Passed before /.

Than from saint Jamys felde the hole Armye passed thorough Holbourne, & so into Chepe; & at Leden Halle seuered & departed / And the last ordeynaunce camme into Chepe ageine abowte fyve of the clokke; so that from .ix. of the clocke in the forenone, tyll fyve at afternoone, this mustre was not ended /.

To see howe full of lordes, ladies & Gentlewomen, the wyndowes in euerye strete were / And howe the stretes of the Cytie were replenysshed with people, many men wolde have thought that thei that mustered had rather bein straungers than Cytezens, consydering that ye stretes euerye where were so full of people, which was to straungers a great marvell.

---

15 Nov. 1547. Vagabonds to be whipt, or pilloried.

(Rep. 11, lf. 388, ink; 364, pencil) Martis, xv^{to} die Nouembris, anno primo Edwardi vj^{ti} [A.D. 1547].

Vagabundes
to be whipt naked at the cart's tail,

Item, it is orderyd & Agreyd that John Launder, James Foster, William Haddok, & John Croydon, valyant & Sturdye beggers,[2] which were apprehended within the Cytie, shall to-morowe be whypped naked att A Cartes Taylle,[3] accordyng to the Lawe / And

[1] Hall prints 'cherefully,' p. 830, ed. 1809.
[2] On Nov. 9, 1518, the Common Council resolvd (*Letter-Book* N, leaf 100) that "Iohn Abbot, peauterer, ys Admytted to be in the stede & place of Henry Barker, for thavoydyng of vagabundes & myghty beggers oute of this Citie; which Henry, for that that he dide not his diligence Aboute the same, & Also for diuerse Consideracions this Court movyng, ys Amoved from' the seid Rome / The seid Abbot to haue lyke wages & lyuery as the said Henry hade."
[3] The Letters Patent of Edward VI, dated June 26, 1553 (just before his death on July 6), which gave Bridewell and its endowment to the City Authorities, bade them take up, and commit to the House of Labour at Bridewell, all 'idle lazy ruffians, haunters of stews, vagabonds and sturdy beggars, or other suspected persons whomsoever, and men and women whomsoever of ill name and fame:' Englishing in the *Memoranda, Royal Hosp.*, 1863, p. 69. And in the Resolution of Common Council, Feb. 29, 1556 (vltimo die Februarii, Annis Regnorum Philippi & Marie, Regis & Regine, &c.), ordering that the money needed for the conversion and fittings of Bridewell should be raisd only from 'the Cheifeste & beste companyes & fellowshippes of the seyde Cytie' (*Letter Book* S, leaf 68, back), and not from poor Citizens, it is recited that Bridewell was given them "to thintente that they shulde, with Convenyente spede, cause the greate number of the vacaboundes, sturdie & valiente Beggers, & Idle maisterles men that the sayde Cytie from tyme to tyme is [leaf 68, back]

App. III. 4. *Vagabonds to be whipt or pilloried.* 175

that Willi*a*m Jakso*n*, Lazarman, who of late hath wrechedly & falsely spoke*n* c*e*rtei*n* slaunderous word*es* against s*i*r Marte*n* Bowes, knyght, maist*er* Barne, Aldrema*n*, & other me*n* of worshyp*e* sytting in the said Courte, shalbe whypped thorrough Chepesyde / And then a*ll* thei .v. to avoyde the Cytie for eu*er*,

and to leave the City.

vppo*n* the paynes in suche case ordeyned & provyded / And that Rob*er*t Shakysberie, being butt A boy, & dysceased w*i*th the palsey, or some other dysease wherew*i*th his bodie shakethe verie sore, sha*ll* lykewyse furthw*i*th dep*ar*te out of y*e* Cytie, vppo*n* payne of whyppyng yf he make defaute /

A palsied boy to leave the City.

Yonge, to sytt vpon the pyllory for his falsehode.

Item, it is ordered & adiuged by the Courte here, that Thom*a*s Yonge, A Sturdy Vagabunde, who was here laufully convycte this daye, aswe*ll* by his ow*n* confessyo*n*, as by good & honest wytnesses, of that / that he doth not onely [*leaf* 388, *back*] Lyve idlely, wythout any maist*er* or s*er*uyce / but also that meny tymes he practyseth & vseth meny false & Craftie meanes wherby he hath dysceaved meny of the kyng*es* leage people, somtyme by forgyng of false tokyns & messages, And sometyme by counterfeityng hym self (stondyng in the hygh weys aboute this Cytie) to be A p*ur*veyo*ur* for the kyng*es* maiestie, allegyng hym self to do yt by Commyssyo*n*, shewyng forth to theim that he p*ar*ceyveth to be vnlerned, A boxe closed, affyrmyng his Commyssyo*n* to be therin / sha*ll* to-morowe, & ij merkett dayes more, in example of oth*e*r offenders, be sett vpo*n* the pyllorye in Chepesyde, with a pap*er* vpo*n* his hed declaryng his seid offenc*es* / And that he sha*ll* stonde there thre houres eu*er*ye of the said Dayes in the m*er*kett tyme / And that, att the Last of those iij dayes, one of his eares shalbe nayled to the pyllorye / And that he, after this his pen*a*unce done, sha*ll* avoyde the Cytie for eu*er* /

He forgd tokens,

and pretended to be a Purveyor for the King.

One of his Ears shall be naild to the Pillory.

muche pesteryd. molested & burdened w*i*thall, their, in some comp*e*tente parte of the sayde howse, to be sett a worke, & be compelled, by some good and necessarye bodely laboures & occupac*i*ons, to gett their owne lyving*es*, & to exchewe and avoyde Idlenes, and theire other lewde and vnlawfull kyndes of lyvinge / ”

See the amusing Letter of the poet Cowper, Nov. 17, 1783 (*Works*, ed. Southey, 1837, vol. xv, p. 134), as to how Molly Boxwell's younger son was whipt at the cart's tail for stealing some iron-work from Griggs the butcher. The Beadle drew his lash thro' his left hand full of red ochre, and left a red stripe on the culprit's back, but didn't hurt him. The Constable thrasht the Beadle with his cane, and a lass pulld the Constable's head back by his club of hair, ‘and slapt his face with a most Amazonian fury.'

App. III. 4. *Street-Cleaning, May Games, Plays.*

1553, June. Streets and Gutters to be daily swept and clenyd with Water.

(Letter-Book R, lf. 256) Barne, Maior.

By the Mayor.

To th[alderma]n) of the warde.

Tell your Scavengers and Rakers to make all Inhabitants sweep the Streets and Gutters before their doors daily at 7 p.m., and flush them with water twice a day.

For Clensinge the Streates and drawing of watter //

Bid the Rakers be ready to carry off the dirt.

We Straightlye Charge and Commaunde yow, that ye call alle the Constables, Skavengers, Bedels, and Rakers of your saide warde, Before yow, and that ye gyve theym) Streightlye in Commaundement, that they from hensforward doo see and cause all the Inhabitauntes of your saide warde, within their Seuerall precinctes, to swepe and clense ye streates & cannelles afore theare dores, every evenynge at vij of the clock Durynge this Somer tyme; And that all the welles & pumpes within the same your warde, euery evenyng and mornynge, at the hower aforsaid, be Drawen withe watter for the better makinge cleane of the same Streates; And that the Rakers of your saide warde, with all dylygence possible, be redye from tyme to tyme to caraye awaye the Sollage[1] of the Clensinge of the saide Stretes. Faile ye not &c /

/ Blackwell / [Town-Clerk]

1554. Order against May Games, Stage Plays, &c. in London Streets.[2]

(Journal 16, leaf 287, back, between 19 April and 22 May, 1 Mary, A.D. 1554.)

No one is henceforth to

set on foot

May Games /

Morris Dances, or Stage Plays, in any open place, or sound a Drum there.

My lorde Mayre, and his brethern the Aldermen of this our moste drade and most benygne souerayn Ladie the Quenes Citie and Chambre[3] of London, on her hignes behalf, do straightlye charge and commande, that no maner of person or persones do in any wyse from hensfurthe make, prepare, or set furthe, or cause to be made or set furthe, eny maner of mayegames or moryce dawnce, or eny enterludes or Stage playes, or sett vpp eny maner of maye pole, or bucler playeng, in any opyn streat or place, or sounde eny drume for the gatheringe of eny people within the said Citie or the lib[er]ties therof /

[1] Soil, refuse.

[2] This Order implies what we know is the fact, that these Games and Plays had gone on in the streets or open places. Vicary must have seen some such. There are many Acts of Common Council against Interludes, Plays, &c.

[3] The Chamberlain's office or Treasury, says Dr. Sharpe: the City of London was cald the King's Chamber.

App. III. 4. *May-Game. Archery-Meeting.*

If any Maypole has been lately put up,

it shall be puld down speedily.

And also, yf any suche maye pole be alredie latelie set vpp in any open place wit*h*in the Citie or lib[er]ties therof, that then the p*a*risheners of the p*a*rishe where eny and eue*r*ye suche maye pole ys set vpp, shaƚƚ cause the same, withe convenient speade, to be taken downe agayne / & no longre suffre them theare to stande, not only vppon payne of ymprisone*m*ent / but also vpon suche further payne as the said lorde Mayor & Aldre*m*en shall thinke meate and convenient /

God save the quene!

1557. 'The xxx day of May was a goly [jolly *or* goodly] May-gam in Fanch-chyrche-strett, with drumes and gunes and pykes; and ix wordes [The Nine Worthies] dyd ryd; and they had speches, evere man; and the morris dansse, and the sauden [Sultan], and a elevant with the castyll; and the sauden and yonge morens [Moors] with targattes and darttes; and the Lord and the Lade of the Maye.'—Machyn's *Diary*, 1550-63, p. 137, ed. 1848.

1557, Aug. 29. An Archery-Meeting in Finsbury Fields, open to all Comers.

(Journal 17, leaf 46, between entries of 4 and 11 Nov. 4 & 5 Philip & Mary, A.D. 1557.)

Offley, Mayor.

A proclamac*i*on for shootinge in Fynnesburye Felde /.

As shooting in the Long Bow has ever defended this Realm, and every good Englishman is bound to uphold it,

the Lord Mayor, &c. appoint a Game of Shooting, on Sunday week, Aug. 29, 1557,

in Finsbury Field at 2 p.m.,

open to all comers

By the Maier.

My Lorde Maier and my m*a*sters the Aldermen) of the Citie of London), callinge to theire remembrance the manyfolde benefit*es* and co*m*modities that haue co*m*men) to this realme by the feate of Archerie and showtinge in the longe bowe, wherby (God be thanked) this saide Realme hathe ever, in tyme heretofore past, ben) defended against the Cruell mallice and daunger of outwarde enymyes / And so from thensfurthe (God willinge) shalbe foreuer / whiche saide feate of showtinge eue*r*ye good true Englisshe man) is naturallie bounden) to maynteyne, supporte and vpholde to the best of his power / And to thintent that the saide feate of archerie shulde be the better maynteyned and vpholden), to incorage the king*es* subiect*es* more and more to vse and exercise the same / My saide Lorde Maior and m*a*sters the Aldermen) haue appointed and fullie concluded, that on sondaie co*m*me sevenight*es*, whiche shalbe the xxix[th] daie of this present monethe of August, shalbe a seuerall game of showtinge, in the felde called Fynnesburie felde, at ij of the clocke at afternone / And who will co*m*me thither and take a longe bowe in his hande, —havinge the standarde therin therefore prouyded,—

VICARY.

178 App. III. 4. *Archery-Meeting in Finsbury Field.*

I. 1st Prize, for the best and longest shot, a Gold Crown or 13s. 4d.;
2nd Prize, a Gold Crown or 10s.;
3rd Prize, a Gold Crown, or 6s. 8d.

II. For the Bearing-Arrow competition, 3 arrows of gold, or money: value
a. 13s. 4d.
b. 10s.
c. 6s. 8d.

III. For Flight Shooting, 3 flights, or cash, value:
d. 10s.
e. 8s.
f. 6s.

When the gamers be assembled togither /.
All men shall keep the peace.

People shall stand out of danger's way,

at least 20 yards off the mark.

At every shot, a Trumpet shall sound, to warn folk.

and fairest drawethe, clenliest delyuerethe, and farthest of grounde shootithe, shall haue for the best game a Crowne of golde of the value of xiij s iiij d̃, or xiij s iiij d̃ in money therefore / And for the seconde game of the saide standarde, he shall haue a Crowne of golde of the value of x s, or x s in money therefore / And for the third game of the saide standarde, he shall haue another Crowne of golde of the value of vj s viij d̃, or vj s viij d̃ in money therefore / And for the best game of the bearinge arrowe, he shall haue an arrowe of golde of the value of xiij s iiij d̃, or xiij s iiij d̃ in money therefore / And for the seconde game of the saide arrowe, he shall haue annother arrowe of golde of the value of x s, or x s in money therefore / And for the thirde game of the saide arrowe, he shall haue one other arrowe of golde of the value of vj s viij d̃, or vj s viij d̃ in money therefore ; And for the best game of the flight, he shall haue a flight of golde of the value of x s, or x s in money therefore / And for the seconde game of the saide flight, he shall haue a flight of golde of the value of viij s, or viij s in redye money therefore / And for the thirde game of the saide flight, he shall haue a flight of golde of the value of vj s, or vj s in money therefore / And god saue the kinge and Quene /.

My Lorde Maier and my masters thaldermen of the Citie of London, on the behalfes of our soueraigne Lorde the kinge, and soueraigne Ladie the Quene, charge and commaunde, That euerye man repayringe to this game of shootinge, kepe the Kinge and Quenes peace in his owne person, vppon the payne of imprysonement; and further to make fyne, by the discression of my saide Lorde and masters / And also that no person approche or comme so neare That he shall stande in daunger of anye Shott, but to be and stande at large, oute of perill and daunger, for his owne ease and others ; and for the good and due orderinge of the same, no person be so hardie to stande within xx yardes of anye of the stakes appointed for a marke, vppon the perill that will fall therof / And to thintent no person shall excuse hym by ignoraunce, there shalbe a trumpett blowen at euerye shott, aswell of the standarde, as of the arrowe or flight / That euerye person maie therby take warnynge to avoide the daunger of euerye of the saide Shottes /.

IV.

VICARY'S BAILIFF'S ACCOUNTS OF BOXLEY MANOR, &c.[1]

Ministers' Accounts, 34-35 Hen. VIII (A.D. 1542-3), No. 127.

Officium Balliuorum Generalium possessionum nuper Monasterij de Boxley } Compotus Thome Vicarye et Willelmi Vicary, Balliuorum Generalium terrarum et possessionum dicti nuper Monasterii, per tempus predictum.

The Account then follows. It shows, first, receipts from various places in Kent and London; then a rent of 15l. 0s. 10½d. received from Thomas Wyat as the tenth part of the clear yearly value of the House and site of the late Monastery, and of the Manors of Boxley, Hoo, and Newenhamme Courte, &c. (except the Rectory of Boxley, &c.), granted in 32 Hen. VIII (1540) to Sir Thomas Wyat at various rents amounting to the sum mentioned.

The grant of the office of Bailiff is recited, and the two annuities mentioned therein are deducted from the receipts.

Certa terre et tenementa in Maydestone } Compotus Thome Vicarye, Collectoris redditus ibidem, per tempus predictum.

* * * * * * * *

Manerium de Chyngley in le Wylde } Compotus Thome Vycarye, Collectoris redditus ibidem, per tempus predictum.

* * * * * * * *

Rumney et Brokelonde } Compotus Thome Vycary, Collectoris redditus ibidem, per tempus predictum.

* * * * * * * *

Redditus in London } Compotus Thome Vycarye, Collectoris Redditus ibidem, per tempus predictum.

* * * * * * * *

[These last four offices were subordinate branches of the bailiwick. Vicary received no extra fees for them. Besides these minor accounts, several receivers in other places accounted to the Vicarys as Bailiffs.]

[1] Extracted by Mr. R. G. Kirk, Record Agent, 27 Chancery Lane, W.C.

180 App. IV. *Vicary's Boxley-Bailiff's Account.*

Ministers' Accounts, 35-36 Hen. VIII (1543-4), No. 150.
Similar accounts to the foregoing.

Ministers' Accounts, 36-37 Hen. VIII (1544-5), No. 146.
Similar accounts to the foregoing.

[This appears to be the last. Two other later rolls have been inspected, one in the reign of Edward VI, and the other in the first year of Q. Mary (1553-4), but the Boxley lands returned are very few, and are not accounted for by Vicary, apparently. In one or two places, however, the name of the accountant is not given.]

Ministers' Accounts, 1 Mary to 1 and 2 Philip and Mary (A.D. 1553-5), No. 17.

m. 71. A few possessions late of the Monastery of Boxley are mentioned, but Vicary is not stated to be bailiff.

m. 89. Possessions of Sir Thomas Wyatt, Kt., attainted of high treason.

Several Manors, with different bailiffs to each.

Manor of Boxley,—John Morse is the Queen's bailiff there.

m. 109 and 109 *d.* A few lands in Boxley.

V.

7 March 1557-8. Mortgage for £100, by Thomas Dunkyn of Shoreditch, of Watsole House and 11 closes of land (60 acres) in Elmsted, Kent, and 3 closes cald 'Wyldes' (18 acres) in Stowting, Kent, to Thomas Vicary, Surgeon, and his nephew Thos. Vicary of Tenterden, clothier (for the behoof of the said nephew): the Mortgage named in Thomas Vicary's Will.

Close Roll, 4 and 5 Philip and Mary, p. 3, membrane 13d.

Indentura inter T Vycary et alium, et T Dunkyn. [May 8, 1566] Wylliam Cordell [Master of the Rolls] Thomas Vycary [Nephew of Thomas Vicary, Surgeon.] [The Mortgage paid off and cancelled.] Vacatur ista Indentura, vnacum irrotulamento eiusdem, pro eo quod infrascriptus Thomas Vycarye Junior, infranominato Thoma Vycary Seniore mortuo iam existente, viij die Maij, anno regni Domine Elizabethe Anglie Regine, quinto, venit coram eadem Domina Regina in Cancellaria sua personaliter, et fatebatur se plenarie fore satisfactum persolutumque, tam de omnibus pecuniarum summis, quam de omnibus aliis articulis, convencionibus et agreamentis, in Indentura ista specificatis, ac pro parte infrascripti Thome Dunkyn perimplendis et observandis, bene et fideliter perimpleri et satisfactum fore,	This Indenture, made the seventh daye of Marche, in the yere of oure Lord God, after the course and rekenynge of the Churche of Englond, a thousand, fyue hundreth, fyftie and seuen: and in the fourth and fyveth yeres of the reignes of oure Soueraigne Lorde and Ladye, Philipp and Marye, by the grace of God, Kynge and Quene of Englond, Spayne, Fraunce, both Sicills, Jerusalem, and Irelond, defendors of the faithe, Archdukes of Austria, Dukes of Burgundie, Myllayne, and Braband, Counties of Haspurge, Flaunders, and Tiroll : Betwene THOMAS VYCARY thelder, of London, Gentleman, seriant of the Kinge and Quenes maiesties Surgions, and THOMAS VYCARY the yonger, of Tenterden in the Countie of Kente, Clothier, one of the sonnes of William Vycary, late of Boxeley in the said Countie of Kente, deceased, on thone partie, And THOMAS DUNKYN, of the paryshe of Saynt Leonard in Shordyche, in the Countie of Middlesex, yoman, on the other partie, WITNESSETH, that the said Thomas Dunkyn,—for and in consideracion of the somme of one hundreth poundes of good and lawfull monye of Englond, to him in hond at thenscaling hereof, by the said Thomas Vycary the elder, and Thomas Vycary the yonger, well and truly contented and paid, (whereof and wherwith the	Indenture dated March 7, 1557-8, (4 and 5 Philip and Mary,) between Thomas Vicary, Surgeon, and his nephew Thos. Vicary, clothier (mortgagees), and Thomas Dunkyn, yeoman, (mortgagor). For £100 lent by the 2 Thomas Vicaries to Thos. Dunkyn,

182 App. V. *Dunkyn's £100 Mortgage to Vicary*, 1558.

	secundum veram intencionem Indenture predicte. Et postulabat Indenturam predictam, unacum *irrotulamento eiusdem*, adnichillari. *Ideo evacuantur, cancellantur, et omnino dampnantur.*
the said Thos. Dunkyn grants	
to the 2 Thomas Vicaries,	
the house *Watsole* in Elmsted, Kent,	
held by Arnold Dunkyn;	
and the 11 Closes of Land belonging to it,	
in Elmstead, about 60 acres,	
also held by Arnold Dunkyn;	
Also 3 Closes cal'd *Wyldes*,	
about 18 acres, in Stowting, Kent, now held by the said Arnold Dunkyn,	
To hold the said house and lands	

said Thomas Dunkyn knowledgeth him selfe well and trulie satisfied, And therof, and of eue*ry* parte and parcell therof, doth clerelie acquite and dyscharge the said Thomas Vycary the elder and Thomas Vycary the yonger, theire heyrs, executors and admynystrators, and eue*ry* of them, by these p*re*sentes,)—hath bargayned, soulde, gyuen and graunted, And by thes p*re*sentes clerely and fully bargayneth, selleth, geueth and graunteth, vnto the said Thomas Vycary the elder and Thomas Vycary the yonger, all and singuler that mesuage or tene*men*te, with thappurtena*un*ces, com*m*onlye called Watsole[1], sett, lying, and being in the paryshe of Elmestede, in the said Countie of Kente / And all and singuler barnes, stables, courtes, yardes, gardens, easementes, co*m*modities and appurtena*un*ces, whatsoeue*r* they be, to the said mesuage or tene*men*te belonging, or in any wise appe*r*teyning, nowe being in the occupac*i*on of Arnould Dunkyn of Elmested aforesaid / And also the said Thomas Dunkyn, for and in considerac*i*on aforesaid, hath bargayned, soulde, geuen and graunted / And by thes p*re*sentes clerelye and fullye bargayneth, selleth, geueth and graunteth, vnto the said Thomas Vycary the elder and Thomas Vycary the yonger, all and singuler those eleuen closes or parcells of pasture grounde, arrable londe, medowe grounde, and wood landes, to the said mesuage or tene*men*te belonging, lying and being in the said paryshe of Elnested [*sic*], conteyning by estimac*i*on three score acres, be it more or lesse, nowe being in the occupac*i*on of the said Arnould Dunkyn; And also three other closes or parcells of pasture grounde, with thappurtena*un*ces, co*m*monlye called Wyldes, conteyning by estimac*i*on eightene acres, be it more or lesse, lying and being in the parysshe of Stowting, in the saide Countie of Kente, nowe in the occupacyon of the saide Arnould Dunkyn, together with all and singuler dedes, charters, wrytinges, te*r*rers, escriptes, and mynimentes, conce*r*nyng the said mesuage and tene*men*te, and all and singuler other the p*re*mysses, with thappurtena*un*ces, or any parte or parcell therof. TO HAUE AND TO HOLDE the said mesuage and tene*men*te, and all and singuler other the p*re*mysses, with thappurtena*un*ces,

[1] Watsole House is not now known (says the Vicar of Elmsted), but Watsoles Street, a road connecting a group of five or six houses in this parish, is well known.—See *Ordnance Survey of Kent.*

App. V. *Dunkyn's* £100 *Mortgage to Vicary.* 183

and euery parte and parcell therof¹, to the said Thomas Vycary the elder and Thomas Vycary the yonger, theyre heyrs and assignes, to thonlye vse and behoufe of the same Thomas Vycarye the yonger, his heyrs and assignes for euer / And the said Thomas Dunkyñ, for him, his heyrs, executors and admynystrators, and euery of them, couenaunteth and graunteth to and with the saide Thomas Vycary thelder and Thomas Vycary the yonger, theyre heyrs, executors and admynistrators, and euery of them, by thes presentes, that he the said Thomas Dunkyñ, the daye of the makyng herof, is lawfully seased in his demeane as of fee, of and in the said mesuage and tenemente, and other the premysses, with thappurtenaunces, withoute eny maner of vse, condicion or dephezaunce ; And that he hath full power and auctorytie, firmly and clerely to bargayne and sell all and singuler the said premysses, with thappurtenaunces, vnto the said Thomas Vycary thelder and Thomas Vycary the yonger, and to the heyrs of the saide Thomas Vycary the yonger, according to the purporte, entente, and trewe meanynge of this Indenture / And that the said mesuage and tenemente, and all other the premysses, with thappurtenaunces, and euery parte and parcell therof, nowe be, and herafter shalbe, clerely discharged, or otherwise saued harmeles, of and frome all maner of former bargaynes, gyftes, alienacions, recoueryes, condempnacions, iudgementes, execucions, leases, grauntes, yssues, liveryes, intrusyons, dowres, joyntours, statutes, recognyzaunces, charges, and encombraunces, whatsoeuer they be, had, made, done or suffered by the said Thomas Dunkyñ or his assignes, or by eny other person or persons by his meanes, consente or procuremente ; The rentes, customes, and seruyces frome hensforth to be due vnto the chief lorde or lordes of the fee or fees therof, and the title of dowry of Jyliañ, nowe the wyf of the said Thomas Dunkyñ, only excepted. And also the said Thomas Dunkyñ, for him, his heyrs, executors and admynystrators, and euery of them, couenaunteth and graunteth to and with the said Thomas Vycary thelder and Thomas Vycary the yonger, theire heyrs, executors and admynystrators, and euery of them, by thes presentes, That the saide mesuage and tenemente, and other the premysses with thappurtenaunces, nowe be, and allwayes herafter shalbe, of the clere yerlye value of syx poundes of lawfull monye of Englond, ouer and aboue all charges and reprises / And further, the said Thomas

to the said 2 Thomas Vicaries

to *the use* of the younger Thos. Vicary in fee.

Covenants for Title by Thomas Dunkyn:

1. that he is seized in fee ot the lands, &c.;

2. that he has full power to grant them to the 2 Vicaries;

free from all encumbrances,

save the chief Lord's dues,

and the dowry of Jylian, the wife of the said Thomas Dunkyn;

3. that the said lands, &c. are worth a clear £6 a year;

184 App. V. *Dunkyn's* £100 *Mortgage to Vicary,* 1558.

Dunkyñ, for him, his heyrs, executors and admynystrators, and euery of them, couenaunteth and graunteth to and with the said Thomas Vycary thelder and Thomas Vycary the yonger, theire heyrs, executors and admynystrators, and euery of them, by thes presentes, **4. that if the said Thos. Dunkyn do** that yf he, the saide Thomas Dunkyñ, his heyrs, executors, admynystrators or assignes, or eny of them, **not pay to the said 2 Vicaries,** do not paye or cause to be paid, to the said Thomas Vycary thelder and Thomas Vycary the yonger, or either of them, their executors, admynystrators or **for the younger of them,** assignes, to the vse of the said Thomas Vycary the yonger, his heyrs or assignes, the somme of one **£100 as hereinafter appointed,** hundreth poundes, of good and lawfull monye of Englond, in maner and forme as herafter followeth, and at suche daye and place as is herafter expressed, That **then the said Thos. Dunkyn, and Jilian his wife,** then he, the saide Thomas Dunkyñ, and the said Jiliañ his wyfe, and eyther of them, and the heyrs of the saide Thomas Dunkyñ, and all and euery other persoñ and persons hauing, or pretendynge to haue, any ryghte, title, vse, interest, or eny parcell therof, **and all other claimants to the said lands, &c.,** by or frome the saide Thomas Dunkyñ, or vnder his title or intereste, of, in, or to, the said mesuage or tenemente, and other the premysses, with thappurtenaunces, or eny parte or parcell therof, shall frome tyme **will, at the request and cost of the 2 Vicaries,** to tyme, and at all tymes, at and vppoñ resonable requeste therof, to be made by the said Thomas Vycary thelder and Thomas Vycary the yonger, or eyther of them, or the heyrs or assignes of the saide Thomas Vycary the yonger, and at the costes and charges in the lawe of the saide Thomas Vycary thelder and Thomas Vycary the yonger, theire heyrs or assignes, make, dooe, and suffer, and cause to be made, done, **make all such further assurances** and suffered, all and euery suche further acte and actes, deuyse and deuyses, conueyaunce and conueyaunces, assuraunce and assuraunces, as (for the better **of the said lands, &c., to them in fee,** assuryng of the same premysses, with thappurtenaunces, and euery parte and parcell therof, to be had in fee symple to the said Thomas Vycary thelder, and Thomas Vycary the yonger,) shalbe, by the saide **as they or their Counsel shall require,** Thomas Vycary thelder and Thomas Vycary the yonger, or th'eyrs or assignes of the said Thomas Vycary the yonger, or by his or theire lerned counsell in the lawe, frome tyme to tyme aduised or deuysed / All which assurances, conueyances, and deuyses shall stonde and **to the use of Thos. Vicary the younger.** be, to the vse of the said Thomas Vycary the yonger, and of his heyrs, according to thintente, purporte, and **Provided always** true menyng of this Indenture / PROUYDED ALWAYES, and it is condiscyoned and agreed betwene the said

App. V. Dunkyn's £100 Mortgage to Vicary. 185

parties to thes presentes, that and yf the said Thomas Dunkyñ, his heyrs, executors, admynystrators or assignes, or eny of them, do paye, or cause to be paide, to the saide Thomas Vycary thelder and Thomas Vycary the yonger, their heyrs, executors, or assignes, the some of one hundreth poundes of good and lawfull monye of Englond, at the place where the founte stone nowe stondeth, within the cathedrall churche of Seynt Paule in London, on the laste daye of the moneth of Marche, the which shalbe in the yeare of oure Lord God, a thousand fyue hundreth threescore and three, betwene the howres of one and fower of the clocke oñ the after none of the same daye, That then and frome thensforth, this presente bargayne and sale to be vtterly voyde and of none effecte[1] / And that theñ, and fromthensforth, all and euery suche assuraunces as shalbe made of the premysses, or eny parcell therof, shall stonde, remayne and be, to the only proper vse and behoufe of the said Thomas Dunkyñ and his heyrs for euer, and to no other vse ne behoufe / Eny couenaunte, graunte, article or agrement before rehersed, to the contrarye in eny wise notwithstanding' / And that theñ the said Thomas Vycary thelder and Thomas Vycary the yonger, or eyther of them, or the heyrs or assignes of the said Thomas Vycary the yonger, receyuyng the said somme of one hundreth poundes, shall make, enseall and delyuer, as his or theire dedes, to the said Thomas Dunkyñ or his heyrs, a sufficiente acquytaunce of the receyte of the said somme of one hundreth poundes, of and for the same / And also shall cause the enrolmente of this Indenture to be cancelled withoute eny delaye, at the costes and charges of the said Thomas Dunkyñ, his heyrs or assignes / And ffurther, the said Thomas Dunkyñ, for him, his heyrs, executors and admynystrators, and euery of them, couenaunteth and graunteth to and with the said Thomas Vycary thelder, and Thomas Vycary the yonger, theire heyrs, executors and admynystrators, and euery of them, by thes presentes, that and yf the said Thomas Dunkyñ, his heyrs, executors, admynystrators or assignes, or eny of them, do not paye, or cause to be paide, the said somme of one hundreth poundes, in maner and forme aforesaid, and at the daye and place aforesaid, that then the said Thomas

Marginal notes: that if the said Thos. Dunkyn shall pay the 2 Thomas Vicaries £100 at the Fontstone of St. Paul's Cathedral, on March 31, 1563, between 1 and 4 p.m., then this Mortgage shall be void, and the lands shall remain the property of the said Thos. Dunkyn: and whichever of the Vicaries receives the £100, shall give a receipt for it, under seal, and shall cause the Enrolment of this Mortgage to be canceld, at the cost of Thos. Dunkyn. And Thos. Dunkyn further covenants with the 2 Vicaries, that if he does not pay them the £100 on 31 March, 1563,

---

[1] The enrolment of the Mortgage was not canceld till May 8, 1566, as noted above.

186 App. V. *Dunkyn's £100 Mortgage to Vicary*, 1558.

he will, at the request of the 2 Vicaries,	Dunkyñ, his heyrs, executors, admynystrators or assignes, at and vppoñ the resonable request of the said Thomas Vycary thelder and Thomas Vycary the yonger, or eyther of them, or the heyrs or assignes of
hand them,	the said Thomas Vycary the yonger, shall delyuer, or cause to be delyuered, vnto the said Thomas Vycary thelder and Thomas Vycary the yonger, or to eyther of them, or the heyrs or assignes of the said Thomas
within 3 months,	Vycary the yonger, within three monethes next after
the Title-Deeds of the said lands, &c.	the said laste daye of Marche, the said dedes, Charters, writynges, terrers, escriptes and mynymentes, before by thes presentes bargayned and soulde¹ / And moreouer,
And will also	the said Thomas Dunkyñ, for him, his heyrs, executors and admynystrators, and euery of them, couenaunteth and graunteth, to and with the said Thomas Vycary thelder and Thomas Vycary the yonger, theire heyrs, executors and admynystrators, and euery of them, by thes presentes, that if he the said Thomas Dunkyñ, his heyrs, executors or assignes, or eny of them, do not
(the said £100 not being duly paid)	paye the said somme of one hundreth poundes in maner and forme aforesaid, and at the daye and place aforesaid / That theñ, he the said Thomas Dunkyñ,
warrant or guarantee, and defend, the possession of the said lands, &c. to the 2 Vicaries,	his heyrs and assignes, and euery of them, all the said mesuage and tenemente, and all other the premysses, with thappurtenaunces, and euery parte and parcell therof, to the said Thomas Vycary thelder and Thomas
to the use of the younger Thos. Vicary, in fee.	Vycary the yonger, theyr heyrs and assignes, to the onlye vse and behoufe of the said Thomas Vycary the yonger, his heyrs and assignes, agaynste all meñ shall warrante, acquite, and defende for euer, by thes presentes. In witnes wherof, the parties aforsaid to theise Indentures enterchaungeablie haue sett theire seals. Yeueñ the daye and yeres fyrst aboue wrytteñ.
March 28, 1558. Thomas Dunkyn acknowledged the above Mortgage in the Court of Chancery at Westminster.	Et memorandum, quod vicesimo octauo die Marcij, et Annis suprascriptis, venit prefatus Thomas Dunkyñ coram dictis Dominis Rege et Regina in Cancellaria sua apud Westmonasterium, et ibidem recognouit Indenturam predictam, ac omnia et singula in eadem contenta, in forma suprascripta.

[This enrolment is crost through with many netlike strokes of the pen, to show its cancellation. To this day, Mortgages are enrold in Chancery on big rolls of parchment like Dunkyn's was, and are canceld in like way.]

---

¹ Now, and for many scores of years past, the Deeds are and have been always delivered over on the completion of the Mortgage.

VI.

WILL OF THOMAS VICARY 1560-1
(1561 NEW STYLE).

[Book *Streate* (Prerogative Court), folio 10, leaf 3.[1]]

In the name of god, amen. The xxvij.th daye of Ianuary in the yere of o*ur* lorde god 1560 / and in the thirde yere of the raigne of o*ur* soueraigne ladie Elizabeth, by the grace of god, quene of englonde, ffraunce and Irelande, deffendo*ur* of the faith, &c. I, Thomas Vicars,[2] Seriante of the Suriant*es* vnto o*ur* saide soueraigne ladie the quenes maiestie, being hole in boddie and in parfecte remembraunce, (thank*es* be giuen to almightie god!) doe ordaine and make this my presente testamente and laste will, in manner and forme followinge. ffirst and principally I bequeath my soule to almightie god, my creator and maker, and to his only sonne, my redemer and sauior, Iesus christe, by the merritt*es* of whose painefull passion, presius[3] deth, glorius resurreccion and blessed assencion, I trust to haue clere[4] remission of all my synnes, humbly beseching the blessed virgin Mary, and all the blessed company of heauen to praye for me,[5] and with me. And my boddie to be buried in *Christ*ian buriall emong those that dye in o*ur* lorde god,[6] wheresoeu*er* it shall pleace god that I shall departe oute of this p*res*ent lief. Also I will that on the daye of my buriall there shalbe made one sermon by some godly and lerned man to preache godes worde, and the declaraci*on* of my faith in the same. / Item I will that the masters of the liuery of my Companie be at my buriall, and they to haue xl^{s.} / for theire dinners, to be deliuered to the wardens at theire commyng to my buriall. And to Johnson, the

Testamentum Thome Vicars.

27 Jan. 1560-1.

Thos. Vicars (or Vicary), Serjeant of the Surgeons to Q. Elizabeth,

leaves his soul to God,

and his body to be buried when he dies.

Directs a Sermon to be preached,

declaring his Protestant Faith; and that the Masters of the Barbers' and Surgeons' Company shall attend his Funeral.

---

[1] Mr. J. Challenor Smith, of the Literary Enquiry Department of the Probate Office at Somerset House, kindly told us of this Will. N.B.—In Will books there are 8 leaves to a folio, so that Vicary's Will is on leaf 83.

[2] He spells it 'Vycary' in the filed copy of his Will.

[3] 'precious' in filed copy.   [4] 'clene' in filed copy.

[5] This survival of Papacy had not died out in the early years of Elizabeth's reign.

[6] no 'god' in filed copy.

188 App. VI. *Will of Th. Vicary*, 27 Jan. 1560-1.

Leaves the poor of St. Bartholomew's Hospital £10;

to the poor of St. Bart.'s the Less, 40s.;

to 5 Hospital Officers 50s. each;

to his Sister, £10;
to Mary Shackston, £10;
H. Picton, his assistant, 20s.;
maid, 20s.;
apprentice, 6s. 8d.;

Clarke of the Company,[1] vj^s· viij^d· And in concideracion of my evell and necligent seruice done to god and to his poore members, the poore of this hospitall of St. Barthelmewes where I now dwell, in recompence whereof, and for the discharge of my concience, I giue and bequeath to thuse[2] of the saide poore, tenne poundes in monney. Item I give and bequeath xl^s· in monney to and amongest[3] fortie poore householders of the saide parish of little sainte Barthelmewes, that is to saye, to euery housholder[4] xij^d· Also I giue and bequeathe l^s· in monney to thospitler, matron, stuarde, Cooke, and porter offecer[5] of the saide hospitall, that is to saye, to euery of them x^s· Item I giue and bequeath to my sister Agnes Osken x^{li}· in monney. Also I giue and bequeath to mary Shackston x^{li}· in monney. Item I giue and bequeath to Henry Picton xx^s·[6] To margaret, now my maide, xx^s· And to Thomas Skair, my ap-

---

[1] John Johnson was elected and sworn Clerk of the Barber-Surgeons' Company on 27 Aug. 1557, 'for so long tyme as he shal behave hymsellfe well and honestly in the saide office.' His salary was £4 a year, with 6s. 8d. extra for paper, ink, and keeping the garden; and 'for wasshinge of the lynen of the howse, iijs. iiijd.'—Sidney Young.

[2] 'the use,' filed copy.

[3] no 'and amongest' in filed copy of the Will.

[4] 'housholder' in filed copy.

[5] In the Hospital, as in early Romances and Ballads, the 'proud Porter' was a person of importance. 'The Ordre of the Hospital of S. Bartholomewes' in 1552, says, 'The officiers are .vii. in nombre, continuable or remouable as the gouernours shall fynde cause, and be thus called: The Hospiteler [Chaplain]. The Renter clerk. The Butler. The Porter. The Matrone. The Sisters .xii. The Byddles .viii. There are also, as in a kynde by them selues .iii. Chirurgiens in the wages of the Hospitall, geuyng daily attendance vpon the cures of the poore.' See below, Appendix XVI.

[6] And a book, *Johannes Vigo*, with half the residue of testator's books and surgical instruments. Henry Picton was not in the Barber-Surgeons' Court, says Mr. Young. He was evidently Vicary's assistant. The Act of 32 Hen. VIII. ch. 42, which made the Barbers and Surgeons one Company, has a last clause enabling any person to keep a Barber or Surgeon as his *Servant*. It enacts 'that it shall be lawfull to any of the Kinges Subiects, not being a Barber or Surgeon, to retaine, have, and keepe in his house, *as his seruant*, any person being a Barber or Surgeon, which shall and may vse and exercise those arts and faculties of Barbery and Surgery, or either of them, in his masters house, or elsewhere by his Masters licence or commandement, any thing in this Act aboue written to the contrary notwithstanding.'—*Statutes*, ed. Pulton, 1636. App. VII.

The 'not being a Barber or Surgeon' in the clause above, was not meant, and would not operate, to prevent Surgeons

App. VI. *Will of Th. Vicary,* 27 Jan. 1560-1.

prentis, vj^{s.} viij^{d.} Also I giue and bequeath to my brother Dunkyn, my gowne furred with white lame,[1] and faced with foyne back*es*,[2] my greate ringe of golde that was master masons,[3] and my veluet bagge with the gilte ring*es* / Item I giue and bequeath to Roberte Baltropp[4] my beste gowne garded[5] with veluet, furred and faced * with Sables, my Cote of braunched[6] veluete, and a sering of siluer, parcell gilte / Also I giue and bequeath to Thomas Bayly[7] my gowne of browne blue lyned and faced with blacke budge,[8] my cassocke of blacke satten fured and garded with veluet, my best plaister box, garnisshed with siluer, my salvitory[9] of siluer, and a sering of siluer, with all other instrument*es* of siluer. Item I bequeathe to Robarte Muddesley[10] my best single gowne faced with blacke satten. Also I giue and bequeath to George Bucke,[11] my best cloke garded with veluet. To George Vaughan,[12] my doblet of crimsen satten. And to master Turke,[13] my Jacket

[margin:] brother Dunkin a gown and ring; R. Baltropp a gown, velvet coat, and syringe; * fol. x, leaf 3, bk. T. Bayly a gown, cassock, plaister box, and silver instruments; R. Muddesley a gown; G. Bucke a cloak; G. Vaughan a doublet;

keeping a Servant or Assistant, but only to enable other men to keep one. See below, Appendix VII.

[1] 'lambe' in filed copy : lambskin.

[2] backs of the Foyne, the wood- or beech-marten (somewhat of the squirrel kind).

[3] 'Massons' in filed copy. 'Probably the Alexander Mason who was Middle Warden of the Barber-Surgeons' Company, 1556 ; Upper Warden, 1561 ; and Master, 1567 and 1573. He died on April 3, 1574.'—S. Young.

[4] Robert Baltrop was admitted to the Freedom of the Barber-Surgeons' Company on 3 March 1545 ; and to the Livery on 20 Oct. 1552. He was Junior Warden in 1560 ; Upper Warden in 1564; and Master in 1565 and 1573.—S. Young.

[5] trimmed, barred.

[6] with branches or any other pattern on it.

[7] Thomas Bayley was Middle Warden of the Barber-Surgeons' Company in 1559.—S. Young.

[8] Lambskin with the wool dressed outwards.

[9] 'a new plaister boxe or salvatory.'—Inventory, 1600 A.D., in South's *Craft of Surgery*, 1886, p. 149. L. *Salvatorium*, a place where things are preserved, a repository.

[10] Robert Muddesley was Junior Warden of the Barber-Surgeons' Company in 1561 ; Middle Warden in 1562; Upper Warden in 1567 ; and Master in 1572 and 1580.—S. Young.

[11] George Bucke is not known at the Barbers' Company.—S. Young. He was probably the brother of Alice Bucke, the second wife, whom Vicary married in 1547.

[12] George Vaughan was admitted to the Freedom of the Barber-Surgeons' Company on 27 June 1536 ; he was Junior Warden in 1558 ; Middle Warden in 1563 ; Upper Warden in 1565 ; and Master in 1569.—S. Young.

[13] We never had a 'Turke' in our Company, that I know of.—S. Young.

190 App. VI. *Will of Th. Vicary*, 27 *Jan.* 1560-1.

Mr. Turke a jacket and doublet;
Rev. R. Wood a gown;
Barber-Surgeons' Hall a Guido,[1] and armour.
Mr. Skinner, some armour.
H. Picton, servant (assistant), a book, J. Vigo;
and all the rest of his surgical stuff to H. Picton and R. Vener.
Nephew[11] Thos. Vicary, junr., when he gets T. V.'s £100,

of russet veluet, and a dublet of blacke satten. Item I giue and bequeath to my louing frende Richarde Wood, clarke, my gowne of london russet, furred with black. Also I giue and bequeath vnto the hawle of my company, one booke called Guido,[1] and ij. billes, .ij. bowes, ij. shefes of Arrowes, ij. bracers,[2] ij. shoting gloves, ij. Scullcs,[3] one handgune, and one Jacke.[4] Item I giue and bequeath to master Skynner,[5] one half hacke,[6] one Jacke, and one murren.[7] And to Henry Picton,[8] my ser*u*aunte aforesaide, one booke called Joh*ann*es Vigo[9] / All the residue of my bookes, stuff and instrument*es* appertaining to surgery, I give and bequeath vnto the same Henry Picton and Richard Vener,[10] equally betwen them to be deuided. ffurthermore my mynde and will is, that as sonne as Thomas Vicary the yonger,[12] (sonne of Willi*a*m Vicary, late of boxley, deceaced,) hath receiued the hundreth poundes that I haue giuen hym, the which I haue putt into

---

[1] Guido de Cauliaco, Guy de Chauliac. His *Cyrurgia* was written in 1363, printed at Venice in 1490, 1497-9, 1500, &c., and other places after. It was translated into French in 1478, Italian in 1493, Spanish in 1498. (See Hain, *Repert. Bibliog.* I. ii. 82-3.) The earliest Englishing in the B. Mus. Catalog is of 1542: 'The Formularye of the aydes of apostemes; of the helps of woundes and Sores,' &c. Guido wrote an Anatomy and other treatises.

[2] Guards for the left arm, in bow-shooting.

[3] Scull-helmets or metal headpieces.

[4] A defensive garment made of small pieces of metal enclosed between two folds of stout canvas or some quilted material,—sometimes costly.—*Fairholt.* '*Bombicinum, anglice* a Iakke.'—Wülker's *Vocab.* 568/29. '*Sarissa, anglice* a materas, *et quoddam genus armorum, anglice* a Jakke of defence.'—*ib.* 609/25.

[5] 'John Skinner' was Vicary's Upper Warden in 1548.—S. Y.

[6] The *demi-hacke* or half-hake was a gun, a smaller kind of 'hackbut,' which was an arquebus with a hooked stock.— Dillon's *Fairholt*. 'Handgonnes or demyhakes.' Inventory of Henry VIII, A.D. 1547.—*Dillon.* Dutch '*een haeck*, a Hooke, or a Claspe. *Haeck, haeck-busse*, an Arque-busse, or a Crock.'—1660. *Hexham.*

[7] A helmet with a projecting rim like a top-hat.

[8] See his bequest of 20s. on page 188.

[9] No doubt his 'Workes of Chirurgerye, Translated by Bartholomew Traherone: London, 1543. folio,' (Lowndes,) or its original.

[10] On 1 Oct. 1566, is translated from the Woodmongers' Company to the Barber-Surgeons, Wm. Slade, "a Surgeon; & learned yt with Ric. Vener & John Hall, at Maydstone." Vener never served as Master or Warden of the Barber-Surgeons' Company.—S. Young.

[11] See *nepoti* in the note of Administration at end.

[12] The filed copy of the Will has the brackets that follow.

App. VI. *Will of Th. Vicary*, 27 Jan. 1560-1.

the handes of my saide brother Thomas Dunkin for hym, that he ymmediatly doe confes the receipte thereof before the master of the Rowles, so that my saide brother Dunkin maye quietly enioye his lande at Elmested,[1] the which standeth bounde for the saide some of one hundreth poundes, by a bargaine of sale, as by writing doth appere, before the saide master of the rowles. And also I giue and bequeath to Steven Vicary,[2] sonne of Williαm Vicary, late of Boxley,[3] in the Countie of Kente, deceaced, all that my house and lande thereto belonginge, set, lieng, and being, next boxeley Churche[4] aforesaid, the which I late purchased of one John Joyce / To haue and to holde the saide[5] house and lande to the saide Steven and to his heires for ever. Item I giue and bequeath to the saide Steven Vicary, all my righte, title,[6] interest and terme of yeres which I haue yet to come, of and in all that leace landes lienge and being in the saide parrish of boxeley / the which I obtained of Sir Thomas wiat, thelder, knighte,[7] for the terme of lx. yeres, as by

to free Dunkyn's land from the charge of it.
(March 7, 1558, in Close Rolls, & Appendix V, p. 181.)

Leaves to nephew Stephen Vicary, his house and land next Boxley Church, Kent,

and his leaseholds in Boxley under Sir Thos. Wyat's Lease of 28 Sept. 1541 for 60 years,

---

[1] Elmsted is 5 miles east from Wye station, 9 north-east from Ashford, and 66 from London. Sir Jn. Wm. Honywood, bart., is now lord of the manor, and lives at Evington-place, about a mile from the Church.

[2] Possibly the 'Stephen Vycary gent.' who was licensed to marry 'Margaret Johnson, spinster,' of the City of London, at St. Margaret, Lothbury, on 23 Jan. 1574-5.—*Chester.*

[3] Boxley is two and a half miles N.E. of Maidstone. As Vicary "was at first a meane practiser in Maidstone . . . untill the King advanced him for curing his sore legge" (Manningham's *Diary*, p. 51), it was but natural that he should buy land close to Maidstone, and also ask the King for part of the Boxley Abbey property, and get it.

[4] Henry VIII's twenty-one years' lease to Thos. Vicary of the tithes and glebe of Boxley Rectory, and the capital messuage and buildings belonging to it, and the monastery's ten pieces of land, was granted in 1537, and therefore expired in 1558.—Hasted's *Kent*, ii. 135. See p. 91, above. [5] No 'saide' in filed copy.

[6] 'title' struck out in the filed copy.

[7] The Poet, born at Allington Castle, Kent, in 1503; died at Sherborne, Dorset, Oct. 1542. He was a great favourite of Henry VIII, though he was twice tried for his life. Had this Lease anything to do with the fact, that on October 5, 1542, Henry VIII granted to Thomas Vicary, and his son William, for the life of the longest liver of them, the office of Bailiff of the Manor of Boxley and all other Manors there belonging to the late Abbey? See Hasted's *Kent*, ii. 125, and p. 93 and 179, above. The Vicarys may have afterwards surrendered this post to the King, as in 1555 it was regranted to Thomas Vicary the father—no doubt after his son's death—by K. Philip and Queen Mary: p. 96, above.

indenture therof made, bering date the xxviij.th daye of September in the xxxiij. yere of the raigne of king henry the eight more plainely appereth / Except and alwaies reserued oute of the same, to thintente and vse hereunder written, that is to saye, the yerely ferme of Polhill¹ feilde, (whiche is xl^{s.} a yere,) now in the tenure and occupacion of Richarde Goldsmyth and Jane his wief, which xl^{s.} a yere I will shalbe distributed and giuen vnto the poore householders dwelling within the same parrish of Boxeley, at ij seuerall tymes in the yere, yerly, during the yeres expressed in the saide leace, that is to saye, xx^{s.} to be giuen in the x.^{th 2} daye of October, and thother xx^{s.} to be giuen in ye xv.th daye of Aprille ; and the saide Richarde and Jane, or either of them, to distribute the foresaide monney by thaduice and discrecion of the vicar and churchewardens of the same parrish churche of Boxeley, yerely, from tyme to tyme. And furthermore, I will that the saide yerely farme of the iiij.^{or} Acres of lande lyeng in Shepelonde, and the ij. Acres lyenge in Bernecrofte, now in the tenure and occupacion of William Boote of the same parrish of Boxeley, (which is xiij^{s.} iiij^{d.} a yere,) I will that the churchewardens of the same parrish for the tyme beinge, shall receiue the saide yerely ferme of xiij^{s.} iiij^{d.}, to be ymploied aboute the moste nedefull reperacions of the same parrish churche of Boxeley. And yf it happen the foresaide Richarde and Janne, theire successors or assignes, to neclecte and not to *giue the saide almes of xl^{s.} a yere at the daies aboue saide, then I will that the vicar and the churche wardens for the tyme being, shall enter in and vppon the saide Polhill feelde and enioye the saide yerely farme of xl^{s.} a yere, and to distribute the same in almes as aboue is mencioned, withoute eyny lett or contradiction of eny person or persons hauing or pretendinge any claime or title in or to the same ; and neuertheles, this exception notwithstandinge, I will that the saide Steven Vicary, or his assignes, shall yerely paye, or cause to be paied, all the rente of xvj^{li.} x^{s.} ij^d yerely, whiche ys reseruid by the saide leace, during all the yeres of the

Marginal notes:
- save 40s. a year for Polhill field held by Rich. and Jane Goldsmith, who shall give this in two sums of 20s. to the poor of Boxley.
- Save also that the rent of 4 acres of Sheepland and 2 a. in Barncroft,
- 13s. 4d. a year,
- shall go to the repair of Boxley parish Church.
- * fol. 10, leaf 4.
- Power of entry to the Vicar, &c., if the 40s. rent is not duly paid.
- Nephew, Stephen Vicary, to pay Sir T. Wyat's heirs their rent of £16 10s. 2d. for their leaseholds.

---

¹ Was this near Poll Mill? In the Certificate of the last Abbot of Boxley Monastery, John Dobbs, dated May, 1535 (27 Hen. VIII), of the yearly value of the Monastery lands, the third entry is "Item, a fullyng [cloth-cleansing mill] called Poll Mill, with th' appurtenaunces, in Boxley foreseid, and in the said diosese [of Caunterbury] ... 3l. 0s. 0d." Dugdale, *Monasticon Anglicanum*, v. 461, col. 2, ed. 1825.

² Better 'xvth' in the filed copy of the Will.

App. VI. *Will of Th. Vicary,* 27 Jan. 1560-1.

saide leace / And as for all other sommes of monney and other thinges by me heretofore bequeathed in my other will[1] to the prison houses and to thother places, I haue alreddie giuen it with my owne handes, requiring my wief to performe the rest. All the residue of my goodes, plate, Juelles, reddie monney, debtes, and all other thinges not bequeathed,—my debtes paied, (yf there be eny at this presente tyme; I know of none,) and my funeralles, my legaces, my will in every pointe and article fulfilled and donne,—I giue and bequeth vnto my welbeloued wief, Alice Vicary,[2] whome I ordaine and make sole executrice of this my presente testamente and laste wiH. And ouerseer of the same, I constitute and ordaine my welbeloued brother, Thomas Dunkyn. In witnes whereof, I haue, vnto this my presente Testamente containing my laste will, subscribed my name with my owne hande, and sette[3] my seale, the daye and yere first aboue written, by me Thomas Vicary. R. Wood / And where I haue giuen vnto Thomas Vicary, sonne of William Vicary, late of boxley, one hundreth poundes, which is deliuered into the handes of my brother Thomas Dunkyn for thonly vse of the saide Thomas Vicary the yonger, wherefore is yerely receiued oute of certaine landes in Elmested[4] in Kent vj^{li} by the yere, as by writing dothe appere, my mynde and will is, that all suche monney as is alreddie receiued of the saide lande, shall stande and be parcell of paimente of the saide hundreth poundes, for the discharge of my concience. And that the saide Thomas Vicary the yonger, ymmediately after the paimente of the rest of the saide C^{li}, shall confes the paimente before the master of the Rowles[5] / Memorand*um*. the very wordes in this Shedule aforesaide was written in paper by the owne hande of the saide Testator, as the[6] persons whose names hereafter followe can testefie and beare witnes, by me Roberte Howell. /

(Margin notes:) Gifts to poor in other Will. Gives all the residue of his personalty (after payment of debts, burial, legacies, &c.) to his wife Alice Vicary, and appoints her sole Executrix, his brother Dunkyn being Overseer. Nephew Thos. Vicary to allow T. Dunkyn the £6 yearly received out of his land at Elmstead, Kent. (Duly done on May 8, 1563: see Close Rolls, & Appendix V, p. 181, 186.) The will was written by Thomas Vicary's own hand.

[1] It was an earlier Will which Vicary had destroyed. The present one, of course, did away with it.
[2] She was his second wife, and once, Alice Bucke of London. Their Marriage-License was granted in Dec. 1547.—*Chester.* Mr. Challenor Smith cannot find her Will. Vicary's son William, by his first marriage (note 7, page 191), no doubt died before him. He was probably the William Vicary admitted to the freedom of the Barber-Surgeons' Company, on July 26, 1547.
[3] 'set to,' affixed. [4] 'Elmysted' in the filed copy of the Will.
[5] 'Masters of the Rolls : 1557, Sir William Cordell ; 1580, Sir Gilbert Gerrard.—Toone, *Chronolog. Hist.* ii. 196, col. 2.'
[6] 'thiese,' filed copy of Will.

194 App. VI. *Will of Th. Vicary*, 27 Jan. 1560-1.

Will proved in the Prerogative Court of Canterbury, 7 April 1562, by Alice Vicary, the widow.

[1]Probatum fuit hujusmodi Testamentum, coram Magistro Waltero Haddon, legum doctore, Curie prerogatiue Cantuariensis Commissario, apud london, septimo die mensis Aprilis, Anno domini millesimo quingentesimo sexagesimo secundo, Juramento, Alicie, Relicte et Executricis in hujusmodi testamento nominato ; Cui comissa fuit administracio et c. de bene, et c. Ac de pleno Inuentario, necnon de vero et plano computo Reddendo. Ad sancta dei Evangelia Iurate[2] /

[from *Probate Act Book*. 1576]

Letters of Administration granted to Thos. Vicary, the nephew, to the goods &c. of Thos. Vicary, dec., which were left unadministered by his widow Alice Vicary.

Thomas Vicary. Quinto die mensis Iulii emanauit com-[3] [5th July] missio Stephano Vicary, nepoti Thome hujusmodi Vicary, nuper perochie Sancti Bartholomei iuxta Smythfild, defuncti / registratum Habentis etc. in Libro Street, Ad administrandum bona, Jura et 10/ credita, eiusdem defuncti per Aliciam Vicary, Relictam et executricem in testamento dicti defuncti, iam defunctam, non administrata. De bene, &c. Ad sancta Dei Euangelia Jurato.

[1] The Proof of the Will is also entered in the Probate Act Book, July 1559 to 1565, with a sidenote as to the Grant of Letters of Administration to Stephen Vicary.
[2] A later sidenote says "v⁺º Julij 1576: emanauit commissio Stephano Vicars, nepoti dicti defuncti, ad administrandum bona et credita eiusdem defuncti per dictam executricem defunctam non administrata, de bene."
[3] The sidenotes are 'Ciuitatis London,' and 'Fedis. / Inventorium exhibitum, primo,' meaning that Stephen Vicary was of the City of London, that he had till the Feast of St. Faith's [October 6] to exhibit his Inventory of the goods administered, and that it was exhibited, and put first in some bundle of like Inventories. The Inventory may be in one of those boxes of such documents in little rolls of parchment which Mr. Challenor Smith and Dr. F. J. F. went through to try to find Shakspere's Inventory. They only got that of Sir Jn. Barnard, who married Shakspere's granddaughter, and found an entry that the 'old goods and Lumber' at (Shakspere's 'New Place' presumably) Stratford-on-Avon in 1674, were worth £4, and the rent of it, £4. See *New Shaksp. Soc.'s Trans*. 1880-6, Appendix II, p. 14†. Lots of the Inventories disappeared at St. Paul's &c., before they came to Somerset House.—*ib.* p. 15†.
[The Register of Burials of St. Bartholomew's the Less commences in 1547; but Vicary's burial is not in it. Dr. Norman Moore has kindly searched for us.]

VII.

STATUTES OF HENRY VIII RELATING TO SURGEONS.

i. A.D. 1511-12. 3 Hen. VIII, ch. 11. The Act stopping the practise of Physic and Surgery by unlicenst folk, and requiring the Examination and Licensing of all Physicians and Surgeons, p. 197 (amended by No. VI, 34 and 35 Hen. VIII, ch. 6).

ii. A.D. 1513-14. 5 Hen. VIII, ch. 6. The Act exempting the Fellowship of Surgeons (12 men), and also the Surgeons of the Barbers' Company, from serving as Constables, Watchmen, Jurymen, &c., p. 198.

iii. A.D. 1530-1. 22 Hen. VIII, ch. 13. The Act providing that Alien Surgeons, Brewers, Bakers, &c. are not to be sued under the Alien-Handicraftsmen's Act, p. 201 (with a Statement showing the cause of it, p. 200).

iv. A.D. 1540. Extract from 32 Hen. VIII, ch. 40, enabling Physicians to practise Surgery, p. 202.

v. A.D. 1540. 32 Hen. VIII, ch. 42. The Act uniting the Barbers and the Surgeons of London into one Company (whereof Vicary was the first Master); and separating the practises of Surgery and Barbery, p. 202.

vi. A.D. 1542-3. 34 and 35 Hen. VIII, ch. 8 (amending No. 1, 3 Hen. VIII, ch. 11). An Act empowering unlicenst folk to cure common ailments and outward wounds by Herbs, Waters, &c. (This, in consequence of licenst Surgeons' greed.) p. 208.

[See VIII, p. 210, &c., the

SUPPLEMENT TO THE STATUTES.

A.D. 1517. Inspeximus, witnest by Letters Patent, of the Act 5 Hen. VIII, ch. 6, with Lists of the 11 Surgeons exempted under it, p. 210.

A.D. 1546. Contract of the Barber-Surgeons with the City of London, varying the Act 32 Hen. VIII, ch. 42, as to serving as Constables, Jurors, Watchmen, &c., p. 215.

with other extracts from the Guildhall Records.]

i.

3 Henry VIII. Chapter XI.[1] (A.D. 1511-12).

An Act concerning Phesicions & Surgeons.

FORASMOCHE as the science and connyng of Physyke [and Surgerie],[2] to the perfecte knowlege wherof bee requisite bothe grete lernyng and ripe experience, ys daily within this Royalme excercised by a grete multitude of ignoraunt persones, of whom the grete partie have no manner of insight in the same, nor in any other kynde of lernyng; some also [can] no lettres on the boke, soofarfurth that common Artificers, as Smythes, Wevers, and Women, boldely and custumably take upon theim grete curis, and thyngys of great difficultie, In the which they partely use socery and whichcrafte, partely applie (p. 32) such [medicynes][4] unto the disease as be verey noyous, and nothyng metely therfore, to the high displeasoure of God, great infamye to the faculties, and the grevous hurte, damage, and distruccion, of many of the Kynges liege people, most specially of them that cannot descerne the uncunnyng from the cunnyng; Be it therfore, to the suertie and comfort of all maner people, by the auctoritie of thys present parliament enacted, that noo person within the Citie of London, nor within vij myles of the same, take upon hym to excercise and occupie as a Phisicion [or Surgion], except he be first examined, approved, and admitted, by the Bisshope of London, or by the Dean of Poules for the tyme beyng, callyng to hym or them iiij Doctours of Phisyk [and for Surgerie, other expert persones in that facultie]; And for the first examyna-

Physic and Surgery are practist by unskilful persons,[3]

Smiths, Weavers, and Women,

who partly use Sorcery and Witchcraft,

to the grievous hurt of the King's liege people.

It is therefore enacted, that none shall practise as a Physician or Surgeon in London,

unless he be examined and approved by the Bishop of London, or Dean of St. Paul's,

---

[1] Two copies of this Act are entered on the Roll, numbers 18 and 22. The Text is printed from the former. *Record Commission Statutes*, iii. 31.

[2] ... To the Original Act a small Schedule is attached ... "Memorandum that Sowrgeons be comprised in this Acte like as Phisicions, for like mischief of ignorant persones presumyng to exercise Sowrgerie." The words relating to Surgery and Surgeons included in Crotchets in the Print, are all interlined in the Original Act.—*Ibid.*

[3] The side-notes being only 18th century ones, we alter and add to them at discretion.

[4] medicyne, nu. 22; medycyns, nu. 18.

198 App. VII. *Licensing Act, 5 Hen. VIII, ch. 6.*

with the aid of 4 Physicians, or Surgeons.

Penalty 5*l.* per Month.

II. In the Country,

Practisers shall be approved by the Bishop of the Diocese, &c., with the aid of Physicians and Surgeons.

Saving the right of Oxford and Cambridge.

cion, such as they shall thynk convenient; And afterward, alway iiij of them that have been soo approved, upon the payn of forfeytour for every moneth that they doo occupie as Phisicions [or Surgeons] not admitted nor examined after the tenour of thys Acte, of vli, to be employed, the oon half therof to thuse of Soveraign Lord the Kyng, and the other half therof to ony person that wyll sue for it by accion of dette, in which no Wageour of Lawe nor proteccion shalbe allowed. And over thys, that noo person out of the seid Citie, and precincte of vij myles of the same, except he have been (as is seid before) approved in the same, take upon hym to exercise and occupie as a Phisicion [or Surgeon] in any Diocesse within thys Royalme, but if he be first examined and approved by the Bisshop of the same Diocesse, or, he beyng out of the Diocesse, by hys Vicar generall; either of them callyng to them such expert persons in the seid faculties as there discrecion shall thynk convenyent, and gyffyng ther letters testimonials under ther sealle, to hym that they shall soo approve, upon like payn to them that occupie [the] contrarie to thys acte, as is above seid, to be levyed and employd after the fourme before expressed. Provided alway, that thys acte, nor any thyng therin conteyned, be prejudiciall to the Universities of Oxford and Cantebrigge, or either of them, or to any privilegys graunted to them.

---

ii.

5 *Hen. VIII. Ch. VI.* A.D. 1513-14 (*Record Stat.* iii. 95).

An Acte that Surgeons be discharged of Constableshipe & other thinges.

The Fellowship of Surgeons,

not above 12 persons, and their predecessors have, time out of mind,

Sheweth unto your discrete wisedomes, your humble oratours the Wardens and felisshippe of the crafte and misterye of Surgeons[1] enfraunchesid in the Citie of London, not passyng in nombre xij persones: That wher-as they and their predecessours from the tyme that noo mynde is to the contrarie, aswell in this noble Citie of London, as in all other Cities and Boroughes within this Realme or ellis wher,—for the contynuall service and attendaunce that they daily and nyghtly

---

[1] See Forewords § 4, and South's *Craft of Surgery* by d'Arcy Power.

App. VII. *Surgeons exempted from Constable duty.* 199

at all houres and tymes gyve to the Kinges liege People, for the relefe of the same according to their science,—have ben exempte and discharged from all offices and besynes wherin they shuld use or bere any maner of armoure or wepyn, And with like privilege have ben entreatid as Herawdes of Armes, aswell in batelles and feldes as other places, ther for to stond unharnessed and unwapenned, according to the lawe of armes, because they be persones that never used feates of warre, nor ought to use, but onely the besynes and exercise of their science, to the helpe and comforth of the Kinges liege people in the tyme of their nede: And in the forsaid Citie of London, from the tyme of their firste Incorporacion when they have ben many moo in nombre then they be nowe, were never called nor charged to be on queste, watche, nor other office wherby they shuld use or occupie any armour, or defencible gere of Warre, Wherthorugh they shuld be unredye and lettid to practice their cure of men beyng in perell : Therfore, for that they be so small nombre of the said felisshepe of the crafte and Misterye of Surgeons, in regarde of the grete multitude of pacientes that be, and daily chaunce and infortune happenyth and encresith in the forsaid Citie of London, And that many of the Kinges liege People sodenly wounded and hurte, for defaute of helpe in tyme to theym to be shewid, perisshe, And so diverse have done, as evidently is knowen, by occasion that your said Suppliauntes have ben com pelled to attende upon such Constableshipe, Watches, and Juries as aforesaid ; Be it enacted and establisshed by the Kinge oure Soveraigne Lorde, and the Lordes spirituall and temporall, and by the Comens in this present Parliament assembled, and by auctoritie of the same, that fromhensforth your said suppliauntes be discharged, and not chargeable, of Constableshippe, Watch, and of almaner of office beryng any armour, and also of all enquestes and juries within the Citie of London ; And also that this Acte in all thynge do extende to all Barbours Surgeons admytted and approved to excercise the said Misterye of Surgeons, according to the fourme of the Statute lately made in that behalfe : So that they excede, ne be, at one tyme above the nombre of xij persons.[1]

attended sick folk night and day,

and have been exempt from bearing arms;

and in war have been treated like Heralds,

because their business was to help the sick.

And in London, from their Incorporation, they've never been called on to serve on quest or watch.

Therefore, since the Surgeons are so few, and London folk fall ill,

while many get wounded,

It is enacted that Members of the Fellowship of Surgeons of London shall be exempt from Constableship, Watch, Juries, &c.

So also shall all Barber-Surgeons duly admitted as Surgeons,

their number being kept to 12.

[1] We suppose the Statute meant only to limit the Fellowship of Surgeons to twelve ; not to say that if it numbered eleven, only one of the many Barber-Surgeons admitted as Surgeons should be entitled to the exemption above-given. Who was to settle which this one was ? See p. 212, below.

200 App. VII. *Alien Surgeons not Handicraftsmen.*

iii.

A. *Statement to show the Cause of the next Statute*, 22 *Henry VIII. Ch. XIII, being passed in* 1531.

<small>Acts on Alien Handicraftsmen.</small>

By the Statutes 1 Ric. III, ch. 9, 10, 12; 1 Hen. VII, ch. 9, 10; and 14-15 Hen. VIII, c. 2, divers enactments were made regulating the trade, work, and status, of Alien and Denizen handicraftsmen in England, restricting their power of taking more than two Apprentices, &c. These enactments having been continually broken by these Aliens, &c., A Decree was, on April 14, 1528 (20 Hen. VIII), made in the Star Chamber " concerninge Straungers Handye-craftesmen inhabitinge this Realm of England" (*Rec. Com. Stat.* iii. 298—301). It recites that the English Artificers and Handicraftsmen complain of the great detriment they suffer from the excessive number and unreasonable behaviour of the said stranger-artificers, who do infringe and break the said Statutes, sell goods at excessive and unreasonable prices, import 'bacon, chese, powdered [salted] beffes, mottons, and other com*m*odytes,' and when they have made money, take it abroad, and settle there, and help the King's enemies, whereby 'our Subjectes handycraftsmen . . . be sore impoverysshed, mynyssed, and almoost utterly decayed and destroyed,' and 'fall to thefte, murder and other great offences :' Considering this, and 'the great scarcyte of grayne and vytell at this present tyme,' It is decreed, this 10th of Febr. 1529, that no Alien shall keep more than two alien Journeymen, though they may have as many English ones and apprentices as they can get; that they shall pay City and Company charges, subsidies, taxes; shall assist in the Searches required by St. 14 and 15 Hen. VIII, ch. 2 ; shall be admitted into Companies on swearing fidelity to the King, and obedience to the Laws; and that Denizens only shall set up new Shops, &c. &c.

<small>Star-Chamber Decree to control them.</small>

<small>They break the Statutes,</small>

<small>and help the King's enemies.</small>

<small>After Feb. 10, 1529, they must obey the Decree,</small>

<small>and the Act confirming it.</small>

This Decree was meant specially to protect the Cordwainers; and it was ratified by the Act 21 Henry VIII, ch. 16 (*Record Stat.* iii. 297), A.D. 1529. But as Surgeons are Handicraftsmen—isn't *Chirurgion* from Greek *cheir* the hand, and *ergon* work ?—and so are Bakers, Brewers, and Scriveners; opportunity was taken by the evil-minded to worry alien Surgeons, Bakers, Brewers and Scriveners under the above-named Act. Consequently Parliament interfered, and by the following Act of 1531, had to class Surgeons with their more lowly brethren, Bakers, Brewers and Scriveners, useful feeders of body and mind.

This Act was unduly turned against Surgeons, &c.

B. 22 *Hen. VIII. Chapter XIII.* A.D. 1530-1.
(*Record Stat.* iii. 332.)

An Acte concernyng Bakers, Bruers, Surgeons & Scryveners.

WHERE dyvers Estatutes penall hertofore have been made ageyn straungers artyfycers for exercysyng of hand craftes within this Realme, and for kepyng of houses, apprentyses, & servauntes estraungers, as by the sayde severall Estatutes more playnly ys rehersed: Sythen the makyng wherof, bere-bruers and bakers whiche bene comon vitaylers, and also surgens and scryveners, beyng straungers inhabyted and dwellyng wythin this realme, hathe bene putte to trouble and great vexacion by occasion of informations brought ageyne them upon the sayde Estatutes, supposyng that Straungers usyng bakyng, bruyng, surgerye, or wrytyng, shulde be hand craftesmen; upon the whiche information greate doutes and ambiguytes have rysen, whether straungers usyng any of the sayde mysteryes or sciences shulde be understande suche handcraftesmen as were entended by any the sayde Estatutes: For playne declaracion wherof [hit is¹] enacted by the Kyng oure Sovereign Lorde, and the Lordes Spirituall and Temporall, and the Commons in this present parliament assembled, and by auctoryty of the same, that no person nor persones straungers, beyng a comon baker, bruer, surgeon or scyvenour, shalbe enterpret or expounded hande craftesmen, in, for, or by reason of usyng any of the sayde mysteryes, or scyens, of bakyng, bruyng, surgery or wrytyng. And that all informations, sutes, accions and processe, had, taken, er herafter to be taken, upon eny of the sayde Estatutes, agayn any suche straunger or straungers beyng bakers, bruers, surgeons or scryveners, shall be, by auctoryte of this present acte, voyde and of none effecte.

Statutes against Alien Artificers for exercising of Handicrafts,

have been wrongly used against Alien Surgeons, &c.

So it is enacted,

that Alien Bakers, Brewers, Surgeons, and Scriveners, shall not be accounted Handicraftsmen.

¹ be it O.

202 App. VII. *The Uniting Act*, 32 *H. VIII, ch.* 40.

iv.

Extract from 32 Hen. VIII, ch. 40, A.D. 1540.

Physicians may practise Surgery.

The Physicians' Act of 1540, 32 Hen. VIII, ch. 40,

32 Hen. VIII, Chapter XL, A.D. 1540 (*Record Stat.* iii. 793), exempts the Physicians in London and its suburbs from serving as Constables, or on watch and ward, as the Surgeons had been exempted by 5 Hen. VIII, ch. 6. It also lays on four Physicians chosen by their Company, the duty of viewing yearly the wares, drugs and stuffs sold by Apothecaries, and ordering the bad ones to be burnt or destroyed. It fines Apothecaries resisting the inspecting Physicians, 100s.; and those inspectors who neglect their duties,

enacts, that as

40s. It then enacts that Physicians may practise Surgery :

Physic includes Surgery,

any Physician may practise Surgery, &c.

"And forasmuche as the science of phisicke dothe comprehend, include, and conteyne, the knowledge of surgery as a speciall membre and parte of the same, therefore be it enacted, that anny of the said companny or felawiship of Phisitions, being hable, chosen, and admitted by the said president and feliship of Phiscians, may from tyme to tyme, aswell within the Citie of London as elsewhere within this Realme, practise and exercise the said science of Phisick in all and every his membres and partes, any acte, statute, or prouision, made to the contrarie notwithstanding."

v.

32 *Hen. VIII. Chapter XLII.* A.D. 1540.
(*Record Stat.* iii. 794.)

Concerning Barbers and Chirurgians.

I.
As it is needful to provide skilful Surgeons for sick men's relief,

THE King our Souveraine Lorde, by thadvise of his Lordis spirituall and temporall, and the Commons in this present parlament assembled, and by auctoritie of the same, by all their common assentis, duely pondering among other thinges necessary for the common welth of this Realme, that it is very expedient and needeful to provide for men experte in the science of fisicke and surgery, and for the helth of man's body whan infirmities and seckness shallhappen ; for the due exercise and maintenaunce wherof, good and necessarie actis be

App. VII. *Act Uniting the Barbers & Surgeons.*

alredy made and provided; yet nevertheles, forasmuche [as][1] within the Citie of London, where men of great experience, aswell in speculation as in practice of the science and [facultye][2] of surgery be abiding and inhabiting, and have more co*m*monly the daily exercise and experience of the same science of surgery then is had or used within other pa*r*tès of this Realme, And by occasion therof ma*n*ny expert pe*r*sonnes be brought up undre them as their serva*u*ntis,[3] apprentices, and other, who by thexercise and diligent information of [the*ir*] said maistres, aswell nowe as herafter, shall exercise the said science within divers other p*a*rtès of this Realme, to the greate relief, comforte, and soccour of muche people, and to the sure savegard of their bodily helth, their lymmes and lyves; And forasmuche as within the said Citie of London there be nowe twoo severall and distincte companyes of surgeons, occupying and exercising the said science and facultie of surgery, thone company being called 'the Barbours of London' and thother company called 'the Surgeons of London,' whiche company of Barbours be incorporated to sue and be sued by the name of 'Maistres or Governours of the mistery and co*m*mynaltie of the Barbours of London,' by vertue and auctoritie of the le*tt*res patentis undre the greate seale of the late King of famous memory, Kinge Edwarde the iiij[th], dated at Westm*inster* the xxiiij[tі] day of February in the first yere of his reigne, whiche afterwarde, aswell by our nowe most dradde Souveraine Lorde, as by the right noble and vertuouse Prince, Kinge Henry the vij[th], father unto the Kinges most excellent Highnes nowe being, were and be confirmed, as by sundry le*tt*res patentis therof made (among other thinges in the same conteynid) more at large may appere; And thother company called 'the Surgeons,' be not incorporate, nor have anny maner of corporation; whiche twoo severall and distincte companyes of surgeons were necessary to be unyted, and made one body incorporate, to thintent that, by their unyon and often assemble to-githers, the good and due ordre, exercise and knowlege of the said science or facultie of surgery shulde be, aswell in speculation as in practise, bothe to them-selfis, and all other their said serva*u*ntes[3] [p. 795] and apprentices, nowe and herafter to be brought up undre them, and, by their larninges

and there are many Surgeons in London

who teach younger ones;

And as two Companies of Surgeons exist in London,

one, Barbers,

incorporated in 1 Edw. III, A.D. 1462,

the other, Surgeons, not incorporated,

and these ought to be united into one body;

---

[1] as O. at, print. [2] facultye O. falcultie, print.
[3] qualified Surgeons, or assistants. See p. 208, below.

and diligent and ripe informations, more perfett, spedy and effectuall remedy shuld be, [then]¹ it hath ben or shulde be if the said twoo companyes of barbours and surgeons shuld contynue severid a-sundre, and not joyned to-gither, as they bifore this tyme have ben and used them selfis, not meddlyng to-gither; Wherefore, in consideration of the premisses, be it enacted by the King our Soveraine Lorde, and by the Lordis spirituall and temporall, and by the Comons in this present parlament assembled, and by thauctoritie of the same, that the said twoo severall and distynct companyes of Surgeons, that is to say, both the Barbours and the Surgeons, and every person of them (being a freeman of either of the said companyes after the custume of the Cittie of London), and their successours, from hensfurth ymmediately be unyted and made one entier and hole body corporate, and one commynaltie perpetuall, whiche at all tymes herafter shalbe called by the name of 'maistres or governours of the mistery and commynaltie of Barbours and Surgeons of London' for ever more, and by none other name; And by the same name to implede and be impleded bifore all maner of Justices in all Courtis, in all maner of actions and sutes, and also to purchace, enjoy and take, to them and to their successours, all maner of landis, tenementis, rentis, and other possessions, whatsoever they be; and also shal have a common seale, to serve for the busynes of the said companye and corporation for ever; And by the same name, peasably, quietly, and indiffeasably, shall have, possesse, and enjoye, to them and to their successours for ever, all such landis and tenementis, and other hereditamentis whatsoever, whiche the said company or cominalty of Barbours have or enjoye, to thuse² of the said mistery and comminalty of Barbours of London; And also shall peasably and quietly have and enjoye, all and singulier benefittes, grauntis, liberties, privileges, [and]³ franchises and free custumes, and also all maner of other thinges at anny time geven or graunted unto the said companyes of Barbours or Surgeons, by whatsoever name or names they or anny of them were callid, and whiche they or anny of them nowe have, or anny or of their predecessours have had, by actes of parlament, lettres patentis of the Kinges Highnes, or other his moost [noble]⁴ progenitours, or

it is enacted that

the said Two Companies are united and incorporated into one Company of Barbers and Surgeons,

with all Privileges, &c. enjoyed by the incorporated Company of Barbers,

with a common seal,

power to hold lands,

and all rights of both the old Companies,

¹ than O. (then = than.) ² the use.
³ O omits 'and.' ⁴ noble O. nobbe, print.

otherwise by anny other laufull meanes have had, at anny tyme afore this present acte, in as large and ample maner and fourme, as they or anny of them have had, might or shulde enjoy the same, this union or conjunction of the said companies togither notwithstanding; And as largely to have and enjoye the premisses, as if the same were, and had ben, specially and particulerly expressid and declared with the best and most clerest wordis and termes in the lawe, to all intentis and purposes: And that all personnes of the said company nowe incorporate by this present acte, and their successours, that shalbe laufully admitted and approved to occupy surgery, after the fourme of the statute in that cace ordeynid and provided, shalbe exempt for[1] bearing of armure, or to be put in anny watchis or inquestis: And that they and their successours shalhave the serche, oversight, punyshement and correction, aswell of freemen as of forreynes, for suche offences as they or anny of them shall committ or doo against the good ordre of Barbery or Surgery, as afore this tyme, amonge the said mistery and company of barbours of London, hath ben used and accustumed, according to the good and politike rules and ordenaunces by them made, and approved by the Lordis Chauncelour, Treasorer, and twoo chief Justices of either benche, or anny three of them, after the fourme of thestatute in that cace ordeynid and provided.

in the fullest manner.

And all Surgeons of the new Company

are exempted from bearing Armour, &c.

They may also punish all Freemen and Foreigners breaking their Rules.

AND further be it enacted by thauctoritie aforesaid, that the said Maistres or Governours of the misterie or comminalty of Barbours and Surgeons of London, and their successours, yerely for ever, aftre their sadd[2] discretions, at their free libertie and pleasure, shall and may have and take, without contradiction, fower personnes, condempned, adjudged, and put to death for felony by the due ordre of the Kinges lawes of this Realme, for anathomyes, without any further sute or labour to be made to the Kinges Highnes, his heires or successours, for the same; and to make incision of the same deade bodies, or otherwise to ordre the same aftre their [said][3] discretions at their pleasures, for their further and better knowlege instruction, insight, lerning, and experience, in the said science or facultie of surgery.

II. Surgeons may yearly take the Bodies of Four Malefactors to anatomize,

and dissect as they like, for their instruction.

SAVING unto all personnes, their heires and successours, all suche right, title, interest and demaunde,

III. General Saving of other folks' Titles

[1] from or against.   [2] well-considered, deliberate.
[3] sadde O.

206 App. VII. *Surgery and Barbery separated.*

In the new Company's Lands.

which they or anny of them might laufully clayme or have, in or to anny of the landis and tenementis, with thappurtenauntes, belonging unto the said company of Barbours and Surgeons, or anny of them, at anny tyme afore the making of this Acte, in as ample maner and fourme as they, or any of them, had or ought to have had heretofore; anny thing in this present acte comprised to the contrary herof, in anny wise notwithstanding.

IV.

And as Surgeons often take diseased persons into their house where they shave men, which

is dangerous to the King's people,

Now, after Christmas next, no Barber in London shall practise Surgery, except Toothdrawing;

And no Surgeon shall be a Barber, or shave any one.

Also, all Surgeons in London, and a mile outside it,

shall have open Shop Signs,

AND forasmuche as suche personnes usyng the mistery or faculty of surgery, often tymes medle and take into their cures, and houses, such [sykke][1] and diseasid personnes as ben infected with the pestilence, great pockes, and such other contagious infirmities, (&) doo use and exercise barbery,[2] as wasshing or shaving, & other feates therunto bilonging, which is very perillous for infecting the Kinges people resorting to their shoppes and houses, there being washed or shaven; Wherefore it is nowe enacted, ordeynid and provided, by thauctoritie aforesaide, that no maner of personne within the Cittie of London, subburbes of the same, and one myle compas of the said Cittie of London, after the feast of the Nativitie of our Lorde God next comyng, using [barbary][3] or shaving, or that herafter shall use any barbary or shaving within the said Citie of London, suburbes, or one myle circuite of the same Citie of London, he nor they, nor none other for them, to his or their use, shall occupy any surgery, letting of bludde, or any other thing belonging to surgery, drawing of teth onelye except; And furthermore, in like maner, who-soever that usith the mystery or crafte of Surgery within the Circuite aforesaid, as longe as he shall fortune to use the said mistery or crafte of Surgery, shall in no wise occupye nor exercise the feate or crafte of barbarye or shaving, neither by himself, nor by none other for him, to his or their use; And moreover, that all maner of personnes using surgery for the tyme being, aswell freemen as forrens, aliens and straungers, within the said Cittie of London, the suburbes therof, and one myle compas of the same Cittie of London, bifore the feast of Sainte Michaell tharchaungell next commyng, shalhave an open signe on the strete side where they shall fortune to dwell,

---

[1] sykke O. like, print.   [2] barbery O. barber .. print.
[3] any barbery O.

App. VII. Wardens of Barbers and Surgeons. 207

that all the Kinges liege people there passing by, may knowe at all tymes [whethir]¹ to resorte for² remedies in tyme of their necessitie. *to let sick folk know where to find them.*

The Record-Commission print of the Statutes from their MS. leaves out the following Sections of this Uniting Act, 32 Hen. VIII, ch. 42, which appear in the black-letter issues of the Statutes. We print them from Pulton's edition of 1636, p. 798.

And further be it enacted by the authority aforesaid, that no manner of person, after the said feast of Saint *Michael* the Archangell next comming, presume to keepe any Shop of Barbery or shauing within the City of London, except he be a Freeman of the same Corporation and Company. *V. None shall be a Barber in London but a Freeman of that Company.*

And furthermore, at such times as haue beene heretofore accustomed, there shall be chosen by the same Company, foure Masters or Gouernors of the same Corporation or Company; of the which foure, two of them shall be expert in Surgery, and the other two in Barbery; which foure Masters, and euery of them, shall haue full power and authority from time to time during their said office, to haue the ouersight, search, punishment, and correction of all such defaults and inconueniences as shall be found among the said Company vsing Barbery or Surgery, as well of freemen, as forcines, aliens and strangers, within the city of London and the circuit aforesaid, after their said discretions. And if any person or persons vsing any Barbery or Surgery, at any time hereafter, offend in any of these articles aforesaid: then for euery moneth, the said persons so offending shall lose, forfeit and pay, fiue pounds: the one moity thereof to the King our Soueraigne Lord, and the other moity to any person that will or shall sue therefore, by action of debt, Bill, plaint, or information, in any the Kings Courts, wherein no wager of law, essoine,³ or protection, shall be admitted or allowed in the same. *VI. Foure Masters or Wardens,— 2 Surgeons, 2 Barbers,— shall be chosen* *to correct all defaults in Surgeons and Barbers,* *native and alien.* *The Forfeitures of the Offenders* *to be £5 a month.*

Prouided that the said Barbers and Surgeons, and euery of them, shall beare and pay lot and scot,⁴ and *VII. All Barbers and Surgeons to pay*

¹ whyther O. whethir = whither. ² for theyre O.
³ See these terms explained in the York Barber Surgeons' Ordinary below.
⁴ *Scot and Lot* (Sax. *Sceat*, pars, and *Hlot*, i. e. *Sors*), Signify a customary Contribution laid upon all Subjects, according to their Ability.—*Spelman.* Nor are these old Words grown obsolete; for whoever in like Manner (though not by equal Portions) are assessed to any Contribution, are generally said to pay *Scot* and *Lot.* Stat. 35 Hen. 8. c. 9.—1744. Giles Jacob, *New Law Dict.* 5th ed.

208   App. VII. *Any one may cure slight Ailments.*

<small>lot and scot in the City.</small>
such other charges as they and their predecessors haue beene accustomed to pay, within the said City of London, this Act, nor any thing therein contained to the contrary hereof, in any wise notwithstanding.

<small>Any person may keep a Barber or Surgeon as his Servant,</small>
Prouided alway, and be it enacted by [the] authority aforesaid, that it shall be lawfull to any of the Kings Subiects, not being Barber or Surgeon, to retaine, haue, and keepe in his house, as his seruant, any person <small>who may practise in his Master's house,</small> being a Barber or Surgeon, which shall and may vse and exercise those arts and faculties of Barbery and Surgery, or either of them, in his masters house, or elsewhere by his Masters licence or commandement, any thing in this Act aboue written, to the contrary notwithstanding.

vi.

34 & 35 *Henry VIII. Chapter VIII.* A.D. 1542-3.
(*Record Stat.* iii. 906.)

AN ACTE that persones being no comen Surgeons maie mynistre medicines owtwarde.

<small>Recital of Stat. 3 H. VIII. c. 11. (no. I, above) for Regulation of Physicians and Surgeons.</small>
WHERE in the parliament holden at Westm*inster* in the thirde yere of the Kinges moste gracious reigne, amongest other thinges for the advoyding of sorceryes, witchecrafte, and other inconveniences, it was enacted, that no per*s*one within the Citie of London, nor within <small>Under it, only examined men were to practise.</small> seven myles of the same, shoulde take upon him to exercyse and occupie as Phisician or Surgeon, except he be first examyned, approved, and admytted by the Bisshopp of London and other, undre and upon certaine peynes and penalties in the same Acte mencioned; <small>But these licenst Surgeons, caring for money only,</small> Sithens the making of whiche saide Acte, the Companie and Felowship of Surgeons of London, mynding oonelie theyre owne lucres, and nothing the profite, or ease of <small>have sued kind folk</small> the diseased or patient, have sued, troubled, and vexed, divers honest per*s*ones, aswell men as woomen, whome God hathe endued with the knowledge of the nature, <small>who have given herbs &c. to people with common ailments, *gratis*,</small> kinde, and operaci*o*n, of certeyne herbes, rotes and waters, and the using and mynistering of them to suche as been pained with customable diseases, as Womens brestes being sore, a Pyn and the Web[1] in the eye, uncomes[2] of hande*s*, scaldinge*s*, burning*es*, sore mouthes,

---

[1] *Web* . . a Pearl or Spot in the Eye.—Kersey.   [2] whitlows or felons.

App. VII. *Any one may treat slight Ills.*

the stone, strangurye,[1] saucelin[2] and morfew,[3] and suche other lyke diseases, and yet the saide persones have not takin any thing for theyre peynes and cooniung,[4] but have mynistered the same to the poore people, oonelie for neighbourhode and Goddes sake, and of pitie and charytie ; and it is nowe well knowen that the surgeons admytted, wooll doo no cure to any persone, but where they shall knowe to be rewarded with a greater soome or rewarde than the cure extendeth unto ; for in cace they wolde mynistre theyre cooning to sore people unrewarded, there shoulde not so manye rotte and perishe to deathe for lacke of helpe of Surgerye as dailie doo ; but the greatest parte of Surgeons admytted, been muche more to be blamed than those persones that they trouble ; for althoughe the most parte of the persones of the saide crafte of Surgeons have small cooning, yet they wooll take greate soomes of money, and doo litle therfore ; and by reasone therof, they doo often tymes impaire and hurte theyre patientes, rather thenne doo them good : In Consideracion wherof, and for the ease, comforte, socour, helpe, relief and healthe of the Kinges poore Subjectes, inhabitauntes of this his Realme, nowe peyned or diseased, or that herafter shalbe peyned or diseased, Be it ordeyned, establisshed, and enacted, by thauctorytie of this present parliament, that at all tymes from hensforthe, it shalbe lefull to everye persone, being the Kinges Subject, having knowledge and experience of the nature of herbes, rotes, and waters, or of the operacion of the same, by speculacion or practyse, within any parte of the Realme of Englande, or within any other the Kinges Domynions, to practyse, use and mynistre, in and to any outwarde sore, uncoom, wounde, appostemacions, outwarde swelling, or disease, any herbe or herbes, oyntementes, bathes, pultes,[5] and emplasters, according to their cooning, experience, and knowlege in any of the diseases, sores, and maladies aforesaide, and all other lyke to the same, or drinkes for the stone, strangurye or agues, without sute, vexacion, trouble, penaltie, or losse of theyre goodes. The foresaide Statute in the foresaide thirde yere of the Kinges most gracious reigne, or any other Acte, ordinaunce or statute, to the contrarye hereof heretofore made, in any wise notwithstanding.

out of pity.

Licenst Surgeons have also askt too high fees,

and have let many folk rot and die,

much to their blame.

Most Surgeons are ignorant too, and often harm their patients.

Therefore good Persons who know the nature of Herbs,

may cure outward Sores by Herbs, Ointments, &c.,

and Stone or Ague by drinks, without being sued under the recited Act of 1511-12.

---

[1] *Strangury* or *Strangullion*, (*Gr.*) a Disease, when the Urine is voided by Drops, with great Difficulty and Pain, and a continual inclination to make Water.—1706. Kersey. [2] Chaucer's sauceflume, salt flegm, a scurvy face, &c.
[3] *Morphew*, a kind of white Scurf upon the Body : from the French Word *Mort-feu, i. e.* dead Fire ; because it looks like the white Sparks that fall from a Brand extinguished.—1706. Kersey. [4] cunning, skill. [5] poultice.

VIII.

SUPPLEMENT TO HENRY VIII's STATUTES,
FROM THE GUILDHALL RECORDS.

The Twelve (pure) Surgeons exempt from Watch, &c., p. 210-12.
Barbers not exempt from Watch, p. 213, 214.
Unlicenst Physicians to be put in prison, p. 213.
Physicians to pay for Exemption, p. 215.
Barber-Surgeons' Statutory Exemption modified, p. 215.

1510. Surgeons to dwell in the City, and serve on Watches, &c.

(Rep. 2, leaf 101, back) xv° die Octobris / Anno regri Regis Henrici viij¹ secundo (1510).

Surgions of London.
Item, At the same Court yt is Agreed, and commaundement gyven) to the Surgeons of this Citie, that they, and euery of them, dwell within the libertie of this Citie / And be obedient to all maner of Somons, watches, and all other charges, as other Citysyns be and ought to be.

1517. Inspeximus of the Act 5 Hen. VIII, ch. 6 (A.D. 1513), with a List of the 11 Surgeons exempted under it from bearing Arms, and serving on Watches, Quests, &c., in 1517 and 1525.

(Guildhall Records, Letter Book N, leaf 44, back: between the 2nd and 14th of July, 9 Henry VIII, A.D. 1517.)

We have inspected the Act 5 Henry VIII ch. 6, past in the Parliament held from Febr. 4, 1512, to Jan. 23, 1517, in these words:
Henricus, dei gratia Rex Anglie & Francie, & dominus Hibernie, Omnibus ad quos presentes¹ litere peruenerint, salutem! Inspeximus quendam Actum in vltimo parliamento nostro apud Westmonasterium quarto die Februarij, Anno Regni nostri tercio inchoato & tento, Ac per diuersas prorogaciones ad & vsque vicesimum tercium Januarij vltimo preteritum Continuato & prorogato, et tunc tento, De Assensu Dominorum spiritualium & temporalium in eodem parliamento existentium, vnacum in-dorsamento per nos superinde facto in hec verba, scilicet: Shewen vnto your discrete

¹ MS. perpresentes.

App. VIII. *Inspeximus of* 5 *Henry* VIII, *ch.* 6.   211

wisdomes, your humble Oratours the wardens and felawshippe of the Crafte and Mistere of Surgeons enfraunchised in the Citie of London, not passyng in Nombre twelue persones, That where-as they & ther predecessours, frome the tyme that noo mynde ys to the Contrarye, Aswell in this noble Citie of London, as in all other Cities and Boroughes within this Realme or elswhere,—for the Contynuall seruice and Attendaunce that they dayly and nyghtly, At All houres & tymes geue to the kynges liege people, for the Relefe of the same Accordyng to the[ir] science—haue been exempte And discharged frome all offices And besynes whereyn they shulde vse or bere Any maner of Armour or Wepyn; And with lyke pr[i]uylage haue been entreated As herawdes of Armes, Aswell in Batelles & Feldes, As other places, there¹ for to stande vnharnesed And vnwapened,² Accordyng to the lawes of Armes, by Cause they be personnes that neuer vsed feates of warre, nor ought to vse, but only the besynes and exercise of the[ir] sciense, to the helpe & comfort of the kynges liege people in the tyme of their nede / And in the forseid Citie of London, frome the tyme of their First in-Corporacion, when they haue been meny moo in nombre than they be nowe, Were neuer Called nor Charged to be on queste, wacche, or other office wherby they shuld vse or occupye Any Armour or defensible G[e]re of warre, where-through they shulbe vnredy And letted to practyse their Cure of Men beyng in perell; Therefore, for that they be so small Nombre of the seid Feaulishhip of the Crafte And Mistere of Surgeons, in Regarde of the Great Multitude of pacientes that been, [&] dayly, Chaunce & infortune happenyth & Encreasyth in the Foreseid Citie of London, And that many of the kynges liege people sodenly wounded and hurte for defaute of helpe in tyme to theym to be shewed, perysshe; And so, dyuerse of theym haue doone, as Evydently is knowen, by occasion that your seid suppliaunttes haue³ been Compelled to Attend vppon shuch Constableshipe, wacches, And Juryes As Aforeseid / Be it enacted & Establisshed by the Kyng, our soueraygne lord, And the lordys spirituall And temporall, And the Comens of thys present parliament Assembled, And by the Auctoritie of the same, that From hensforth your seid suppliaunctes be dischargied,

As the Fellowship of Surgeons not exceding 12 men, have, they and their Foregoers,

always been exempt from bearing Arms

both in the battlefield and elsewhere,

because they seek only to help the King's folk;

And as, in London

they have never been required to serve on Quest or Watch,

since they are few, and Patients many,

and yet some hurt folk have died for want of timely help,

because Surgeons have been forst to serve as Constables and Watchmen, It is enacted

¹ MS. there there.
² See the description of them 'without harnes' in the Citizens' March-past before Henry VIII in 1539, p. 173 above.   ³ MS. haue haue.

P 2

212 App. VIII. *List of Surgeons exempt from Watch.*

that Members of the Fellowship of Surgeons shall not be liable to serve as Constables, Watchmen, Inquestors or Jurors, nor shall any licenst Barber-Surgeon be so.	And not Chargeable, of Constableshipe, wacch, And of All maner off office beryng Any Armore / And Also of All enquestes & Juryes within the Citie of London) / And Also that this Acte in All thyng do extend to All Barbours Surgeons Admytted & Approued to exercise the seid Mistere of Surgeons Accordyng to the fourme of the Statute latly made in that behalfe / Nos
This Act, we have, at the request of the late Wardens of the Fellowship of Surgeons, directed to be verified, as we testify by these our Letters Patent.	autem tenorem Actus predicti, Ad Requisicionem Johannis Hart & Ricardi Hogekyn), tunc Gardianorum Societatis Artis siue Mistere De le Surgeons Ciuitatis nostre Londonie, duximus Exemplificandum per presentes. In cuius rei testimonium, has literas nostras fieri fecimus patentes. Teste meipso apud Westmonas-

Witness Ourself, this 10th of March 1517.	terium, decimo die Marcij, anno regni nostri quinto / /porter/ Ingeramum Bydell ⎫
	   Examinatur per ⎬ Clericos
	          Willelmum Porter ⎭

The names of the 11 Members of the Fellowship of Surgeons in March, 1517.	Hereafter folowyth the Names of those Surgeons which[1] be exempt from Almaner offices, enquestes & wacches, accordyng to the Acte of parliament heretofore made, enacted, & presented by Doctour Yakesley:

Thomas Thornton) ⎫	Richard Hockekyns ⎫	James Monford ⎫
Thomas Rosse ⎬ wardens ·	Robert Marshall ⎬	Thomas Palley ⎬
Robert Beuerley ⎫	John Rutter ⎭	Edward Holway ⎭
Christofer Turner ⎭	Garet Fereys	

In Journal 11, lf. 296, back, is the following list of exempt Surgeons in 1525, enterd on a blank page left during the Mayoralty of Sir Jn. Rest, 1516-17. The first names are those above given. On the deaths of Beverley and Turner,[2] 2 fresh Surgeons were added in 1525. All follow a copy of an Inspeximus like that above printed, which Inspeximus is dated March 10, 1514 (5 Hen. VIII).

Intratur	Heraftur folowen) the Names of those Surgeons whiche be exempt from almaner enquestes and watches accordyng to the Acte of parliament heretofore made, enacted and presented by / Doctour Yakesley.
mortuus	Thomas Thornton) † ⎫ Wardeyns Thomas Rosse ⎭
mortuus	Robert Beverley
mortuus	Christofer Turner †
	Richard Hochekyns †
mortuus	Robert Marshall †

[1] MS. which which.
[2] The later deaths of Thornton, Hochekyns, and Marshall, are enterd in another hand.

App. VIII. *Barbers not exempt from Juries, &c.* 213

John Rutter
Garet Fereys
Jamys Momford
Thomas Palley
Edward Holway
Edward Clache
Cristofer Dyxson
} impositi ad instanciam Gardianorum & aliorum Mistere predicte, 6. 12. Anno 17 (6 Dec. 1525).

1520. The Barber-Surgeons' claim for Exemption from Juries, &c., not allowd by the City.

(Rep. 5, lf. 29) Jouis, 15 die Marcij (? 11 Hen. VIII, A.D. 1520).

[*Present*] Maior [Sir Jas. Yardford], Recorder, Aylmer, Boteler, Exmewe, Brugge,[1] Milburn, Feure, Aldernes, Mundy, Baldry, Bayly, Aleyn, Seymer, Spencer, Kyme, & Ambo vicecomites [Jn. Wilkinson, Nicholas Partrich]
Barbitonsores Isto die, lecta fuit Supplicacio Barbitonsorum excercentium Misteram de Surgeons, essendis exemptis[2] ab omnibus Juratis &c : Et dictum fuit per Magistrum Recordatorem, quod omnes Concessiones facte per Edwardum 4, Resumpte fuerunt per Dominum Henricum 7 ; Et nulla prouis[i]o facta fuit.

(Repertory 4, lf. 62) Martis, 28 die Augusti (? an. 12 Hen. VIII, 1520)

At this Courte camme Pereson & Bankes, Wardens of the Barbours Surgeons, & Showed forthe their Graunte of Kyng Edward iiijth, wherby they Claymed to be dyscharged & exempte of all maner of Juries & other Inquisicions &c. *Et* non allocatur. Whereuppon they hadde in Commaundement to geve warnyng to all theyre Company tappere as others do, vppon theire perell, & cⁱ [This entry is repeated in Repertory 5, leaf 64.]

1525. Unlicenst Physicians to be put in Prison : All Prescriptions to be filed.

At a Common Council held on Thursday the 18th of April, 16 Henry VIII, A.D. 1525 [*leaf* 280], the following Resolutions were past :—

---
[1] John Brug or Bruges.
[2] We suppose the ablative, and not the genitive, is the proper case.

214 App. VIII. *Unlicenst Physicians to be imprisond.*

(Journal 12, lf. 281, bk.)

Phisicions

Unlicenst Physicians may be imprisond for 20 days as often as they practise till they are licenst.

As licenst Physicians won't sell medicines that can be got of Apothecaries, so Apothecaries shall not make up unlicenst Physicians' prescriptions.

Apothecaries shall file all Recipes, to show whether they were good, or hurtful.

[*Journal* 12, *leaf* 282 *or* 292]
Physicians to be registerd in the City.

Item, at this Comen Counsell it ys agreed & decreed, that suche as occupie phisike within the liberties of this Citie, not beyng' examynede & approuyd by the Collegge accordyng' to the statute in that behalf ordeyned & prouided, may be, at the Requeste of the College, commaundyd & compelled vppon the payne of imprisonament of xx days, tociens quociens, that they shall no more occupie phisike till they be examyned. Item, where-as all the College & those whom they admytte, be swore that they shall sell no medicynes theym self, yf they may haue the same of the apothecaries, so that it be prouydyd that thapothecaries may be swore, and vppon) a payne commaundid, that they shall not serue eny byll of eny physicions not examyned & approved. Item, that thapothecaries shall kepe the billis that they serue, vpon) a fyle, to thentent that, if the pacyent myscary / it may be by the College considerid whether the bill were medecynall, or hurtfull, to the siknes.

Item, that when) eny persone ys admytted by the seyd College to occupie phisike, that then) they shall, from tyme to tyme, Certifie the same to my lord Mayer' for the tyme beyng', to thentent that it may here Remayne of Recorde ∴

These entries are also in *Letter-Book N, leaf* 262.

1525. Barbers to serve on Inquests in the City.

Bayley [Mayor].

Intratur

The King's letter asking that the Barbers may not serve on Inquests

is utterly denied.

Commune consilium tentum xx die Julij, Anno Regni Regis Henrici viij^{ui} Decimo Septimo [A.D. 1525], in presencia Willelmi Bayly, Militis, Maioris, Georgii Monoux, Willelmi Boteler', Thome Exmewe, Johannis Brugge, Johannis Milbourne, Johannis Mundy, Militis, Johannis Aleyn), Johannis Rudston), Nicholai Lamberd, Johannis Caunton), Johannis Hardy, Stephani Pecok & Christoferi Ascus [? Ascue], Aldermannorum, & diuersorum aliorum Cominariorum &c. c' . . . . .

Also the kynges lettre sent to this Comen Counsell, in the Favour' of the Barbours of this Citie to be discharged of goyng' in enquestes, in like wyse was Redde at length, & well vnderstande / and for asmoche as it ys expressely ageynst the kynges lawes, and also ageynst the liberties of this Citie, it ys therfore vtterly denyed &c.

App. VIII. *Composition as to Watch-Duty.* 215

1538. The Physicians' Composition with the City as to Constableship and Watches, &c.

(Repertory 10, leaf 27 back) Jouis, xxviij die marcij anno 29 H. 8 [A.D. 1538].

Gresham [Mayor]. Intratur. Phisycyans. Item, my lorde Mayer moved, that phisicyans shall pay xx s to the vse of ye parysshe where he ys elect constable, & as longe as he remayneth wit*h*yn the same parysshe; & iij d̛ for a man to watche¹ whan hys tourne commyth nyghtly; and also all phisycyans shall pay clerkes wages & all other duetyes to the chyrche, &c.

1538 (Repert. 10, lf. 35). Jouis, vj Junij, anno 30, H. 8 [A.D. 1538].

Phisicians to pay 20s. for every exemption from Constableship, and 3d. from Watch. Item, that the phisicians Inhabyttyng wit*h*in the Citie of london shalbe constables, & shall pay xx s / & whan ye tourne for watche, shall pay iij d̛ for euery tyme.

(Repert. 10, leaf 50 back) Martis, viij die septembris, anno 30, H. 8 [A.D. 1538].

Phisicyans. 14 Physicians. (3, K. Henry's) agree to the City's terms. Item, Master Yaxley, Master Bartlet, Master Bentley, Master Clement, Master Wotton, Master Freman, Master Gwyn, Master Nycholas, Master Cromer, Master Fryar, Master Burges, Master Pyerson, Master Owen, Master Augustyn, phisicyans, have agreed to doo theyre duetyes accordyng to an Act of comon cownseyll therof made / And it ys agreed that the persones aforesayd shall enjoy the benefytt of the same.

1544-6. The Barber-Surgeons' Contract with the City of London as to Inquest-Duty, Contributions, Constableship and Watches, varying their Statute of 1540, 32 Hen. VIII, ch. 42.

1544. (Rep. 11, lf. 73 ink, 71 pencil) Sabba*ti*, xxiiij^{to} die Maij, Anno xxxvj^{to} Henri*ci* viij (A.D. 1544).

Waren [Mayor] [*Present*] Mayor, Recorder, Gresham, Forman, Dormer, Cotes, Laxton, Hoberthorn, Amcottes, Wylforde, Judde, Hyll, Jervys, Rede, Ac Tolos & Dobbys, vicecomites /.

(lf. 73, bk.) Barbours Item, yt is Agreyd that the Wardeyns of the Barbours shalbe warnyd to be here next Court day, for the mater here meved this day by Master Tolos, Shreve, for that, that they refuse to apere & passe vpon Enquestes, &c /

¹ serve on the Watch.

216 App. VIII. *Contract as to Inquests, &c.*

1545. (Rep. 11, 1f. 175 ink, 153 pencil)   Jouis, xij° die Marcij, Anno xxxvj^to Henr*ici* viij¹ (A.D. 1545).

(lf. 176, or 154 pencil)

Laxton)
Mayor.
Barbour-
surgeons

Item, the petyc*io*n of the Wardeyns of the barbour-surgeons to be dyscharged of Constableshipe, Watche, & all enquest*es* savyng⁺ the Enquest*es* of Wardemote onys in the yere, was red ; And aunswer made vnto theym) by the mouthe of M*aste*r Recorder, that theyr seyd Offer, mencyoned in theyr seyd petyc*io*n, to go Apon) enquest*es* of wardemote, shulde be Allowed & entred of Recorde, And that for the resydue of the mater of the seyd byll, the Court wolde be further Advysed, &c /

(Repertory 11, lf. 187 ink, 185 pencil.   Guildhall Records.)

1546. M*artis* xxviij° die Ap*rilis*, A*n*no xxxvij° Henr*ici* viij° (A.D. 1546).

[leaf 187, back]
[L]axton)
M*ayor*
[Ba]rbo*urs*

Item, the Court, At the petyc*io*n) of dyue*r*se of the barbou*r*s & surgeons, made vnto theym) in the name of theyr hole Felowshipe, is contentyd that theyr Offer her-tofore made to the sayd Court, to go vpon) the Warde-mote enquest At Crystmas, shall so be pennyd that yt shall not be preiudycyall or hurtefull to theyr graunte, made vnto theym) by acte of p*ar*liament

(Repertory 11, lf. 229 bk., ink ; 206 bk., pencil.)

1546. Jouis, viij die Octobris, A*n*no 37 H 8 (A.D. 1546).

Barbo*ur*s

Aylyff

Item, the Barbo*ur*surgeons haue   day oue*r* vntyll this day seve*n*nyght, for theyr olde matter of dyscharge from) Offices & other charg*es* / And Are wyllyd to send M*aister* Aylyffe worde to be here vpon Tuysday next, for the fyndyng⁺ of suertyes for thoffyce of Blakwell hall, wherof he hath the reue*r*syon).

(Repertory 11, lf. 231 ink, 208 pencil.)

Jouis, xv° die Octobr*is*, A*n*no 37 Henr*ici* viij¹ (A.D 1546).

[leaf 232 or 209]

The Barbers and Surgeons are to embody their Proposals in Articles.

Item, the petyc*io*n of the Barbours & Surgeons to be dyscharged of bering⁺ of Armoure & other charg*es*, Accordyng⁺ to the tenou*r* of thacte of p*ar*lyame*n*t A*n*no 32 / Henr*ici* viij¹, Cap*itulum* / 52¹ / was this day redd, & by the Court well debatyd ; And Fynally Agreyd, that they shall drawe the hole effect*es* of the same theyr byll in Artycles, Ageynst the next Court day ; And that then), the same beyng⁺ reasonable, shalbe Allowed vnto theym), & entred here of Record.

¹ That is, chapter 42.

App. VIII. *Contract as to Inquests, &c.* 217

(Repertory 11, lf. 234 ink, 211 pencil.)

Jouis, xxij° die Octobris, Anno 37 Henrici viij¹ (A.D. 1546).

Laxton) (Mayor) — [*Present*] Mayor, Recorder, [R.] Gresham, Hoberthorn), Amcottes, Tolos, Wylford, Lewen), J. Gresham, Judde, Dobbys, Hyll, Whyte, Chertsey, Lok; ac Berne & Aleyn), vice*comites* [sheriffs].

Barbours & Surgeons

Their Articles being reasonable, are agreed to by the Court.

Att this Court, the boke conteynyng the Artycles of certeyn) charges, & thexercyse of certeyn) Offyces to be bourne from) hensforward & exercysed by the Barbours & Surgeons of this Cytie, grauntyd & Agreyd vnto the seyd Barbours & Surgeons, was redde; And the same, by the Courte well perceyvyd & vnderstondyn), thought good & reasonable, And therupon) grauntyd by the same Court, & Agreyd, that the same Artycles shalbe entryd here of Recorde, Att All tymes herafter to be iustely obserued & kepte, & putt in due execucion from) tyme to tyme for euermore; the true tenour of whiche boke herafter ensuyth in these wordes:

[*The Barber-Surgeons' Agreement with the Corporation of London for varying the Statute* 32 *Hen. VIII, ch.* 42.]

To the ryght honourable sir Wylliam Laxton), knyght, lorde Mayer of the Citie of London), & his ryght Worshipfull Brethern), thaldermen) of the same /

Intratur

Forasmuch as some Citizens grudge the Barbers and Surgeons being exempted by Parliament from Services that other Citizens perform,

In theyr moste humble wyse, shewen) vnto your good lordeshipe & Maistershipes, your humble besechers, the maysters or gouernours of the mystery or cominaltye of the Barbours & Surgeons of the seyd Citie: that for asmoche As some grugge & dyspleasure ys lately, syth the vnyon) & Coniuncyon) of theyr seyd Felowshipe [felt] by dyuerse of theyr' neighbours, being Citezeins of this Citie / As they¹ be, by reason) that they, your seyd Supplyantes & theyr seyd Felowshipe Are nowe of late, for sundry good & reasonable cawses & Consyderacions (As yt hath semyd unto the kynges highnes & his graces moste high Court of parlyament,) sumwhat Allevyatyd, exoneratyd & dyscharged,—Aswell by vertue of sun)dry lettres patentes of his graces moste noble progenytours by his maiestie most gracyously Confyrmed, As Also by Auctoryte of dyuerse Actes of parlyament hertofore made & establyshed in that behalfe /—of & from) certeyn) Offyces & other charges that other the Citizeins of this Cytie Are elygyble & lyable vnto, for the whiche

¹ That is, the Barbers and Surgeons are also Citizens.

[If. 234 or 211, bk.] gruge & displeasure, your' besechers Are nott A lytle sorye / for the playn declaracion wherof, & for the eschuyng' & Avoydyng' & vtter extingguysshement of the seyd grugge & dyspleasure from) hensforwarde / They, for & in the name of theyr hole Felowshipe Aforeseyd, Are now Agreable & contentyd that yt may be ordeyned, enactyd, & decreyd by your' good lordshipe & Maistershipes, by the Auctoryte of this honourable Court, that they, your' seyd Supplyantes, shaH & may be from) hensforwarde, charged & Chargeable with other the Citezens of this Citie, in aH the affayers of the same, Accordyng' to the Tenour, true meanyng', purporte & effecte of the Artycles herunder wrytten), And no further, nor in eny otherwyse / And that the same Artycles may here be entred of Record / Att AH tymes herafter perpetually to be Obserued & kepte.

the Company of Barbers and Surgeons are willing to do such Services as follow:

1. That all Barbers and Surgeons shall serve on the Wardmote Quests,

[1] Fyrst, that the seyd hole felowshipe of Barbours & Surgeons shhaH, for euer (As theyr Course & turne shaH happen)) be sworne, go & passe, vpon) the Wardemote enquestes of this Citie from) tyme to tyme, in lyke maner as aH other the Citezens of this Citie, for theyr partes, do go & passe vpon) the same; So alwayes that they & euery of theym) may clerely be dyscharged of Almaner of Sumons & passyng' vpon any maner of Jurye or enquest Att AH tymes herafter within the seyd Citie bytwene party & partye, or otherwyse to be taken), Accordyng' to theyr lybertyes & privyleges to them) hertofore grauntyd, AsweH by Acte of parlyament / As other wyse.

but not on any other Jury or Quest.

2. That all Freemen Barbers and Surgeons not practising as licenst Surgeons,

[2] Item, that AH & euery person) & persones that nowe Are, or that herafter shalbe, free of this Citie, of & in the seyd Company of Barbours & Surgeons, nott vsyng', practysing', or occupying' the Facultye & Scyens of Surgerye, laufully therunto Admytted & approvyd, shalbe Contrybutorye to the charges of this Citie Att eny tyme herafter growyng' or arysing' for the affayers of the same Citye, after theyr rate & substance; And also be Constables, & kepe Almaner of Watches, as theyr turne & Course [leaf 235 or 212] shaH duely yt requyre, As other the Cytezens of the seyd Citie shaH do / eny graunte, lybertye or privylege to theym) or eny of theym) Att eny tyme hertofore, by eny maner weyes or meanys made or grauntyd to the contrary, in eny wyse nott withstondyng' /

shall pay all City dues,

and serve as Constables and Watchmen, like other Citizens.

3. That all practising Surgeons shall pay and do all City dues and services,

[3] Item, that AH & euery person & persones fre, & that herafter shalbe free of the Mysterye & felowshipe of Barbours & Surgeons, vsing', exercysing' & practys-

ingᵗ the Faculty & Scyense of Surgerye, shalbe Contrybutorye to all maner of charges, paymentes, & imposycions / other then the seyd offices of Constableshipe & Watchingᵗ / that Att eny tyme herafter shull fortune / to be bourne, payed & Susteyned by the Citizens of this Citie, for the honour, welth & necessarye Affayers of the same Citie, in lyke maner & fourme in euery poynt, After theyr substance & value / As other the Citezens of this Citie shall bere, susteyn & paye / eny lawe, Acte, Ordenaunce, graunte, vsage & privylege, Att eny tyme hertofore to theym made, grauntyd or obteyned to the contrary, in eny wyse notwithstondyngᵗ.

[4] Item, that Aswell those xij persons[1] free of the seyd Mysterye of barbours & Surgeons, that hertofore haue bene named & presentyd to this Court, to haue & enioye suche lybertyes & privyleges as the Surgeons of this Citie hitherto haue had, obteyned & enioyed, & yett do enioye / As also all & euery other person & persones of the seyd Felowshipe & Mysterye of Barbours & Surgeons that herafter shalbe named & presentyd to the seyd Court to be of the seyd number of xij / And lykewyse all & euery other person & persones that nowe are, & that herafter shalbe Freemen of this Citie of & in the seyd Company of Barbours & Surgeons vsyngᵗ & exercysingᵗ the facultye & Scyence of Surgerye, shalbe clerely exoneratt & dyscharged of beryngᵗ eny maner of Armour wythin the seyd Cytie, & of & from the offyce of Constableshipe & kepyngᵗ eny maner of Watche Att eny tyme herafter wythin the seyd Citie / eny lawe, Acte, Ordenaunce, vse or custome, Att eny tyme hertofore made, provyded, Allowed or vsed to the contrary, notwithstondyngᵗ /

save Constableship and Watching.

4. That the present and future 12 Surgeons priviledgd under the Act 5 Hen. VIII, ch. 6, shall enjoy all their old exemptions,

and shall be free from bearing Arms and serving as Constables and Watchmen.

---

[1] See the Act 5 Henry VIII, ch. 6, p. 198 above, and the Iuspeximus or Letters Patent of March 10, 1517, p. 210-212.

IX.

TEN RECIPES[1] BY HENRY VIII AND HIS PHYSICIANS, Dr. AUGUSTYNE, Dr. BUTTS, AND Dr. CROMER.

From the Sloane MS. 1047.

WITH A POEM "WHAT VEINS TO BLEED IN."

I.

[leaf 1]   The Kinges Maiesties owne plastre.

Take the rootes of marche mallowes; washe and pike them cleane; then slytte them, and take owt the Inner pythe, and cast it awaye, and take the vttre parte that is faire and white, and cutt them in small peces, and brysse them a lytle in a mortre; And take of them half a pounde, and putt them in a newe erthen panne: Then putt therto, of linesede, and fenigrec,[2] of eche ij vnces, a lytle bryssed in a mortre. Then take malvesie and white wyne, of eche a pynte, and styrre all these to-guether, and lett them stande infuse two or thre dayes. Then sett them over a softe fyre, and styrre it well, till it waxe thick, and lyke a slyme: then take it from the fyre, and strayne it thorough a pece of newe canvas. [leaf 1, back] Thus haue yow the [mu]scellage redye to make the plastre with. Then take fyne oyle of rosys, a quarte, and washe it well with rose-water and whyte wyne; then take the oyle cleane awaye from the wyne and the water, and sett it over the fyre in a brasse panne, allwaies stirring it; and put therto the pouldre of lytherge, of golde, and of silver, of eche of them viij vnces; ceruse, vj vnces; redd corall, ij vnces; bole armoniac,[3]

---

[1] They are taken at random, by their titles.

[2] *Fœnum Græcum, Carphos, siliqua Columellæ* . . Fenugreek . . outwardly it helps all inflammations, and alleviates paines in raw and excoriated places, Imposthumes, Ulcers, &c. (p. 57) . . . The Meal is Emollient and Emplastick; and boyled with Mead, and applied, it helps all inflammations, and dissolves hard swelling . . It discusses, and is Anodyne, insomuch that its mucilage (made by decoction in water) is put into most Cataplasms for those intentions.— 1678. W. Salmon, *London Dispensatory*, p. 147.

[3] *Bolus Armenus* . . *Bole Armoniack.* It is so called because it comes from Armenia; but it is also found in Germany. Schroder saith, it is a pale red Earth, impregnated chiefly with Iron Vapours. It is very dry, Astringent and strengthening . . often used outwardly in strengthning Cataplasms and binding pouders.—1678. Salmon, *Lond. Disp.* p. 433.

App. IX. *Recipes by Hen. VIII & his Physicians.* 221

sanguinis draconis,[1] of eyther of them, one vnce : And in any wyse lett them be fynely [p]uldered and cersed [sifted]. Then putt them into the oyle over the fyre, allwaies styrring; and lett not the fyre be to bigge, for burnyng of the stuff. And when it begynneth to waxe [leaf 2] thicke, then put in x. vnces of the saide muscellage, by a lytle at ones, or elles it woH boyle over the panne. And when it is boyled ynough, ye shall perceaue by thardenes or softenes thereof, when ye droppe a lytle of it vppon the botom of a dysshe, or a sawcer, or on a colde stonne / Then take it frome the fyre; and when it is nere colde, make yt in rolles, and wrappe them in parchement, and kepe them to your vse. This plastre resolveth humours where as is swelling in the legges.

II.

[leaf 5] A blacke plastre devised by the kinges hieghnes.

Take ˙gummi armoniaci .ʒ.iiij. olei omphacini[2] ʒ.iij, fyne therebinthine .ʒ.vj. gummi Elennij[3] .ʒ.j., Resun [leaf 5, back] pini ʒ.x. Boyle [them] to-guether strongly on a softe fyre of coolys in a faire laten basyn, allwayes styrring it vntiH it be plaster-wyse; and so make it vppe in rolles, and kepe it to your vse.

III.

[leaf 8, back] A plastre devised by the kinges Maiestie at G[r]enewich, and made at Westminstre, to take awaye inflammacions, and cease pay[n]e, and heale excoriacions.

Take of plantaigne leaues, violett leaues, honye-suckle leaues, con-

[1] It is the Tear of a Tree, red like blood, the Fruit of which is like to a Cherry, whose skin being taken off is like a Dragons, from whence came that name. It comes from *Portus Sanctus* in America . . . It is temperate, drying and binding . . . Outwardly, it heals Wounds, stops Bleeding, fastens Teeth, dries up Catarrhs, and laid to the Navel, stops Dysenteries.—1678. Salmon, *Lond. Disp.* p. 172.

[2] MS. omphatini. *Omphacinum Oleum*, Oil made of unripe Olives.—1706 Kersey's Phillipps. It is cooling, drying and binding, and strengthens the Stomach, heals exulcerations, cools the heat of burning Ulcers, repercusses Tumors in the beginning.—1678. Salmon, *Lond. Disp.* p. 167.

[3] *Elemi*, a sort of transparent Gum or Rosin, which issues from a Cedar-tree in Ethiopia; being of a whitish Colour, and mix'd with Yellow Specks.—1706. Kersey. *Elemi Gummi*, Gum Elemni . . . It dissolves in oyly bodies, heals Wounds and Ulcers in the Head, . . ripens and eases pain. It is mild and agreeable with the Body, and gently cleanses and fills Ulcers up with flesh.— 1678. Salmon, *Lond. Disp.* p. 161.

222 App. IX. *Recipes by Hen. VIII & his Physicians.*

solide¹ maior' and minor', solatr',² the budde*s* of rosys ³of eche one hanfule.³ Beate all these to-guether, and strayne hem̅. Take, the fatte of capons or hennys ℥.xij. Boyle hem̅ wi*th* yo*ur* Iuces, vntyll the Iuce*s* be consumed : then̅ strayne it; and putto, these thinges folowing : lytherge of silver⁴ ℥.iiij., redde corall .℥.ij., cornu cerui vsti .℥.j., cornu vnicorum⁵ ℥.ij., margaritar*um* ℥β [½ oz.]. Preparate and pouldre [leaf 9] all these fynely, and putt them to your fatt*es*, and boyle them̅ all to-guether over a softe fyre, styll styreing it vntyll it be [plaster] lyke : then putt therto thiese muscellages following : Take of quynsede,⁶ of linesede, an*a*, ℥.j. Drawe the muscellage of them̅ wi*th* rose-water and white wyne, wherin therebintyne hath lyen̅ iiij daye*s* infuse, being oftymes moved ones or ij in an howre. And take of that .℥.ij. and putto the other, and make thereof a plaster, or a spasmadrappe.⁷

IV.

[leaf 15, or fo. 17, back] Jacobbe*s* Plaster.

Take lapidis colaminaris,⁸ terre sigillate,⁹ lapidis lazulj,¹⁰ lapidis

¹ See notes to Recipes VII, VIII, below.
² *Solanum, Solatrum* . . Nightshade . . The Essence helps St. Anthony's fire, the Shingles, pain of the Head, Gout, Sciatica, pains caused by hot, sharp and biting Humours, heart burning, heat of the Stomach, and hot Inflammations : it is to be used with caution, yet is not so dangerous as Opium.—1678. Salmon, *Lond. Disp.* p. 104.
³—³ In a corrector's hand, over the line.
⁴ *Silver* . . To purge it from other more imperfect Metalls. This is done . . . By melting of it with Lead, continuing the fire till the imperfect Metalls with the lead turn to fume, or come off like froth or dross, which is called *Litharge* of silver.—1678. Salmon, *Lond. Disp.* p. 277.
⁵ See Salmon's *London Dispensatory*, p. 207, and 220.   ⁶ quince seed ?
⁷ See The Table of Spasmadraps, or dipt plasters, on leaf 32.
⁸ *Calaminaris lapis* . . Caliminare. It is a yellow stone, not hard, which when burning, gives a Yellow fume : found in Metallick mines : Of this, Copper-smiths make Brass . . This stone dries, cleanses, binds, cicatrizes and incarnates ; fills Ulcers with flesh ; and made into pouder, and sprinkled upon gald places in Children, drys and heals them suddenly.—1678. Salmon, *Lond. Disp.* p. 407.
⁹ *Terra Sigillata, Silesiaca* . . . Sealed Earth. There are several sorts . . as the . . Turkish, which is properly so called, and that which is intended here, viz. that from Constantinople, which is of an ash-colour, and indeed the best of all Earths which are known to us . . . *Terra Sigillata* is drying, binding, sudorifick, and alexipharmick, resisting Plague, Poyson, Putrefaction, and all kinds of Malignity and Venom . . . Outwardly, it cures the bitings of Venemous Beasts, and cleanses malignant Wounds.—1678. Salmon, *Lond. Disp.* p. 436.
¹⁰ *Lazuli lapis* . . the Azure Stone . . Of this stone is made that glorious colour called *Ultramarine* . . It is a wonderful thing (according to the Opinion

App. IX. *Recipes by Hen. VIII & his Physicians.* 223

sanguinarij,¹ lapidis emerj, of eche two vnces; sang[u]inis draconis, boli armenj,² of eche .j. vnce; lytherge of golde, ceruse, of eche one vnce; lett all these be pouldered small, and cersed [sifted] fynely. Then take oyle of rosys a pynte, and sett it over a softe fyre, and putt therto white waxe small cut, half a pounde; deres suett, iiij vnces; And when they be relented, put therto all the poulders, and styrre them well, and [leaf 16, or fo. 18] lett them boyle a lytle while; and then take it from the fyre, and putt therto mastique and olibanum,³ of eche one vnce fynely [pouldered]; And when it is almost colde, putt therto ij vnces of therebintyne, and ij drammes of camphere in fyne pouldre, and make it vppe in rolles, and kepe it in lether. This plaster is goode for all maner of olde sores.

V.

[leaf 26, back] An other plaster deuised by Master Chambre, Doctour Buttes, Doctour Augustyne and Doctour Cromer, the which doith both consolidate and comforte the membre, and temperately heate, and healeth the Vlcer.

Take oyle of rosys, ℥.viij., succorum plantaginis,⁴ centinodij,⁵ burse pastoris,⁶ foliorum rubei, ana, ℥.i. : boyle the oyle to the consump-

of Fioravantus) in the Cure of Malignant Feavers, and the worst of Ulcers.—Salmon, *Lond. Disp.* p. 413.

¹ We don't see this in *Salmon*, unless it means Coral (p. 422-4, 'the Tincture of the Coral like blood) or Ruby, *Pyropus*, p. 417.

² See note 3, p. 220, above.

³ *Olibanum*, Gum of the Male Frankincense-tree (p. 167). *Thus*, Frankincense. It is a native Rosin from an Arabian Tree called *Lovan*, which we call the Frankincense Tree. It is Male or Female : the Male is called *Olibanum*, which is a Rosin, hard, clear, of a yellowish white within, fat, and round like drops. The female is softer . . . *Olibanum* is the best of the two (being from Trees which grow on Mountains) . . It is Pectoral, Cephalick, Stomatick, Anodyne, and Vulnerary. It cleanses, fills Ulcers with flesh and heals them, cures green Wounds, chiefly of the Head ; is good against Kibes and Chilblains, and helps Ulcers in the Fundament.—1678. Salmon, *London Disp.* p. 179.

⁴ *Plantago* . . Plantain . . it cures old Ulcers, Issues, Rheums . . heals Ulcers, and soreness of the mouth and Privy parts.—1678. Salmon, *Lond. Disp.* p. 89.

⁵ *Centinody*, an Herb having as it were a hundred Knots, Knot-Grass. *Knot-grass*, an Herb lying on the Ground, with long narrow Leaves like a Bird's Tongue. It is good against the Stone, Strangurys, Bloody-flux, hot Swellings, fistulous Cancers, &c.—1706. Kersey.

⁶ *Bursa pastoris* . . sheppards Purse . . It binds and astringeth, is good in

224 App. IX. *Recipes by Hen. VIII & his Physicians.*

tion of the Iuces; then putt therto myrtylles, hipocistidos,[1] galles brusyd̛ [leaf 27], of eche ℥.ᶜ [½ oz.], plantaigne water, rosewater, water of honye-suckle flowres, of eche ℥.ᶜ. Boyle aƚƚ thiese to-guether with the oyle to the consumption of the waters; then straygne them thorougħ a fayre clothe into a clayne vesseƚƚ, and putt therto lytherge of golde and syluer, and ceruse, and redde coraƚƚ combusted. Aƚƚ these weƚƚ preparated, of eche one vnce : lapidis Ematitis,[2] tutie,[3] cornu cerui vsti, perlys; of eche of thiese finely pouldered, half an vnce. Boyle aƚƚ thiese to-guether over a softe fyre, tyƚƚ it be almoost plaster wyse : then putt therto of muscellage seminis consilij[4] draweñ with rose water, ℥.ij. And when yt is [leaf 27, back] boyled ynougħ, take it besyde the fyre, allway stirring it; and in the cooling, putt therto half añ vnce of fyne pouldre of redde dammaske rosys, and ℥.ij. of fyne pouldre of camphore; And so make it vppe in rolles, and kepe it for your vse.

VI.

[leaf 64] An Oyntement devised by D. Chambre, D. Buttes, D. Cromer, and D. Augustyne, against the eville complexione of hoote cawses of Vlcers in the legges, and partes that be soore.

Take lytherge of golde,[5] lytherge of silver, anɑ, ʒiij., Tutie[3] prepa-

---

bleeding at the Nose, spitting of Blood, pissing of blood, bloody flux, and the flux of Womens courses; it stops a looseness, cures Wounds, and stops bleeding in any part of the Body.—1678. Salmon, *Lond. Disp.* p. 39.

[1] *Hypocistis.* It is the juyce of the root of the Shrub *Cistis* or Holly Rose, dried in the Sun . . Is binding, stops all fluxes . . It strengthens parts debilitated through superfluous moisture, stops vomiting and spitting of Blood, binds violently, and is Vulnerary.—1678. Salmon, *Lond. Disp.* p. 164.

[2] Hematite, the sesqui-oxide of iron, red, from *haima* blood.

[3] *Tutty, Pompholix* or *Spodium*, is a thin Volatile Ash, which sticks to the upper part of the Furnace when brass is melted; looking almost like flocks of Wool, and falling down when touched. It is also made of *Cadmia*, by calcining of it with a violent fire to Ashes. But that is properly *Spodium* which is heavy, and falls down to the bottom, called *Nil, Nihili Gryscum*, or Greek *Spodium*. Being washed, it drys without sharpness, and is the best of all drying medicines, exceeding good in all malignant and cancerous Ulcers, and other old and running Sores which abound with moisture.—1678. Salmon, *Lond. Disp.* p. 355.

[4] We can't find *Consilium*, but suppose it is Consound, like *Consolida:* Fr. '*Consire, Consolde, Consoulde:* f. The hearbe Comfrey, Consound, Asse-eare, Knitbacke, Backwort.'—1611. Cotgrave.

[5] *Of Vnprepared Litharge.* It is an Excrement arising from the refining of Silver or Gold with Lead : it is twofold, either white or yellow, (called Litharge

rate, calcis nonies lote,[1] cerusse, ana ℥.ʆ. (½ oz.): make these in verray fyne pouldre. Take the Iuce of nightshade, the Iuce of plantaigne, the Iuce of Rubee,[2] ana ℥.i.; worke them in a leaden morter with the poulders. Take oyle of rosys, ℥.i.; washe it well in [leaf 64, back] rose water, and so make vppe your oyntement therewith. Et fiat.

VII.

[leaf 82]   A Decoccioun devised by the Kinges Maiestie.

Take of Rose water, honysuckle flowres, ana ℔i .iiij; mallowes, nightshade, consolida maior[3], consolida media[4], plantaigne, sage, holyockes, chamomell flowres, dammaske rose leaves, ana, M.j. Take beane flowre, M.ʆ, and boyle all these to-guether over a softe fyre, tyll the thirde parte be consumed; then strayne it, and putt it in a fayre glasse, and take such quantitie thereof as shall suffise, and warme yt a lytle, and wasshe the membre therwith; and weete your [leaf 82, back] clothes therin, and wrappe them abowt, and so rowle it vpon.

VIII.

[leaf 83] A Water made and devised by the Kinges Maiestie.

Take the flowres of rosys, the flowres of [leaf 83, back] barberies, the

of Silver and Gold,) but they only differ in boyling; for the yellow is that which is most boyled or burnt, and is indeed only Lead half calcined. It drys, cools, bindes, repels, generates flesh, fills up hollow Ulcers, cleanses, cicatrizes raw places . . . —1678. Salmon, *Lond. Disp.* p. 354/1. See note 4, p. 222, above.

[1] Chalk washt nine times to purify it.
[2] *Rubie maieur, ou des taincturiers.* The hearb Madder, red Madder. *Rubie mineur.* Clauer, Loue-man, Goose-share, Goose-grasse.—1611. Cotgrave.
[3] *Consolidæ,* (*Lat.*) the Herb Consound, or Comfrey, of great Virtue for curing Wounds, looseness of the Belly, Sharpness of Humours, Consumptions, &c.—1706. Kersey. 45. *Consolide maioris, Symphiti,* of Comfry, cold in 1°, dry in 2°. It is mucilaginous, Vulnery and Conglutinative. It heals all wounds external and internal, stops fluxes of blood in wounds, helps spitting of blood, and Ulcers in the Lungs: It is good against Ruptures and pains in the back: It Cures broken bones and dislocations, and very powerfully stops the Terms, Whites, and running of the Reins: It may be used in powder, but a Mucilage is best; otherwise a Decoction in strong Ale will serve the turn: The bruised root applied, immediately easeth the Gout.—1678. Salmon. *London Dispensatory,* 6/1.
[4] 46. *Consolidæ mediæ, Bugulæ,* of Bugle; Temperate and dry in 1°. It is an exceeding good Vulnerary, both inwardly and outwardly, healing Ruptures, bruises, and the like: Inwardly it helps the Jaundice, and opens obstructions.— 1678. Salmon. *ib.* 6/2.

226 App. IX. *Recipes by Hen. VIII & his Physicians.*

flowres of po*m*me granate, the flowres of honye suckle, an*a* equaliter. Boyle aH these to-guether ; And in the boyling, putt to these poulders folowing : the rootes of consolida maio*r* and minor¹, Cincquefoile, water lyllie : Boyle them aH to-guether a goode space, and straigne them, and putt therto of mirobolane cytrine² pouldered, one vnce, and of met ros*arum*³.℥.ij, and boyle a decoccion.

IX.

[leaf 89] A Cataplasme made vngtment-lyke of the King*es* Ma*ies*ties devise, made at West-m*inster*.

Take a quarte of mylke, a fyne manchett⁴, a handfuH of mallowes, a handfuH of rose leaves : Boyle these to-guether tyH they be softe ; then strayne them, and drawe the pulpe of them, and putt therto the muscellage of p*ar*sly, ℥.j., the yolk*es* of ij newe layd egg*es*, the pouldre of long wormes weH washed and dryed ℥.ß [½ oz.], the pouldre of mellilote flowres⁵, and chamomel flowres, of eche, half an vnce, oyle of gardeyn lyllies⁶ as much as shaH suffise : Et fiat.

¹ 47. *Consolidæ minoris, Prunellæ,* of Self-heal : Temperate [&] dry in 1°. It is, like the former, a good Vulnerary, and has all the same Virtues.—1678. Salmon. *London Disp.,* 6/2. (For *Consolida Regalis,* Larks-spur, see 48/1.)
² 63. *Myrobolani Bellericæ, Chebulæ, Citrinæ, Emblicæ, Indicæ.* The five sorts of Myrobolans. The *Bellerick* purge Flegm : The *Chebulæ* first purge Flegm, then Choler : The *Citrine* or *yellow* purge Choler : the *Emblick* purge Flegm and Water : The *Indian* or *black* purge Melancholy, Dose *à* 3vj *ad* ʒj ss. The Bellerick are round ; the *Chebulæ,* long, with corners ; the Citrine are round like the Bellerick ; the Indian black, and eight-cornered. *Horstius* saith that they are *Prunorum quædam genera,* a kind of Prunes found growing in the Kingdom of *Cambaia,* which the Arabians call *delegi. Sala* makes an Extract of them (being stoned) by beating the pulpy part, and steeping it in water for some days, then straining and inspissating ; others add juyce of Pearmains, and then inspissate.—1678. Salmon. *London Dispens.,* 136/2. See also p. 79, col. 2, no. 429. *Myrobolanus,* Myrobolan Tree, a kind of Outlandish Prune, not known to the Greeks, but found out by the Arabians. . . They grow in the East-Indies, and are found wild in *Goa,* being a Fruit sharp in taste, much like to Service-berries.
³ 75. *Mel Rosarum commune,* sive *Foliatum,* Honey of Roses. Colledg.) *Recipe Red Roses not quite opened* lb:ij. *Honey* lb.vj. *set them in the Sun according to Art. Salmon.*) It strengthens the Stomach, and heals Ulcers of the Mouth and Throat.—1678. Salmon. *London Dispensatory,* 605/2.
⁴ *Manchet* or *Manchet-Bread,* the finest and smallest sort of Wheaten Bread. —1736. Kersey.
⁵ 404. *Melilotus, Corona Regis.* . . . Melilot is a kind of strong-scented Trefoil : It is Emollient, Discussive, Anodyne, Traumatick, Vulnerary, wasting,
⁶ See next page.

X.

[leaf 93, back] An other pultes devised by M*aster* Chambre, Doc*tour* Buttes, Doc*tour* Cromer, and Doctour Augustyne.

Take a gallon of milke, and a quarte of faire water, and the herbes folowyng: of nightshade leaves, lactuce leaves, henbayne leaves, howseleke leaves, plantaigne leaves, mallowe leaves, violett leaves, thre [leaf 94] swete appuls : Boyle all these to-guether tyll the moysture be consumed, and that it be thick. Then drawe the pulpe of them thorough a strayner, and putt therto these thinges folowing: of barlye meale, beane meale, Ote meale, ana, ʒ.j. Mixe all these to-guether, and boyle them on a softe fyre, tyll it be somwhat thicke. And in the coolyng, putt in thiese thinges folowing: the yolkes of thre egges, of the pouldre of rose leaves, of chamomell flowres, of mellilote flowres, [leaf 94, back] ana ʒ.ß [½ oz.], oyle of rosys[1], quantum sufficit. Worke all well to-guether, and [sprede] on a faire clothe, and vse it warme to the membre.

[*End of the MS.*]

ripening, Diaphoretick, Diuretick, Lithontriptick, and an Opener of Obstructions : the Juyce or Essence dropt into the Eyes, clears the Sight, consumes the Pin and web (see note 1, p. 208), and dissolves the Pearl and other Spots which offend them. See our *Synopsis Medicina, lib.* 3, *cap.* 22, *Sect.* 198 ; and *cap.* 59, *Sect.* 3.—1678. Salmon. *London Disp.*, 76/1.

[6] *Oleum Liliorum*, Oyl of Lillies.
Colledge.) *It is made in the same manner as Oyl of Roses.* [See next note.]
Salmon.) It eases pain, and ripens Tumors: It was much used in Pestilential Bubo's.—1678. Salmon. *Lond. Disp.*, 728/2.

[1] *Oleum*, seu *Pinguedo Rosarum*, vulgo *Spiritus Rosarum*, Oil, fat, or spirit of Roses.
Colledge.) *Recipe as many fresh Damask Roses as you will; steep them 24 hours in a sufficient quantity of warm water; press them out, and repeat the infusion certain times, till the liquor is sufficiently strong, which destill in an Alembick with its Refrigeratory, or a Copper with its Worm: separate the Spirit from the water, and keep the water for another infusion : you may also do the same being pickled with Salt* (as is taught, Chap. 2, Sect. 9, of this Book). *And in the same manner you may draw* Oleum, seu Spiritus Rosarum rubrarum, *Oyl or Spirit of red Roses.*
Salmon.) It is a great Cephalick and Cordial ; It chears and recreates the Animal and Vital Spirits, quickens the Senses, and revives the heart, exhilerates the mind, expells Melancholy, is wonderful against all fainting and swooning fits, and, in a word, performs whatever any Cordial can do. Dose *à* gut. ij. *ad* vj.—1678. Salmon. *London Dispensatory* . . Lib. IV. Cap. 3, p. 465, col. 1. See too the Oyls of Roses, Omphacine and Compleat, p. 726.

What beins to bleed in.

Egerton MS 2572 (Statutes of the Company of Barbers and Surgeons of York), leaf 69.

To knawe the vaynes to let blode one.

	ȝe that[1] wyll lette gude men blode,
	And vaynes wyth all ȝowre liues fode,
Only certain veins should be bled from.	Some vaynes, vse ȝe,
	And mony other lette ȝe be. 4
	Therefore nowe wyll I. them schawe,[1]
	And tell ȝowe them apone a rawe,
	And where they lye, euer ylke ane,[2]
	And for what thynge they shall be tane.[2] 8
Every man has 33 veins:	Iilke a mane hath xxx and thre :
	Lythe[3] and I shall tell them the ;
	Some er abowne, and some benethe ;
	Lithe,[3] and thowe shall knawe them ethe[4] : 12
2 behind the ears,	Behynde the heres, fyndes thowe twa ;
	If thowe lett blode of tha,[5]
	His syght shall neuer fale,
	And heles of[1] torne-seke, and of scale. 16
2 at the temples,	Two at the templys shall noght be leuyde, [leaf 69, back]
	For werke and stangynge of the he[ue]de.
1 amid the forehead,	In the myddis the forehede, fyndis thowe ane,[2]
	For lepir and sausfleme shall be tane.[2] 20
1 under the nose,	Vndir the nose lyes a wayne,
	There-wythe shall the frensi[6] be sclayne,
	And the gome rosage alswa[7] ;
	And when the eiien[8] tholis wa,[7] 24
1 on each side the nose, by the eye; (2)	Apone the nose, fast by thy ne,[9]
	Schall thowe lete blode, if thowe be sle[10] ;
	For yll blode and the scome,
	Then shall thowe hele them all and some. 28
2 in the neck-holes,	Two in the neke holes shall thowe fynde,
	For lepir and for stratnes of wynde.
2 in each lip, (4)	Two vaynes er in ether lippe ;
	Those wyll I noght thowe ouer lyppe 32

---

[1] MS schewe. The copier has altered the dialect forms in many words. We don't change all back.
[2] MS one, tone, altering the dialect ; *tane* is 'taken.'
[3] listen   [4] easily
[5] MS *two, thoo*, altering the dialect: see *tha*, l. 38.
[6] MS sreusi
[7] MS alswo, wo, changing the dialect. See *wa* in l. 52. *tholis* is 'suffers.'
[8] MS euen.   [9] thyn e, thine eye   [10] sly, clever.

App. IX. *What Veins to bleed in.*

Tyll oppyne¹ whene the mouthe is flane,²
And other euels euer ilke ane.
Vndir the tonge, two, seys³ thowe lye,
For euyll of tong*is* and swynaysy.⁴   36   ² under the tongue,
   Nowe benethe⁵ wyll I ga,
So that thowe may knawe all tha⁶;
Ilke man that is on life,
In his arme hath vaynes fyfe :   40   5 in each arm, (10)
Abowne the hede he behovis them blede,
Whene the hede hath ony nede;
For all thy body, in myddis the Arme;
Beneth, when yᵉ leu*er* tak*is* harme.   44
Aboue yᵉ thovme is the make;
That' shall thowe take for the cardiake.
Thy ryght' hande has I. wane,⁷ in fay,   [leaf 70]   1 in the right hand,
Thy litill fynger hath yt aye.   48
When the leuer hath ony qwyke,
In the left hande for the mylte;   1 in the left hand,
Wythin the Ankeles, domistica,   1 inside each
When the bledir hath ony wa;   52   ankle, (2)
Wythout the Ankeles, Siatica,   1 outside. (2)
For siatica, that shall thowe ta;
And wemen that' hath tynt' ther floures,   1 in women's bowers.
Lete them blede in there bowres.   56   [33 in all.]

[? *poem incomplete*]

These lines are a metrical version of the prose descriptions (in circles) of the drawing of *Homo Venorum* on leaf 50, a naked man with vermilion direction-lines running from his bleeding-points. These lines—each with its circular label—start from the head:

(1, 2) Be-hynde þᵉ eres er twa vayns þat' er gude to be opynd for t*u*rnseke and for scall, & alsso for euyll sight'.

(3, 4) Þe vayns i*n* þe tempyls of' þe hede, for warkyng & stangyng' i*n* þe hede; & alsso it' wyll lett' þe sheddyng of þe schett'.

(5) Þe vayn i*n* þe forhed is calde 'ariote,' to opyn for þe fransy & sauce-flemy*n*g' in the face, and alsso for þe emoraudes & for lunatikus.

(6, 7) Opyn þe vayn on þe nese, fast' by the eghe, for bleryd eghen, & for þe scome of' mense eghen, & dymnes of' þame.

(8) Vndyr þe nese, on þe end þerof, lyggys a vayn þat is gud to opyne for þe gut' roset', & for þe fransy in þe hefd.

¹ overleap to open, omit to bleed from.   ² ? MS slane.   ³ seest
⁴ quinsy.   ⁵ MS beneth benethe   ⁶ MS thay   ⁷ One vein.

230 App. IX. *What Veins to bleed in.*

(9, 10) Twa vayns er in þe lippis,[1] þat er gude to be opynd when þe mouth es flayne wyth abundans of¹ blude. [*left col.*]

(11) It es gude for to blede on þe tonge for þe sqvnesy, and for bolny[n]g [swelling] of¹ þe tonge. [*right col.*]

(12, 13) In þe nek hole er ij vayns þat¹ er gude to opyne for leper and for straytnes of wynde. [*right col.*] (*See Poem*, 1. 29.)

(14, 15) Opyn þe hed vayns þat es called cyphalica, and lyggis hyest¹ in þe arme, for clensyng¹ of þe hede and of þe brayne. [*left col.*]

(16) þe vayne of þe hert¹ es callyd cardiaca,[2] for rysyng at þe hert, & for þe impostoum [?] of¹ spirituale membris. [*left col.*]

(17, 18) þe vayn of þe lyuer þat lyggis beneth in þe arme, & es called basilica, for yuell of¹ þe lyuer and splene. [*left col.*]

(19, 20) It es better to blede on þe purpur vayn in þe left¹ arme in wynter, þan on þe right arme, and eyuer so.

(21) þe vayne in þe bake, it es gud to be opynd for þe purgyeng¹ of melancolye. [*right col.*] (*Not in the Poem.*)

(22, 23) It es gud to blede on þe left¹ hande for þe passyone and deses of þe mylt & oder membris. [*right col.*]

(24, 25) þe vayns betwix þe lityll fynger & þe next¹ fynger es gude to opyn for þe litarge and for ylle eghen). [*left col.*]

(26, 27) þe vayne betwyx þe fyngere & þe thombe es gud to be opyd for het of warke in þe swldyrs & migram in þe heue[de].

(28) þe vayne on þe pyntyl es gude to blede for hete & scaldyng þerof¹, and for bolny[n]g or bryssyng þerof¹. [*middle.*]

(29, 30) þe vayn vnder þe ankle within þe fute, þat¹ es called domestica, for þe bledder, and for yuelle humors.

(31, 32) Opyn þe vayn vnder þe ankylle with-owten), þat¹ es callyd saluatica, for þe sciatike and for þe emorodys.

---

[1] The Poem above puts 2 veins in each lip (line 31), and two under the tongue (l. 38); but has only 1 temple vein, and no back vein.

[2] See the Poem, l. 46.

X.

PAYMENTS BY HENRY VIII AND PRINCESS MARY, TO DOCTORS, &c. OTHER THAN THOS. VICARY, IN 1517—1543.

WE could not find Vicary's name in the Harl. MS. 21,481 (Henry VIII's Accounts 1509-1518), leaf 257, at foot.

ib. leaf 263 [July 1] a° ix^{no} (1517). Wedenysday at Grenewyche.

Item to Docto*u*r Vernando de Victoria, phe- sic*i*on with the quenes grace, for his half yeres wag*es*, due vnto hym at his mydsom*er* last passe*d* ... ... } xxxiij ti. vj s. viij d

leaf 269.   Quarter Wag*es* due at Michell*mas*, a*n*no ix° (1517).

Item for Pyers, barbour, wag*es*[1]	lxvj s. viij d.
Item for Pero, the frenshe coke, wag*es*	lxvj s. viij d.
Item for Massy, barbour, wag*es*	lxvj s. viij d.

leaf 271.   A*n*no ix^{no}, xxv^{to} die Octo*br*is (1517).

[back] Item to the P*r*ior of saint bartilmewe, opo*n*) a warrante toward*es* the making of the manou*r* of Newe Hall in Essex[2] ... ... } ^t_M ti

[1] The December (1517) wages are on leaf 276, back. The Easter (1518) ones on leaf 286.

[2] New Hall is 1½ miles N.W. of Boreham (which is 3¾ miles N.E. from Chelmsford), and stands a mile back from the road. Its fine old avenue of trees, nearly a mile long, is now much curtaild. It was probably built about 1500, was soon after ownd by Sir Thos. Boleyn, Q. Anne B.'s father, and past from him to Henry VIII about 1517. He made it a Royal Residence—one of the grandest in the kingdom—cald it *Beaulieu*, and in 1524 celebrated the Feast of St. George there (Hall's *Chronicle*, The .xvi. yere, p. 677, ed. 1809). He enlarged the building. His arms are still over a door at the back of the Hall, with a Latin inscription saying that ' K. Hen. VIII, renownd in arms, executed this sumptuous building.' Q. Mary livd there several years before her accession.   Q. Eliz. also enlarged New Hall: her arms, with an Italian inscription, are still over the entrance door. The Palace consisted of 2 large quadrangles, with all necessary offices. It had a most splendid chapel, with a grand East window, which is now in St. Margaret's, Westminster. This window was originally meant as a present from the magistrates of Dordt in Holland, to Hen. VII. Perhaps about a fifth of the original building is left

232 App. X. *Payments to Doctors, &c.; not to Vicary.*

Henry VIII's New Year's Gifts in 1518.
Harl. MS. 21,481, leaf 279.

Fryday, Newyeres day, primo die Januarij, anno ix° (1518).

Item to Doctor taillour seruaunt	xx. s.
Item to master Chambre [Henry's physician] seruaunt	xx. s.
Item to Doctor Fairfax, for a pricksonge boke	xx. ti.
[back] Item to the blynde poyete	C s.

leaf 283. Tewesday at Wyndesore, Candelmas Day (2 Febr. 1518).

Item to Doctor Vernando, þᵉ quenes Fesicion, opon) a Warrante for transporting his wyf ⎱lxvj ti. xiij s. iiij d. oute of Spaigne into England ... ...

leaf 284, back. Primo die Marcij a° ixⁿᵒ at Wyndesore (1518).

Item to Doctor Farnando, the quenes phisicion, for his half yeres wages due primo die marcij, ⎱xxxiij li. vj s. viij d. anno ixⁿᵒ ... ...

X° die Maij anno ixᵐᵒ (1517) Sonday at Richemounte.

Item to Richard Pynson)[1] opon) a warrant for prentyng of certan) bokes concernyng the ⎱ti xiij s iiij d. kinges subsidye ... ...

[1] Vicary's name does not occur in *The Privy Purse Expenses of King Henry the Eighth*, from Nov. 1529 to Dec. 1532, ed. (Sir) N. Harris Nicolas, 1827, though those of Henry's Physicians and Apothecary do. See for Dr. Chambers, p. 194, 243; for Dr. Butts, p. 262, 305; for Dr. Bartelot, p. 146; ? Dr. Goodryke, p. 8; Dr. Nicholas (who attended Wolsey in his last illness), p. 192.

For payments of the bills of Cuthberd, the king's apothecary, see p. 44, 124, 165, 203, 251. See also Master John, the apothecary, p. 147; and the Sergeant Apothecary, p. 79, and 146 (July 11, 1531: 'paied to Jacson for certeyne gloves fetched by the sergeant Apoticary, iiij s. x d.').

in the present large mansion, a red brick building in the Tudor style, with stone facings. The old hall is still intact, and is used as a Chapel. It measures 50 ft. by 20, and is 45 ft. high. New Hall is now a Roman-Catholic school or training-college, founded by some nuns of the Order of the Holy Sepulchre, who took refuge there when driven from Liège by the first Revolution in 1793.—Durrant's *Handbook for Essex*, ed. W. H. Utley, p. 48-50.

[1] 'William Copland of London, merchaunt,' gets £380 at Christmas 1517, leaf 277, back, 'for certan) stuf by him provided for the manour of New-Hall, & also for certan) Iuelles by hym delyuerd to the kinges grace.' Was he any relation of the printers, Robert and William Copland?

App. X. *Payments to Doctors, &c.; not to Vicary.* 233

The Surgery entries are only :

p. 67. Aug. 19, 1530. "Item the same daye to the frenche fletcher in Rewarde towardes his Surgery ... ... xl. s."

p. 128. April 15, 1531. "Item the same daye paied to a surgeoñ that heled litle guilliam [one of the King's crossbow makers] xl. s."

p. 245. 17 Aug. 1532. "Item the same daye paied to graunde guilliam[1] [another cross-bow maker] by the kinges commaundement, for his surgery, when he was syke at Londoñ ... xxx s."

In Madden's *Privy Purse Expenses of the Princess* (afterwards Queen) *Mary*, Dec. 1536-44 (London, 1831), Dr. Owen appears as Physician both to her and Prince Edward:

p. 52. Jan. 1537-8. "Item to Doctour Owen, the Prince phesition, in likewise [a Dublet clothe of Satten]" xxiiij s."

Then on p. 114, in April 1543, " Item, payed to Doctour owen, x ħi;" and afterwards, 3 entries of payments to messengers sent for him :

(p. 129. Sept. 1543) "Item to crabtre for goyng to Doctour owin, from grafton to Dunstable ... ... xiij d."

(p. 133. Oct. 1543) " Item paid to Crabbetre for his Costes, sent vnto Doctour owen ... ... xij d."

(p. 134. Oct. 1543) "Item geuen to nycholas, grome of the Stable, sent from grafton to Doctour owen ... ... ij s."

And on p. 164, Sept. 1544, Mrs. Owen's servant gets 5s. for bringing the Princess a present.

Dr. Michael[2] (? Delasco) was another Physician of the Princess, and there are several entries relating to him, and gifts of money to (?) his wife, Mrs. Mary.

[1] Item, for Guilliam le Craunt, crosbowmaker, x s. Payments on 1 April, Anno xxxj° Hen. VIII [A.D. 1539], Arundel MS. 97, leaf 72, at foot. Another payment to him of x s. iiij d. in May, anno xxxj° (1539), lf. 75 ; others elsewhere, and another of ix s. viij d. in Feb. 1540, lf. 118.

[2] 'The same instrument which appoints John de Sodo apothecary to the Princess (*Rymer*, xiv. p. 578), dated 29th Jan. 1537, also nominates *Michael Delasco*, "in Medicinis Doctorem " to be her Physician, with a salary of 100 marks sterling per annum ; and in the "Book of Payments" his name occurs in Midsummer, 1539, as "phesicion to the Lady Marye," with the quarterly allowance of 16*l*. 13*s*. 4*d*. Mrs. Mary Mychaell is presumed to be his wife ; and it is probably her picture that occurs in the list of those at Westminster (*MS. Harl.* 1419, A). She appears in the roll of New Year's gifts, 1556, and presents "twelve pistyllets," which are valued at 3*l*. 14*s*., and receives in return a gilt jug. Quære, whether the above Michael Delasco be the same with Michael de Securis, a physician " in partibus Normanniæ oriundus," who receives letters of naturalization, dated 28th Nov., 25 Hen. VIII, 1533.—*Rymer's Coll.*, vol. iv. MS. *Addit.* 4622.'—*Madden*, p. 249, col. 2.

234 App. X. *Payments to Doctors, &c.; not to Vicary.*

p. 28. May, 1537. Item, for j hoggeshed wyne for Doctour mighell ... ... xxxj s. viij d.
p. 30. 1537, June 30. Item, payd for the hyre of a Barge for Doctour mychaell, and m^r Iohn) poticary, commyng to my ladys grace, beyng sicke ... ... vij s. vj d.
p. 36. Aug. 1537. Item, geuen) to *Cristofer* Wright, sent vnto Doctour michaell ... ... v s.
p. 37. Aug. 1537. Item, geuen) to Thom*as* guye, sent vnto Doctour michaell ... ... v s.
p. 45. Nov. 1537. Item geuen) at the Cristenyng of Doctour mychaell Childe, a Salt, silu*er* and gilt, my lad*ies* [grace] being godmother to the same : price ... ... lxvj s. viij d
Item geuen) to the mydwyfe and the norce ... ... xij s. vj d.

(There are many payments (as in Hen. VIII's book) to midwives and nurses.)

Dr. Nicholas, who attended Henry VIII and Wolsey (see above), is another Physician who, in April 1543, bleeds the Princess Mary, as 'one Harry does her women and her: p. 113—

Item, geuen) to Doctour nicholas, letting my lad*ies* grace Blode xx s.
Item, geuen) to one Harry, letting my lad*ies* women) Blode [1] x s.
p. 123. July 1543. Item, to Harry, surgion), for letting of hir grace blood ... ... xx s.
Item, p*ai*d to ferrys,[2] the king*es* surgion) ... ... x s.

Dr. Nicholas was also sent for to the Princess in 1543; and he attended the laundress at Greenwich.:

p. 107. Jan. 1543. " Item, p*ai*d to Crabtre, sent vpon) my lad*ies* busynes for Doctou*r* Nycholas ... ... iij s."
p. 121. June 1543. " Item, to Doctou*r* Nicolas for comyng to the Launder, beyng seek at grenew*i*ch ... ... x s."
p. 121. June 1543. "Item, to one of the gromes, for goying for Doctou*r* Nicholas ... ... xx d."

In July 1526, Dr. Wootton was Dean of the Princess's Chapel, and her Physician (Harl. MS. 6807, leaf 3); and at a later period Dr. Fynch is her Physician, when she is in the Marches of Wales (MS. Cott. Appendix xxix, leaf 51).—*Madden*, p. xxxix, xl.

---

[1] Below is Item, p*ai*d for a payr of Shoes for Jane the fole ... vj d.
Item, to the Barbour for shaving hir hed ... ... iiij d.
and on p. 111, March 1543. ' Item to [the] Barbo*ur* for shaving of Janys hed, iiij d.'

[2] See him in Holbein's picture, no. 8, the right-hand head in the lower row of the kneelers. ' He receives C s. per quarter in the King's Household Book, 1542-4, in Sir Tho. Phillipps's collection.'—*Madden.*

App. X. *Payments to a Surgeon and Dr. Huyck.* 235

Christopher the Surgeon[1] (*not* Christopher Bradley, keeper of the Princess's greyhounds,) is paid four times for bleeding her:

p. 30. June 1537. " Item, payed to Cristofer, who dyd let my lad*ies* grace Bludde ... ... xx s."
p. 74. July 1538. " Item, geuen) to one Cristofer, a s*u*rgion), letting my lad*ies* g*r*ace Blood ... ... xxij s. vj d."
p. 89. April 1540. " Item, geuen) to C*ris*tofer the Surgion, letting my lady maryes g*r*ace blode ... ... xxij s. vj d."
p. 90. May 1540. " Item, geuen to C*ris*tofer the Surgion, com*m*yng from) Londoñ to tittonhanger[2], to lett my lad*ies* g*r*ace Bloode ... ... xxij s. vj d."

Exch. Q. R. Anc. Misc. $\frac{56}{9}$, (? 1 Mary, A.D. 1553-4,) lf. 16, in a List of the Members of the Household, are

Phisicians
Thom*a*s Hues
George Owen
Thomas Wendie
                † Rowland
Potecary
Iohn Savarye ./

Dr. Robert Huyck's Annuities of £50 and £100.
Tellers' Roll, Nº. 110.
Mich. 4-5 Elizabeth (1562).

m. 46[d] Rober*t*o Huyck, Docto*ri* Medicine, de an*nu*i*ta*te sua ad 1 li. pe*r* annu*m*, sibi debit*a* p*ro* trib*us* quarter*iis* anni finit*is* ad festu*m* S*a*ncti Mich*ae*lis Arch*a*ngeli, A*n*no iiij° Elizabethe Regine, den*ariis* rec*eptis* per Marke Steward   xxxvij li. x s.

m. 51. Also another quarter's payment to him at Christmas 12¹ 10ˢ

ib. He also had another annuity of £100, a quarter's payment of which was made at Christmas.

---

[1] ? The Christopher Samon of Holbein's Picture.
[2] The hamlet of Tittenhanger in Hertfordshire is 3 miles South of St. Alban's, and 17½ from London. Henry VIII and Queen Katherine stayd at Tittenhanger Park in 1528, during the sweating sickness in London. The Colne flows along its western boundary.—Thorne, *Environs of London*, under 'London Colney.'

XI.

PAY OF ARMY AND NAVY SURGEONS TO HEN. VIII.

(*t. Hen. VIII. Royal MS.* 7 *F XIV, art.* 24, *leaf* 138 *bk.*)

A Declaracio*u*n made by Io*h*n Ienyns, of a*l*t the Charg*es* of the Kyng*es* Armye Roia*l*t nowe beyng o*n* the See, Aswell in his Navye and Fleete Roiall, beyng then in the Retynue of my Lor*d* Admyra*l*t, As in the Retynue of S*ir* William Fitz-William, knyg*h*t, Vice-Admyra*l*t; that is to wete, for oone hoole monet*h*, Accomptyng xxviij daies for the monet*h*, as here after foloweth /

[*in margin*] The Henry grace de dieu, M¹ and VC to*n*.

Sir Io*h*n Walloppe and s*ir* Io*h*n ⎫
Wysema*n*, Capitaynes, for theire ⎬ iiij *t*i iiij s.
Dyett*es* for the sai*d* monet*h* ⎭
Robert Basfor*d* and Isley, pety Capi- ⎫
teynes, either of theym at xij d by ⎬ lvj s.   Som*m*e
the day   ⎭   of men ix C vij

Souldiours —— CCCL ———— iiij vij *t*i x s
Thomas Spert, Maister ———— v s
Maryners ——— VC ———— Cxxv *t*i   of money CCxlvj *t*i
Gonners ———— 1 ———— xij *t*i x s   xiij s ij d
Dedesharys ——— xlvj ———— xj *t*i x s
Rewardes to gonners ———— iiij *t*i x s x d
Surgions ———— ij ———— xxiij s iiij d

The other crews are given at length; but we just state the number of tons, Surgeons—2 at 23*s*. 4*d*. a month, and 19 at 10*s*.—and men (soldiers, mariners, &c.) in each:

'The new Spanyar*d*' 260 ton, 1 Surgeon at 10*s*., 182 men.
'The Mary Roose' 600 ton, 2 Surgeons at 23*s*. 4*d*., 405 men.
'The great Galey' 700 ton, 2 Surgeons at 23*s*. 4*d*., 454 men.
'The Peter Pomegarnade' 400 ton, 1 Surgeon at 10*s*., 304 men.
'The Barbara' 400 ton, 1 Surgeon at 10*s*., 303 men.
'The Io*h*n Baptist' 400 ton, 1 Surgeon at 10*s*., 303 men.
'The greate Nicholas' 400 ton, 1 Surgeon at 10*s*., 303 men.
'The Mary Jamys' 300 ton, 1 Surgeon at 10*s*., 253 men.

App. XI. H. VIII's Army & Navy Surgeons' pay.

'The Mary George' 300 ton, 1 Surgeon at 10s., 193 men.
'The great Barke' 400 ton, 1 Surgeon at 10s., 213 men.
'The lesse Barke' 240 ton, 1 Surgeon at 10s., 193 men.
'The new Barke callid the Mynyon)' 160 ton, 1 Surgeon at 10s., 133 men.
'The Swepestake' 80 ton, 1 Surgeon at 10s., 60 men.
'The Swalowe' 80 tons, 1 Surgeon at 10s., 60 men.
'The Kateryne Galey' 80 ton, 1 Surgeon at 10s., 63 men.
'The Galey Foyste': no tonnage or Surgeon named; 62 men.
'The Mary Gonson' 460 ton, 1 Surgeon at 10s., 303 men.
'The Nicholas Draper' 180 ton, 1 Surgeon at 10s., 123 men.
'The Margarete Bonaventure' 180 ton, no Surgeon, 122 men.
'The Mighell Fowler' 40 ton, no Surgeon, 41 men.
'The Cryste' 180 ton, 1 Surgeon at 10s., 123 men.
'Sir Robert Iohns Shippe' 160 ton, 1 Surgeon at 10s., 143 men.
'The Mary Harper' 80 ton, 1 Surgeon at 10s., 73 men.
'My Lord Admyralles Bark' 80 ton, 1 Surgeon at 10s., 53 men.

Elizabethan Ships, Whale, and Dolphin from Christopher Saxton's Maps, 1573-9.
(From the *Pall Mall Gazette* blocks.)

XII.

HENRY VIII's PAYMENTS TO HANS HOLBEIN,[1] 1538-1541, AND TO PLAYERS, MUSICIANS, &c.

From the Arundel MS. 97, in the British Museum.

Quarter's Wages, Lady Day, 1538.

(lf. 11) Item, for Hans Holben, paynter[2] ... ... vij ħ x s

(lf. 26, bk.) Yet quart*er* Wag*is* at Midsom*er* a° xxx° (1538)
Item, for Hans Holbyn, paynter, for one hole yeres annuitie
aduaunced to him beforehand, the same yere to be accompt- } xxx ħ
edde from o*ur* ladye dey last past [1538], the somme of /

Yet paymentes in Decembre, anno xxx° (1538).

(lf. 48) Item, payde to Hans Holbyn, one of the kingis paynters, by the king*is* commaundeme*n*t, certefyed by my lorde
pryviseales le*tt*re, x ħ for his cost*es* & charg*es* at this tyme } x ħ
sent abowte certeyn his grac*es* affares into the p*ar*ties of
High Burgony, by way of his graces rewarde

(lf. 67, bk.) Yet Quart*er* Wag*is* at o*ur* Lady day a° xxx° (1539)
Item, for Hans Holbyn, paynter ... nihil, q*uia* p*ri*us per warr*an*to

(lf. 81, bk.) Yet quart*er* wag*is* in June A° xxxj° (1539)
Item, for Hans Holbyn, paynter ... ... vij ħ x s

Michaelmas, 1539.

(lf. 90) Item, Paide by the Kyngis highnes commaundement,
certified by my Lord Pryviseales le*tt*res, to Hans Hol-
benne, paynter, in the advauncement of his hole yeres
wagis beforehande, after the rate of xxx ħ by yere, which } xxx li
yeres ad*u*auncement is to be accompted from this present
Michaelmas [1539], and shall ende vltimo Septembris
next commynge, the somme of

---

[1] These have been printed before (we find) in the *Archæologia* and in Wornum's *Life of Holbein:* perhaps elsewhere too.

[2] See the payments to Anthony Toto and Bartilmewe Penn, payn*t*ers, xij li, x s (lf. 51, bk.), &c., in note 3 on p. 101, 117, above.

App. XII. *Henry VIII's Payments to Holbein.* 239

(lf. 93, bk.) Yet quarter Wagis [in Septembre] Anno xxxj (1539)
Item, for Hans Holben, paynter ... ... vij ti x s

(lf. 107) Yet quarter Wagis at Cristmas A° xxxj° (1539)
Item, for Hans Holbyn, paynter, ... vij ti x s

(lf. 125) Yet quarter Wagis at our lady day A° xxxj° (1540)
Item, for Hans Holben, paynter ... ... vij ti x s

Yet quarter wagis, at midsomer, A° xxxij° (1540.)
(lf. 137, bk.) Item, for Hans Holben, paynter ... ... vij ti x s

Yet paymentes, in Septembre, A° xxxij° (1540.)
(lf. 147) Item, paid to Hans Holbyn, the kinges paynter, in aduauncement of his wagis for one half yere beforehande; the same half yere accompted and reconned, fromme Michaelmas last paste [1540], the somme of } xv ti

Yet quarter wagis at michaelmas, a° xxxij° (1540.)
(lf. 151) Item, for Hans Holbyn, paynter,—nil, quia prius, per warranto
Quarter's Wages, Christmas, 1540.

(lf. 163) Item, for Hans Holbyn, paynter, wagis—nil, quia prius per manibus

Yet paymentis in Marche, Anno xxxij° (1541)
(lf. 179) Item, paied to Hans Holben, the kinges painter, in aduauncement of his half yeres wagis before-hande, after the rate of xxx ti by yere, which half yere is accompted to beginne primo Aprilis, anno xxxij° [1541] domini Regis nunc / and shall ende vltimo Septembris then next ensuynge, the somme of } xv ti

Yet quarter Wagis at our Ladyday, A° xxxij° (1541.)
(lf. 181, bk.) Item, for Hans Holben, paynter, wagis—nil, quia prius [per] manibus

Yet quarter Wagis at midsomer, A° xxxiij° (1541.)
(lf. 195) Item, for Hans Holbyn, paynter—nihil, quia prius.

[Mr. Fenwick says there are no payments to Holbein in the Phillipps MS, A.D. 1542-3, at Cheltenham.]

240 App. XII. *Hen. VIII's Payments to Players, &c.*

Some Payments to Players, &c.

(lf. 53) Rewardes geuen on Wensday New Yeres day at Grenewiche, aº vt supra (xxxº, 1539).

(lf. 55, bk.) Item, to yᵉ kinges pleyers for pleying before yᵉ king this Christemas ... ... vj ɫi xiij s iiij d
(lf. 56) Item, to yᵉ quenes pleyers for pleyng before yᵉ king this Cristemas ... ... iiij ɫi
Item, to the Princes pleyours for pleynge before the kinge this Christemas by yᵉ kinges commaundement ... ... iiij ɫi
(lf. 68, bk.) Item, for Iohn Slye, pleyour (½ years wages) xxxiij s iiij d

Rewardes geuen on Thursday, Newyeres day, at Grenewiche, as hathe be accustumed. Anno tricesimo primo (1540).

[*Arundel MS.* 97, *Brit. Mus.*]

(lf. 108) Item, to master Crane, for playinge before yᵉ king with the children ... ... vj ɫi xiij s iiij d
(lf. 110, bk.) Item to yᵉ kingis pleyers, for playng before yᵉ king this Cristmas [1539] ... ... vj ɫi xiij s iiij d
(lf. 111) Item, to the Quenes pleyers, for playing before yᵉ kinge iiij ɫi
Item, to the Princis pleyers, for playinge before yᵉ kinge ... iiij ɫi
(lf. 125, bk. : 25 March, 1540) Item for Iohn Slye, pleyour
xxxiij s iiij d

Rewardes geuen on Saterday, Newyeres day, at Hamptoncourte, Anno xxxijº (1541).

(lf. 164, bk.) Item, to Master Crane, for playinge before the king with the children of the chappell, in rewarde ... vj ɫi xiij s iiij d
(lf. 167, bk.) Item, to the kingis pleyers, in rewarde vj ɫi xiij s iiij d
Item, to the Quenes pleyers, in rewarde ... ... iiij ɫi
Item, for the princes pleyers, in rewarde ... ... iiij ɫi
(lf. 181, bk. : Lady Day, 1541) Item, for Robert Hinscot,[1] George Birche, & Richard Parloo, pleyers ... ... xxxiij s iiij d
(lf. 194, bk. : Midsr. 1541) Item, for Robert Hinscot,[1] George Birche, & Richard Parow, pleyers ... ... xxxiij s iiij d

Some New Year's Gifts to Minstrells, &c. 1540-1.

1 Jan. 1540.

(lf. 108) Item, to Thomas Evans / Thomas. Bowmān & Andrewe Newmān / the Quenes minstrelles, in rewarde ... ... xl s

1 Jan. 1541.

(lf. 164, bk.) Item, to Thomas Evans, William More, and Andrewe Newmān the Queen's minstrellis, in rewarde ... ... xxx s

---

[1] This may be Hinscoɔc.

App. XII. *Henry VIII's private Band,* 1540-1. 241

(lf. 164, bk.) Item, to Lewes de Basson, Anthony de Basson, & Baptist de Basson, Jasper de Basson, John de Basson, the king*is* minstrell*is*, by the king*is* co*m*maundement certified by maister Charles Hawarde } iiij ti

(lf. 167, bk.) Item, to Guilliam de Trosshes, Guilliam dufaite, and Petie Johīn, minstrell*is*, in rewarde¹ } iiij ti

Henry VIII's private Band in 1540-1.

[As a sample of the Monthly Payments to the Band all thro' the Arundel MS. 97, we take those of March, an. xxxj, 1540; and as a specimen of the New Year's Gifts to them, those of Jan. 1, 1541. Note Anthony ' Mary,' the sackbut-player; and the Italian fiddlers or violists at the end.]

(lf. 122, bk.) Yet Paymen*tes* in Marche, A*n*no xxxj° (1540).

Item, for xij Trumpetters, wag*is* in xvj d a dey, eu*er*yon.	xxiiij ti
Item, for fyve other Trumpeters, in viij d a dey, eu*er*yon ...	v ti
Item, for Philip Welder, luter, wag*is*	lxvj s viij d
Item, for Petir Welder, luter, wag*is*	xxxj s
Item, for Iohīn Seue*r*nake, Rebeke, wag*is*	nil
Item, for Thomas Evans, Rebeke, wag*is*	xx s viij d
Item, for Will*ia*m More, Harper, wag*is*	xxxj s
Item, for Thomas Bowman, minstreH	xx s viij d
Item, for Andrewe Newman, the wayte	x s iiij d
Item, for Arthur Dewes, luter, wag*is*	x s iiij d
Item, for Hans Highorne, ViaH, wag*is*	xxxiij s iiij d
Item, for Hans Hosenet, ViaH, wag*is*	xxxiij s iiij d
Item, for Marke Anthony, Sagbut	xl s
Item, for Pilligrine, sagbut, wag*is*	xl s
Item, for Nicholas Vorcifall, sagbut	lv s vj d
Item, for Guilli*a*m Duwayte, minstreH	liij s iiij d
Item, for Guiliam de Trosshes, minstreH	liij s iiij d
Item, for Iohīn Buntanus, tabret	xlj s iiij d
Item, for the Children of the Chapell, bordwag*is* ...	xxvj s viij d

¹ Item, to a wom*ā*n that gave a booke [tablet] of wax ... ... x s
Item, to diu*er*se po*re* men, women and children, that brought capons, hennes, egges, bookes of waxe, and other triffelles: in rewarde } lxiiij s iiij d
Item, to Robert Morehous, that gave the kinge a purse withe bottonnes of golde } vj s viij d
Item, to Francis, a straung*er* that gave y° king p*er*fumed gloves and other perfumes /. } xl s
Item, to Cornelis Smith, *that* gave a basket of Iron ... vj s viij d
VICARY. R

1 7 *

242 App. XII. *Henry VIII's Musicians*, 1540-1.

(lf. 123) Item, for Burtill and Hans, dromslades ... xxxiij s iiij d
Item, for Hans quere, dromslade ... ... xx s viij d
Item, for Iohn Pretre, fyfer, wagis ... ... xx s viij d
Item, for Nicholas Andrewe, Sagbut ... ... xx s viij d
Item, for Anthony Symon, Sagbut ... ... xx s viij d
Item, for Anthony Mary, Sagbut[1] ... ... xli s iiij d

(lf. 164, bk.) Rewardes geuen on Saterday, Newyeres day, at Hamptoncourte, Anno xxxij° (1541).

Item, to the Kinges Trumpeters, in rewarde ... ... v li
Item, to the Sagbuttes, in rewarde ... ... l s
Item, to the Kinges Drumslades, in rewarde ... ... xx s
Item, to the stille minstrelles,[2] in rewarde ... ... iiij li
Item, to the newe Sagbuttes, in rewarde[3] ... ... iiij li

(lf. 165) Item, to Vincent da Venitia, Alexandro da Venitia, Ambroso da Milano, Albertus da Venitia, Ivam Maria da Cramona, and Anthony de Romano, the Kinges Vialles, by like commaundement, certified by maister Charles Hawarde } iiij li

[1] Item, for sir Iohn Wolf, prest, devisour of herbers ... ... xx s
Item, for Mathewe de Iohna, caster of the barr ... xx s viij s
[2] ? What was a still Minstrel? Surely not one who didn't sing.
[3] See the Queen's and King's Minstrels, above.

[*From Andrew Boorde*, p. 125.]

XIII.

THE 185 FREEMEN OF THE BARBER-SURGEONS' COMPANY,

THE MOST NUMEROUS IN LONDON, IN 1537,

WITH THE NUMBERS OF THE OTHER 38 CITY COMPANIES.

IN order to show that the Barbers' (or Barber-Surgeons') Company was—even before its statutory union with the Surgeons—the strongest Livery Company in the City of London, the following list of its 185 Members has been copied from the Return (in the Record Office) of all the Companies' members, in 1537, the year in which Thomas Lewyn was sheriff, with Sir John Gresham, while Sir Richard Gresham was Lord Mayor (Stow, *Survay*, p. 445, ed. 1598; p. 532, ed. 1603).

The Barber-Surgeons are 185 strong. Then come the Skinners, 151; the Haberdashers, 120; the Merchant-Tailors are 7th, with their 96; the Tilers (a Rafe Burbage among them) have 90; while the others dwindle away so that the Barbers make half-a-dozen (or more) of them. The point of numbers is of moment, not only as witnessing the importance of the Company to which Vicary belonged, but also the share which the Barbers took in the civic processions, and the number of armed men they could produce when called on.

[A.D. 1537.[1]] Chapter-House Books B ⅟₁.

The seuerall companyes of all the Mysteryes, Craftes and occupaciones within the Cytie of London, with the names of euery free mañ[2] beyng householder within the same / first / Mercers...

---

[1] 'Thomas Lewyn, Shiref of London,' is 2nd of the Yrenmongers, on lf. 13 of the MS.
[2] Among the Freemen of 'the Paynter Stayners,' is 'Agnes Best, widowe.'

244 App. XIII. *Freemen of the Barber-Surgeons*, 1537.

[leaf 21]   Barber Surgeons.

Nicholas Symson		Cristofer Samond		Thomas Mede	[lf. 22, bk.]
William Kyrckby		Robert Waterford	46	Ioĥn Anger	
*Thomas Vycars		Henry Atkyn		Thomas Worseley	
Ioĥn Bankes	4	Christofer Bolling		Ioĥn Gilberd	92
Ioĥn Potter		Robert Stocdale	[lf. 22]	Cristofer Haynes	
Thomas Twyn		Mathewe Ioĥnson	50	William Smythe	
Ioĥn Ioĥnson	7	Davy Sambroke		Ioĥn Mosseley	
Ioĥn Holland	[lf. 21, bk.]	Ioĥn Atkynson		William Hill	96
William Rewe		Thomas Waryn		George Wenyard	
Ioĥn Aylyff		Robert Grove	54	Ioĥn Barker	
Edmond Harman		Robert Brownhill		William Barker	
Ioĥn Peñ	12	William Spencer		Iames Wod	100
Richard Tayler		Thomas Butfilane		Ioĥn Stere	
Harry Carrier		Robert Forster	58	William Hetherley	
Rauf Garland		Edmond Tyrell		Olyver Wilson	
Ioĥn Enderbye	16	Ioĥn Philpott		William Grene	104
Peter Devismand		Ioĥn Thowlmod		Henry Rawshold	
Robert Postell		Edward Ingalby	62	Bartilmewe Dobynson	
Ioĥn Bird		Richard Elyott		Henry Patterson	
Iames Tomson	20	Thomas Wilson		Philip Pegott	108
William Kydd		Ioĥn Smythe		Robert Downys	
Ioĥn Yong		William Hiller	66	Antony Barowes	
Thomas Sutton		Richard Tholmod		Iamés Hogeson	
Charles Wyght	24	Ioĥn Awcetter	68	Robert Wevir	112
Ioĥn Newmañ				Ioĥn Surbut	
Thomas Crome	26	Richard Sermond	69	William Sewell	
		Hugĥ Lymcocke		Ioĥn Denys	
William Higges	27	Ioĥn Bordman		Ioĥn Page	116
Ioĥn Dene		Rauf Stek	72	Robert Todwell	
Thomas Surbutt		Henry Hogekynson		Ioĥn Cutberd	
William Billing	30	Ioĥn Tomson		Ioĥn Gray	
William Lyghthed		Hugĥ Dier		William Dauntese	
Ioĥn Raven		Edward Freman	76	Thomas Appilton	121
Robert Hutton		Thomas Mone		Ioĥn Crayell	
Henry Pemberton	34	William Yenson		Thomas Arundell	
William Shirbourne		Ioĥn Banester		William Ioĥnson	124
George Genne		William Trewise	80	Henre Adam	
Thomas Iohnson		Christofer Hungate		William Downham	
Robert Spignall	38	Ioĥn Hutton		Rogier Skynner	
Richard Boll		Ioĥn Browne		Ioĥn Gerard	128
Nicholas Alcoke		Ioĥn Grene	84	Richard Rogiers	
William Tylley		Ioĥn Tymber		Thomas Dicson	
Ioĥn Northcote	42	Ioĥn Shreue		Thomas Gylman	
William Wetyngton		Thomas Staynton		Thomas Dester	132
Henry Yong		Thomas Pays	88	Edward Hewett	[lf. 23]

App. XIII. Barber-Surgeons & other Companies. 245

Iohn Dormot		Iohn Robynson	152	Iohn Edlyn	
George Batman		Richard Coley		Iohn Samond	
Thomas Vivian	136	Iohn West		Henry Bodeley	172
George Brightwelton		William Welfed		Thomas Stanbrige	
Iohn Waren		Iohn Smerthwaite		William Borrell	
Iohn Grenway		Iohn Lybbe	157	Richard Nicols [lf. 23, bk.]	
Iohn Bell	140	George More		Edward Hughbank	
Laurens Mollyners		Thomas Burnett		Iohn Charterane	177
Iohn Cobbold		Iohn Hamlyn	160	Henry Wotton	
William Draper		Richard Child		Robert Hastynges	
Richard Smythe	144	Thomas Baily		Alexander Mason	180
Robert Ledes		George Vaughan	163	Thomas Darker	
Iohn Gamlyn		Thomas Wetyngham		Thomas Fyshe	
Thomas Cutbert		Iohn Bonair		Edward Rollesley	
Robert Chamber	148	Richard Cokerell		Iohn Braswell	
Lewis Bromefeld		William Walton		William Symsyn	185
Richard Worseley		Geferey Fraunceis	168		
Iohn Oskyn		Thomas Fayles			

It will interest some Readers to see the comparative and actual strength of the City Companies and Trades in 1537. The first column below shows how they rank in point of numbers; the second, their rank in the City. The Stationers and other trades are left out (we assume) because they were not then incorporated.

A.D. 1537.

Order by number of Members.

1. 185 Barber-Surgeons
2. 151 Skinners
3. 120 Haberdashers
4. 113 Leather Sellers
5. 109 Fishmongers
6.   99 Tallow-Chandlers
7.   96 Merchant Tailors
8.   90 Tilers
9.   89 Brewers
10. 77 Drapers
11. 69 Cloth Workers
12. 65 Cutlers ⎫
13. 65 Founders ⎬
14. 65 Bakers ⎭

Order in the MS. and City.

1. Mercers, 55
2. Drapers, 77
3. Merchant Tailors, 96
4. Fishmongers, 109
5. Goldsmiths, 52
6. Grocers, 59
7. Salters, 40
8. Vintners, 33
   (1. Sir James Spencer, knight
    2. Mr. Carter, King at Armes)
9. Haberdashers, 120
10. The Broiderers, 33
11. The Paynter Stayners, 53
12. Bakers, 65

246 App. XIII. London City-Companies in 1537.

Order by number of Members.	Order in the MS.
15. 63 Coopers	13. Ironmongers, 59
16. 60 Sadlers	(1. William Denton, Alderman
17. 59 Grocers	Thomas Lewyn, Shiref of
18. 59 Ironmongers	London.)
19. 56 Cordwainers	14. Skinners, 151
20. 55 Mercers	15. Brewers, 89
21. 53 Painter-Stainers	16. Waxchandlers, 45
22. 52 Joiners	17. Cloth Workers, 69
23. 52 Goldsmiths	18. Leather Sellers, 113
24. 48 Armourers	(leaf 18 back, foot. Lawrence
25. 47 Pastelers	Cornewe, sergeant)
26. 45 Wax-Chandlers	19. Innholders, 43
27. 44 Fletchers	20. Bowyers, 19
28. 43 Innholders	21. Fletchers, 44
29. 40 Salters	22. Barber-Surgeons, 184
30. 39 Fruiterers	23. Plumbers, 25
31. 38 Curriers	24. Weavers, 30
32. 37 Freemasons	25. Cutlers, 65
33. 33 Broiderers	26. Sadlers, 60[1]
34. 33 Vintners	27. Cordwainers, 56
35. 30 Weavers	28. Curriers, 38
36. 25 Plumbers	29. Tallow-Chandlers, 99
37. 25 Blacksmiths	30. Freemasons, 37
38. 20 Spurriers	31. Armourers, 48
39. 19 Bowyers	32. Pastelers, 47
	33. Fruiterers, 39
	34. Coopers, 63
	35. Founders, 65
	36. Blacksmiths, 25
	37. Spurriers, 20
	38. Tilers, 90
	39. Joiners, 52
	[No Stationers, &c.]

Readers will note that the Barber-Surgeons have only one Lighthead among them (no. 31). Let us hope that their one Well-fed (no. 155) showed the condition of Vicary and all his mates, Surgeons and Barbers alike.

[1] The 3 last Sadlers are 'The good wife Pounde, The good wif Coupir, The good wif Yong.' The Company still has Women as Freemen.

XIV.

ORDINANCES

OF THE

BARBER-SURGEONS' COMPANY OF LONDON,

SEPT. 1529,

as approvd by the City's Committee, and submitted (on Oct. 20, 1529) to the Chancellor and Treasurer of England (Sir Thomas More, and Thomas, Duke of Norfolk), and the Chief Justices of the King's and Common Benches (Sir Jn. FitzJames and Sir Robert Norwich), and by them revised into, and ratified as, the Company's Ordinances from May 14, 1530.

(From the Guildhall *Letter-Book* O, leaves 114 back, to 118.)

WITH

LISTS OF THE WARDENS OF THE SURGEONS AND BARBER-SURGEONS 1488—91 (p. 260)

AND

ACTS OF THE COMMON-COUNCIL RESTORING TO THE BARBER-SURGEONS THEIR OLD PLACE AS 17TH IN THE RANK OF CITY-COMPANIES (p. 261).

248 App. XIV. *Barber-Surgeons' Draft-Rules*, 1529.

The late Mr. John Flint South, or his Guildhall copiers, seem to have mist the following Document, which is described in the 1530 Revision of it printed in South's *Craft of Surgery*, p. 339—350, as

'a Boke conteyning dyuers Statutes, actes and Ordyn*au*nces, heretofore devysed, ordeyned and made, for the Fellowship of Barbours Surgeons, and their Successors, and for the Common weale and conservac*i*on of the good estate of the sayd Crafte and Mysterye of Barbors Surgeons aforesayd, and for the better Rules and ordyn*au*nces of the same Fellowship, establysshed, ordeyned and vsed.' —*Ib.* p. 340.

As the Act 19 Henry VII, chapter 7 (A.D. 1503),[1] requird all Ordinances of London Gilds or Fraternities to be examind and approvd by the Chancellor and Treasurer of England, and the Chief Justices of the King's Bench and Common Bench, or three of them, Vicary and his Brethren, on Oct. 20, 1529,[2] duly submitted the Barber-Surgeons' proposed Rules to these Officials, and on May 14, 1530, had them returnd, revised and duly ratified, with a change of the order of Clauses and of some words, a Prolog reciting the Act 19 Hen. VII, ch. 7, and the 'Boke' following, &c., and an Epilog saving the King's rights, and adding the Proviso on p. 254, below, that no Freeman of the Company might 'open any Shoppe of Barbarye' till he ownd goods of the value of 10 Marks sterling, £6 13s. 4d.

[1] Statutes, ed. Pulton, p. 434-5. He notes references to 28 Hen. VIII, ch. 5, and 31 Hen. VIII, ch. 41.
[2] The MS. and South's print give the date a year later, making the Revised Ordinances of May, 1530, recite these Draft ones as sent-in in Oct. 1530. It is plain to us that the Draft Ordinances were submitted to More and his Colleagues directly they were clear of the City Committee in Sept. 1529.

1528, Dec. 17. A Committee appointed to revise the
City Companies' Ordinances.

(Letter-Book O, leaf 131, back.)

Comune Consilium Tentum die Jouis, videlicet, xvij°
die Decembris, Anno regni Regis Henrici octaui vicesimo
[1528], in presencia Johannis Rudstone, Maioris, Brugge,
Mylbourne, Mundy, Baldry, Seymer, Spencer, Englishe,
Dodmer, Hardy, Pecok, Askue, Champneys, Hollys,
Pergetour, & Waren vicecomitis[1] . . . .

A Committee of 6 and the Common Clerk appointed to look over the

Johannes Clarke, Draper ⎫   with the Comen
BenJamyn Dygby, mercer ⎪   Clarke Attendaunt
Ricardus Fermour, Grocer ⎪   vppon theym/named
Poule Wythypolle, merchaunt- ⎬ and appoynted to
    taillour   ⎪   pervse and oversee
Olyver Leder, Fishemonger   ⎪   suche Bookes of Actes
William Hampton, Skynner ⎭   & ordynaunces as

Ordinances granted by the City to City Companies,

heretofore were given and graunted by the Maier and
Aldremen to dyuers Felishippes of this Citie / whether
that they be good and Resonable, and ougĥt to be con-
fermyde by Auctoritie of Comen Counsell or not, & c'. /

with power to

That they, or the more parte of theym, haue full power
and Auctorite to peruse, oversee, examyne, Refourme,

revise them,

& correcte suche Bookes and ordynaunces as heretofore
were gevyn and graunted by the Maier and Aldermen
then for the tyme beynge, to dyuerse Felishippes of this

and to authorise all such Ordinances as they think reasonable.

Citie / And alle suche of the saide Actes and Ordynaunces As vppon the examinacion and Reformacion of
theym as they shalle thynke to be good and Resonable,
and ougĥt to be conferrmed by Auctorite of Comen
Counsell, They soo to allowe & admytte & c'./.

/ finis.

1529, Feb. 3. The Ordinances of the Mystery of Barber-
Surgeons of London.

(Letter-Book O, leaf 114, back.)

Where at A Comen Counselle holden yn the Guy-
hatt of the Cytie of London, the xvijth daye of December

---

[1] Raphe Waren and John Long were the Sheriffs.

250   App. XIV. *Barber-Surgeons' Draft-Rules,* 1529.

Recites the above Appointment of the Ordinance-Revision Committee on Dec. 17, 1528,	yn the xxth yere of the Reigne of o*u*r sou*e*raigne lorde, Kyng Henry the viijth, thiese p*er*sons foloyng, that ys to say / Joħn Clerk, drap*er*, BenJamyɴ Dygby, Mercer, Rychard Fermour, Grocer, Pauħ Wythypoll*e*, m*er*chau*n*t Tayllour / Olyuer Leder, Fysshemonger, and Wylli*a*m Hamptoɴ, Skynner, w*i*th the co*m*en Clerke, wer named, appoyntid and Auctorysed, by auctorytie of the same, that they, or the more p*a*rte of theym), shulde have
with power to amend all Companies' Rules,	fuħ power and auctorytie to p*e*ruse, oue*r*se, examyn*e*, Refo*u*rme and correcte suche book*es*, Act*es* and orden-au*n*c*es* as heretofore wér geueɴ and graunted by the Mayre and Aldremeɴ theɴ for the tyme beyng, to
and pass such as they think reasonable;	dyu*er*s Felyshipps of this Cytie / And aħ suche of the sayde Act*es* and ordena*u*nc*es* As vpoɴ thexamynacyoɴ and Reformacyoɴ of theym), they shaħ thynke to be good and Reasonable, and ougħt to be conferm*e*d by Auctorytie of co*m*en Counseyħ, they so to Allowe and
and that Thos. Vicary, and other Wardens of the Barber-Surgeons, on Feb. 3, 1529, showd the Committee	Admytt; Whervpon, Walter Kelett, Thomas Vycar, Joħn Potter and Thomas Suttoɴ, Wardeyns of the Crafte or Mystere of Barbour Surgeons, Afterward, that ys to say, the iij^{de} day of February, the xxth yere afo*re*sayd, exibyted to the sayde p*er*sones so named and
a Book of the Ordinances of their Company; and the Committee have revised and past these, in the form following:	Appoynted, A certeyɴ booke or volume concernyng dyu*er*s Articles for the good ordre of the sayd Mistere, whiche booke they have, by good delib*er*acyoɴ perused & oue*r*seeɴ / & dyu*er*s of the sayd Articles they have corrected & [*word rubd out*] yn maner and fourme ensuyng':
	To the Right honou*r*able and their Singu*l*er good lorde and Maisters, my lord Mayre and his worshipfuħ Bretherne, Thaldermeɴ of the Citie of' Londoɴ
The Supplicacyoɴ	Mekelye besechen yo*u*r good Lordshippe and Maister-ships, the Maisters or Goue*r*ners and Co*m*i*n*altie of the Mystere of Barbo*u*rs Surgeons of Londoɴ, That for the better Rule & more quyete ordre hereafter to be had and vsed yn the sayde Mystere / It maye please you to graunte vnto theyɴ the Articles, ordyn*a*u*n*c*es* and othes ensuyng, whiche they, by yo*u*r Favours, suppose to be verye necessarye and behouefuħ for theyɴ to haue & execute; And they shaħ praye to god for yo*u*r good contynewaunce and p*r*osperous p*re*seruacyons. /
(1) Paying of qu*a*rterages. Liverymen shall pay 6d. a quarter,	Firste it ys enacted and ordeyned that eu*er*y manɴ yn the Clothing or lyuere of the sayd Mistere shaħ paye quarterly to the mayntenau*n*ce of the Co*m*en charg*es* of the same / vj d / and eu*er*y manɴ oute of the Clothyng,

App. XIV. Barber-Surgeons' Draft-Rules, 1529. 251

other Freemen, and widows, 3d.,

under penalty of 3s. 4d.

and euery wydowe kepyng an open Shoppe / iij d / And this to be payde quarterly, vpon payne and Forfeyture at euery tyme offendyng or dooyng the contrary / iij s iiij d / the oon halfe thereof to be Applyed to thuse of the Chambre of London, And the other halfe to the Almes of the sayde Felishippe / So Alweys that the sayde quarterage be lawfullye demaunded

[¹ leaf 115]
(2) All Sommons to be obserued

under penalty of 3s. 4d.

¹Also it ys ordeyned that euery persone enfraunchesed yn the same Crafte, shalbe redye at aH maner of Sommons of the Maysters or Gouerners of the sayde Crafte for the tyme beyng / And yf any suche persone Absent hym from any suche sommons wythoute cause Reasonable, to be tryed by his othe before the Maisters or Gouerners, yf they thinke yt necessarye / Than he to paye for euery so doyng¹ / iij s iiij d, ²to be deuyded and Applyed yn maner and fourme Aforesayde /—/²

(3) The howre of Sommons to be kepte

under a fine of 2d. for the Alms-fund,

and 3s. 4d. penalty for disobedience.

³Also that euery man enfraunchesed yn the sayde Crafte, beyng duely warned or sommoned, that kepeth not his howre accordyng to his Sommons, withoute cause reasonable, to be tryed yn maner⁴ aforesayde, for euery tyme so doyng¹, shall paye to the Almes of the sayde Crafte / ij d. And he or they that disobeyeth this ordenaunce, shall paye for his or their disobedyence yn that behalf, for euery tyme so offendyng / iij s iiij d / to be deuyded yn fourme Aforesayde /

(4) To Auoyde discorde amonges theym of the company.

No Freeman shall sue another till he has first complaind to the Masters of the Company. They shall try and settle the matter. If they can't, in 14 days,

⁵Also, yf any mater of Stryff¹ or debate hereafter be betwene eny persones⁶ of¹ the sayde Crafte (as god forfende !) That noon of theym shaH make any pursute yn the comen lawe; butt that he whiche fyndeth hym Agreved, shall Fyrste make his complaynte to the Maysters or Gouerners of the sayde Crafte for the tyme beyng, to thentent that they⁷ shaH ordre the sayde Matier or cause of complaynt so made, yf they can / And yf it fortune that they can nott, or⁸ doo nott, ordre & Appese the same matier withyn xiiij dayes than next ensuyng, That than yt shalbe lyefuH to the partye Aggrevyd, to take hys Aduauntage at the Comen lawe / So Alweys that the partye Ayenst

---

¹ South, p. 342.
²–² the one half to the Chamber of London, and the other to the Almes of the Crafte.—Sir Thos. More's Statutes, in South, p. 342.
³ amalgamated with the preceding article, in South, p. 342.
⁴ to be fixed in the maner and forme.—More.
⁵ More's Ordinances put this after No. 7, p. 252, below, that no Freeman shall teach any one but his apprentice.—South. p. 345.
⁶ person, M.   ⁷ he, M.   ⁸ nor, M.

App. XIV. Barber-Surgeons' Draft-Rules, 1529.

the plaintiff may go to law; and the defendant mustn't bolt. Penalty, 13s. 4d.

whome the compleynt ys made, be nott fugitife / And who so doith the contrary herof, shall paye for euery tyme so dooyng¹ / xiij s iiij d / to be deuyded and Applyed yn fourme aforesayde /—//

(5) No man to Reuyle Another.

¹Also, that no person of the sayde Felyshippe shall Reuyle, Rebuke, nor Reproue an other of the same Felyshippe by eny vnsittyng,² opprobryous, cedicyous,³ or dishonest wordes, yn the presence of the Maysters or gouerners, or eny of theym, nor before eny other persones yn eny other places / And he that offendyth yn this behalfe, & due profe thereof had, shall paye for euery suche defaulte, vj s viij d, to be deuyded and Applyed yn fourme Aforesayde.

Penalty, 6s. 8d.

(6) A Remedye agaynst theym that wyll not be of the lyuerey, nor bere offyce.

⁴Also, that no person of the sayde Crafte shall Refuse to be of the Clothyng of the sayde Mystere, or to bere office yn the same, at any tyme whan he, by the Maysters or gouerners & Assistentes of the sayde Mystere, or the more parte of theym, shalbe Abled therto, vpon payne to pay xl s., to be Applyed yn fourme aforesayd. And that the Maysters or gouerners of the sayde Mystere for the tyme beyng, shall nott take nor Admytt any person ynto the Clothyng or lyuerye of the same Mystere, withoute the comen Assent of⁵ xxiiij^ti Assistentes of the same, or the more parte of theym, vpon lyke payn as ys aforsayd for euery tyme so dooyng, to be deuyded & Applyed yn fourme aforesayd.

Liverymen to be elected by a majority of the 24 Assistants.

[leaf 115, back]

(7) Ayeynst theym that techen Forrens.

⁶Also yt ys ordeyned that no persone enfraunchesed yn the sayde Mystere, shall enfourme or teche⁷ eny Foren, other than hys Apprentyce, eny poynte of his Crafte belongyng to Barbery or Surgery, vpon payn, for euery tyme so doyng, xl s / to be Applyed yn fourme Aforesayd.

(8) No Apprentice to be taken but he be Fyrst presented to the Maysters.

⁸Also yt ys ordeyned that no persone enfraunchesed yn the sayde Crafte, shall take any Apprentyce vnto⁹ the tyme that he Fyrst present the same person before the Maysters or Gouerners for the tyme beyng, that they maye see he be clene, withoute contynuell¹⁰ Diseases or grevous Infyrmyties, wherby the Kynges lyege people myght take hurte, vpon payne for euery tyme so doyng, of xl s / to be Applyed yn maner Aforesayde.

---

¹ South, p. 345.   ² vnfything, More : South ; (vnsittyng is unsuitable).
³ condycions, M.   ⁴ South's *Craft of Surgery*, p. 346, line 1.
⁵ of the, M.   ⁶ South, p. 345.   ⁷ charge, M.
⁸ South, p. 343.   ⁹ until.   ¹⁰ chronic, permanent.

App. XIV. Barber-Surgeons' Draft-Rules, 1529.

(9) What shalbe payde at the takyng of Apprentice.
3s. 4d.

Also yt ys ordeyned that euery persone of the sayde Felyshippe shall pay towardes theyr [1]comen Charges, for euery Apprentice that he taketh / iij s iiij d / To be payde / xx d / at his presentacyon, & the other xx d withyn the same yere / And yf it fortune the sayde Apprentice to dye or avoyde Awey withyn the Fyrste yere, wherthorow hys Mayster taketh noon Aduauntage of hym / That than the sayde iij s iiij d to stonde for the payment of hys next Apprentyce, So that he brynge ynto theyr hall the Indenture of the sayde Apprentice so ded or gon Awey / And he or she Refusyng this to doo, shall forfeyte & paye / x s /, to be Applyed & deuyded yn fourme Aforesayde /—//[1]

Penalty 10s.

(10) None yn the lyuerey to have aboue iiij Apprentices & seruauntes [Assistants] togyder at ons.

[2]Also yt ys ordeyned that no persone of the sayde Felyshippe, beyng yn the clothing or lyuerye, shall have any mo seruauntes, Apprentice or couenaunte,[3] vsyng the facultye or mysterye of Barberye or Surgerye togyder at ons, aboue the nomber of iiij persones / Prouyded Alweys that withyn halfe A yere of the goyng oute or endyng of the terme of oon of the sayde iiij persons, yt shalbe lyefull to euery suche persone to take and[4] have an other Apprentice or seruaunte, the sayde Acte not withstondyng. And he that offendyth yn brekyng this[5] Acte, shall forfeyte and paye / xl s / to be deuyded and Applyed yn fourme Aforesayde /—/

(11) None oute of the lyuerey to have above iij Apprentices & seruauntes togyder at ons.

[6]Also yt ys ordeyned, that no maner persone[7] of the same Felyshippe, beyng oute of the Clothyng, shall have togyders at oons aboue the nombre of Three Apprentices or seruauntes to occupye the sayde Mystere and Faculte / Prouyded [8]as yt ys prouyded aforesayde yn[8] the later Article, and vpon lyke payn).

(12) For Takyng seruauntes Alowes [hired] or Alyauntes,

[lf. 116] Also yt ys ordeyned that no persone of the sayde Felyshippe shall take to hys seruyce as seruaunte Allowes [hired], any Englyssheman[9] Forren, or Alyaunt Straunger, to occupye the facultie of Barbery or Surgery / But Fyrste the[10] sayde persone shall present the same seruant[11] withyn iij dayes next after hys commyng to the sayde person, to and before the Maysters or[12] gouerners of the sayde Felyshippe for the tyme

---

[1–1] charge for every Apprentice that he taketh, ij s vj d, to be payed at the presentacion and allowyng of euery Apprentice.—More, in South, p. 343.
[2] South, p. 343.   [3] apprentices or Foreins, M., p. 343.
[4] or, M.   [5] of this, M.   [6] South, p. 344.   [7] of parson, M.
[8–8] as ys prouyded in, M.   [9] Englishe, M.   South, p. 344.
[10] but the, M. (but = except).   [11] person, seruaunt, M.   [12] and, M.

254 App. XIV. Barber-Surgeons' Draft-Rules, 1529.

and Ratyng of theyr wages

beyng, to thentent that he, before theym̃, maye be Sessed, what wages he shall take / And yf he be An Alyaunt Straunger borne, he[1] to paye yerely of hys wages, to the Almes of the sayd⁣[1] Felyshippe, iij s iiij d̃ / And that money to be taken quarterly, of the Mayster of the same straunger, and of his wages / And who that doyth contrary to this Rule, shall forfeyt, at euery tyme so dooyng, xl s / to be deuyded and Applyed as ys aforesayde /—// . .

Penalty, 40s.

(13) None that ys made Free, shall open his Shoppe tyll hee have doon his duetye at theyr hall

Paid 6s. 8d. to the Company, and 4d. to the Clerk.

Penalty, 40s.

[2]Also yt ys ordeyned, that no persone of the same Feliship*e*, after that he be admytted and sworne Freman of this Citie afore the Chamberleyn, presume to opyn his Shoppe wyndowes before he hath presented hymself to & before the Maysters or Gouerners of the sayde Mystere for the tyme beyng, and with theym have Agreed yn paying hys dutye Accustumed, that ys to saye, to the vse of the Companye vj s viij d[3], & to the Clerk⁣ iiij d[4], to the mayntenaunce of their comen charges, And yn takyng his othe afore theym, accordyng to the lawdable custome & ordre, yn the same Mistere of olde tyme vsed, vpon payne to lose, forfeyte, & pay xl s / to be deuyded and Applyed yn fourme aforesayde.[5]

(14) For entisyng of seruauntes, & takyng of Foreyns.

Penalty, 13s. 4d.

[6]Also yt ys ordeyned that no persone of the sayde Crafte shall entice or desire eny seruaunte from his Maister, nor shall take any Forren ynto his seruyce for lesse terme than for oon yere ; and he to be cessed or Rated for his wages, by the Maysters or gouerners of⁣ the same Mystere : And this to be doon yerly euery yere, vpon payne for euery tyme doyng the contrary, of xiij s iiij d̃ ; The oon halfe to be Applyed to thuse of the Chambre of london, And the other halfe to thuse of the Almes of the sayde Felyshippe.

---

[1] More leaves out 'he' and 'sayd.' [2] South, p. 344.
[3] iij s iiij d, M. [4] xij d, M.
[5] The Revised Ordinances of May 14, 1530, add the following :—
Provyded alwayes, that for dyvers consyderacions, as well for the welthe of the kinges leige people, as for the honestye of the sayde Crafte, yt is now condescended and agreed that, from hensforthe, no parsons of Felayship, after he or they be made Free of the sayd Companye, shall presume to sett open any Shoppe of Barborye, unto suche tyme as he or they be abled by the sayd Maister or gouernors, without he be of the clere value, of his owne proper goods, to the value of Tenne markes sterlinge, upon payne of Forfayture of xl s, the one half to the Chamber of London, and the other half to the Almesse of the sayd Crafte.—South, *Craft of Surgery* (1886), p. 349.
[6] South, p. 344.

App. XIV. *Barber-Surgeons' Draft-Rules*, 1529.   255

(15) A penaltye of xl s for shavyng on the Sonday.

[leaf 116, back]
(16) For takyng of Syke or hurte persones vnto theyr Cure.

(17) For the lecture of surgery wekely at their hall.

[1]And Where, by dyuers and[2] high Auctoryties for the honour & Reuerence of the Sondaye, yt is ordeyned[3] of olde Antiquytie, that no barbour dwelling withyn this Citie, or Suburbs of the same, nor elleswhere,[4] shaH occupye shavyng on the Sondayes, neyther withyn theyr hous nor withoute, pryvely nor Appertly / It ys nowe therfore ordeyned and enacted, that no persons free of the sayde company, fromhensforth occupye [5]eny maner Shavyng, priuy or peirt,[5] [on the Sondayes,][6] withyn this Citie nor liberties of the same,[7] vpon payne and forfeyture for euery tyme so doyng, of xl s / The oon halfe therof to the Chambre of London, And the other half therof to the Almes of the seyd Crafte

[8]Also yt ys ordeyned that no maner persone beyng Free of the sayde Felishippe, shall take any seke or hurte persone or persones to hys cure, whiche ys in pereH of deth or mayne, but yf he shewe the same seke or hurte persone, by hym receyved, to the Maysters or gouerners of the sayde Mystere, or twoo of theym for the tyme beyng,[9] for savegard of the kynges people[10] / And that withyn iiij dayes next after the Receyvyng of the sayde seeke or hurte persone; vpon payne for euery tyme doyng the contrary, of xx s ; [11]The one half thereof to the Chambre of London, And the other half therof to the Almes of the sayde Felyshippe.[11]

[12]Also yt ys ordeyned, that euery man enfraunchesed yn the sayd Felishippe, occupying Surgery, shall comme to theyr haH to the Redyng of the lecture concernyng Surgery, euery Courte daye[13]; And euery man, after his Course, shaH Rede the lecture hymself, or elles fynde An Able man of the sayde Felyshippe to Rede for hym, And nott to Absent hymself at hys daye of the same Redyng withoute cause Reasonable, And withoute he gyve lawfuH warnyng therof before the daye, vpon the payne to forfeyte and loose for euery tyme

---

[1] South, p. 346.   [2] More leaves out 'and.'
[3] ordeyned and enacted, M.   [4] ells who, M.
[5-5] any Shaving, M. (peirt = appert, open, public).
[6] on the Sondayes, M.   not in Letter-Book O.
[7] M. puts in 'prevely nor apertlye.'   [8] South, p. 346.
[9] See earlier provisions to this effect in South's *Craft of Surgery*, p. 17 (A.D. 1369), p. 19 (1390), p. 25 (1416), &c.   Also in Riley's *Memorials*, 337, 393, 519, &c.   M. leaves out 'for the tyme beyng.'
[10] Liege people, M.
[11-11] to be devyded and applyed in maner and forme aforesayed, M.
[12] South, p. 347.   [13] Daye of assemble therof.

(18) No man) to supplant Another yn takyng from) hym his Cure.	¹Also yt ys ordeyned that no man) of the saydc Felyshippe shall take eny Cure from) Another of the same Felishippe, nor supplant oon) Another, nor geve or speke any Slaunderus wordes yn disablyng hym) of hys science or connyng / but be rather yn a Redynes to geve good Counseyll to helpe the Kynges people : And every man) offendyng yn this behalf, to pay at every
Penalty, 13s. 4d.	tyme so offendyng, xiij s iiij d / the oon) half therof to the Chambre of London), And the other half to the Almes of the sayde Felyshippe / Provyded Alwey that
But a Patient may change to a 2nd Surgeon, after paying the 1st.	yf the pacyent fynde hymself Aggreved with his surgeon), That than) the same pacyent, paying to hys Fyrst Surgeon) Reasonably for hys labour, shall and maye take and have eny other Surgeon), at his libertie and pleasure.
(19) What every man) shall paye for his Dyner.	²And where, of olde custume, yerely vpon) the Sondaye next ensuyng the Feast of Seynt Bartylmewe Thappostell [Aug. 24], A dyner ys kepte & prouyded for theym) of the lyverye of the sayde company yn theyr comen hall called Barbours hall, And on the ³morwe foloyng³ A dyner for theym) of the same Company beyng oute of the lyverye / It ys ordeyned and enacted that every man) that hath been) vpper Mayster or vpper Gouerner of the said company, shall paye at and for
12d.; and 8d. for his wife,	the same dyner, xij d for hymself, and viij d for his wif, yf she⁴ com) ; And every other man) beyng of the lyverye of the same company, shall paye yn lykewyse for hymself viij d, and for his wyf, yf she com), iiij d ; Prouyded Alwey that the Maisters or Gouerners of the sayde com-
[* leaf 117] unless she helps prepare the Dinner.	pany *for the tyme beyng, shall paye nothyng for their wyfes commyng to the dyner for that yere, Forasmoche as theyr wyfes muste of necessitie be there to helpe that every thing there be sett yn ordre⁵ ; And that every man) of the sayd Company beyng oute of the lyverye, shall pay at and for his dyner on the sayde morowe, iiij d,⁶ And for his wyf, yf she com), ij d—//
(20) ⁷The othe of every man) of the Companye.	Ye shalle swere that ye shalbe good and true vnto our liege lorde the Kyng, and to his heyres, Kynges of Englond, and obedyent to the Mayre, and his Brethern) the Aldermen) of the Citie of London) ; And also ye

¹ South, p. 347. ² Ibid., p. 347.
³⁻³ daye of Saynt Cosme and Damian, yf it be not on the Satterdays.—More.
⁴ they, M. ⁵ Lady Aylyf once gave a table-cloth. ⁶ viij d, M.
⁷ M. puts this and the next oath first, after the Proem.—South, p. 340-2.

App. XIV. Barber-Surgeons' Draft-Rules, 1529.

Swear to obey the King, the Mayor and Aldermen, the Governors of your Company, and its Rules, present and future.

shalbe obedyent to the Maysters or gouerners that nowe be, & herafter shalbe, of the Crafte of Barbour Surgeons, wherof ye be nowe made Free / ye shall Also obey, kepe, & obserue all the good orders, Rules, and ordynaunces of the said Crafte heretofore made and not Repelled, and hereafter to be made, So helpe you god and all seyntes, and by this Booke /—//

(21) The othe of the Maisters or Gouernours.

Ye shalle swere that ye shall obserue, kepe, & maynteigne the worshippe, profyte, and comen wele of the Crafte of Barbour Surgeons, yn all poyntes lawfull and lyefull¹, as good and profytable Maisters or Gouerners and Rulers ought to doo, after your Connyng,² good diligence, and power / Also ye shall kepe and maynteyne, and doo to be kepte and maynteyned duryng your tyme, asferforth as ye lawfully maye / Aswell all suche good vsages, custumes, liberties and ordynaunces of this same Crafte, and at this day vsed, Approved and contynued / And alle and singuler poyntes conteyned yn the premysses, duely and truly³ ye shall putt yn execucyon, whan) & As often as the caas shall Requyre duryng your tyme / And also ye shall duely

To maintain the well-being of the Company, and its good old Customs;

to make Searches thro' the Craft,

and truely make your Serches thorough all the company of the same Crafte withyn the Citie of London) and Suburbs of the same; And thervpon), as the caas shall Require, alle the defaultes and neclygences, concilementes⁴ and inconuenyences that may hapne or fall to be founde yn the Crafte of Barbery or yn Surgery⁵ yn your tyme, ye diligently shall Refourme and sett yn good Rule, And truly correcte and punysshe, accordyng to the power and Rules for the Reformacyon) had and made for the same yn the sayde Crafte / And for and duryng your tyme, correcte and lawfully punysshe, after the qualyties and Gravyties of & vpon) the demerytes & defaultes founden) yn the same, after your connyng and power / Also ye shall not Admytte any Forreyn) to be of this Misterie,—whiche herafter shall sue to be A free man)⁶ of this Citie by Redempcyon), and to be enfraunchesed yn this Mistere,—withoute thassent of the xxiiij^{ti} Assystentes of the same Crafte, or the more parte of theym) / And ouerthat, ye shall not charge the hole bodye of this Felyshippe by puttyng the comen Sealle of the same Mystere to any maner wrytyng, *cause or matere, wherby the same Company yn any wyse may be charged, hurte,⁷ or hyndred /

reform defaults,

punish offences,

and not admit Aliens,

save by consent of the majority of the 24 Assistants.

Not to misapply the Company's Seal.

[* leaf 117, back]

¹ leafull.—More. ² good connynge, M. ³ when ye, M (wrongly).
⁴ of comytementes, M (wrongly, for 'concylementes' of the MS).
⁵ or Surgerye, M. ⁶ be ffreman, M. ⁷ hurted, M.

VICARY.

258 App. XIV. Barber-Surgeons' Draft-Rules, 1529.

In all things

to behave uprightly,

not heeding prejudice, &c.

And to administer this Oath to your successors.

Also yn alle the premysses, and other thinges necessarye concernyng the weale & profytt of the sayde Crafte, ye[1] shall truly, lawfully, dilygentlye, and Indifferently behaue yourself, after your connyng and power; and neyther for nede, love,[2] Fauour, Affeccyon, nor for drede, malyce, hatred or enuye, otherwyse procede, Rule, or conclude, to or with any persone or persones with whiche ye shall haue to doo, by Reason of your sayde office / Than the good vsages, Rules, liberties and ordinaunces for the good ordre of the same Crafte heretofore made, and nott Repelled, and hereafter to be made / Also, at thende of your office, ye shall geue vnto the Maisters or Gouerners that shall succede you nexte yn the same occupacyon, this present othe, So that they shall duely and truelye in all thynges duryng the tyme that they shalbe yn lyke office, perfourme & fulfyll the same othe; So god you helpe, and all Seyntes, & by this boke.

(22) Howe euery man shall behave hymself yn the Courte tyme.

No one to talk more than is necessary; and to stop when he's told to.

Penalty, 20d.

[3]Also yt is ordeyned, that at euery Courte[4] holden yn the comen hall of the sayde Mystere, no man beyng there present, shall multiplye langage yn the Courte[5] tyme, that ys to saye / yf any man there[6] speke mo wordes, or multiplye more langage yn the Courte,[5] then the Maisters or Gouerners for the tyme beyng there[7] present, thinke to be good and necessarye / That than, yf[1] they or oon of theym commaunde hym to keepe cylence, that than he shall so doo, yn kepyng his obedyence / [8]And also no man commyng to eny of the sayde Courtes,[9] shall departe from thens duryng the Courte[10] tyme, withoute licence of the Maisters or gouerners there[11] present, or oon of theym / And the Offender yn eny of the sayde / ij / poyntes or cases, to forfeytt and paye at euery tyme so offendyng, xx d, to be deuyded and Applied yn fourme aforesayde.[12]

---

[1] that, M. (misread or miswritten for 'ye').   [2] Love, meede.—More.
[3] South, p. 348.   [4] assemblie, M.   [5] assemble, M.
[6] mans othere, M.   [7] then, M.
[8] This is a separate article in South, p. 348.   [9] Assembles, M.
[10] Assemble, M.   [11] then, M.
[12] After this, and before the final clause of the Barber-Surgeons' 'Boke' in Letter-Book O, comes the following repetition of a general Act of 1364:

A generall Acte for all the Occupacyons and Mysterees of London.

Be yt remembred that the thursdaye next before the Feaste of Seynt Thomas Thappostell [Dec. 21], the yere of the Reigne of Kyng Edward the iij[de] after the conquest, the xxxviij [A.D. 1364], in the presence of Adam A Bery, than Mayre of the Citie of London / John Louekyn / Adam Franceys / Stephyn Cauendisshe / John Noot / Thomas Ludlowe / Wylliam Holbech / Wyllyam Tuden-

App. XIV. Barber-Surgeons' Draft-Rules, 1529. 259

(23) All Livery-men are to walk and sit by order of Seniority in their Company,

according to their Beadle's Roll,

under a Penalty of 12d.

¹Also yt ys ordeyned that no man of the Clothyng or lyverye² of the said Company, presume to go, oon Afore Another of theym, yn processions, buryalles, or Anniuersaries, nor yn sittyng yn their Courtes,³ Assemblees, or yn their hall at dyner or other Repastes there, or yn any other honest place, to be hadde otherwyse than he ys yn Auncyentie yn the same companye, And Accordyng to the true entraunce therof yn theyr bedylles Rolle⁴ / Nor that eny of theym, of eny scrupulositie, Frowardenes, follye, or⁵ pusillanimytie, Refuse to take hys owne Rowme or place Accordyng to the ordre aforeseyd / Butt that euery man yn thiese ij Cases kepe and occupye his owne Rowme and place, yn fourme aforesayd (wyll he, nyll he) yn good and⁶ obedyent maner / And he of theym that offendyth yn brekyng the ordre yn any of the sayde ij Cases, shall forfeytt and paye at euery tyme so offendyng, xij d, to be Applyed and deuyded yn fourme aforesayde⁷—//—//

All the City Crafts shall be so ruled that no false work be done in them.

Each shall be governd by 4 or 6 (or more or less) persons.

Rebellious Members shall be fined and imprisond more heavily for successive offences.

ham / John Biernes / John A Chichester / Wylliam Welde / Water Forester / Symon Worsted / John of Seynt Albones / James of Thame / Thomas Pykenham / James Andrewe / Bartholomeu Frestelyn and John Litle, Aldremen of the same Citie, this ordinaunce ensuyng was made (amonges other) for the profytt of the comons of alle mysteres of the Citie of London, that ys to saye / It is ordeyned that alle the craftes and occupacyons of the Citie of London shalbe lawfully Ruled and Gouerned, eueriche of theym yn his nature, yn due maner, So that no falsed, nor false worke ne deceyte, be founde yn nowyse yn the sayde Craftes or occupacyons, for the honour of the good people of the sayde Craftes, And for the comen profyte of the Kynges liege people / And that of euery occupacyon be chosen and [leaf 118] sworne, iiij or vj, or mo or lesse, after the busynesse of the occupacyon; whiche persones so chosen and sworne shall haue full power of the Mayre, the sayde occupacyon welle and lawfully to Rule and Gouerne / And yf eny persones of the sayde occupacyons be Rebelle, contraryous or disturbyng, So that the sayde persones chosen and Sworne can nott duely perfourme & execute their office, And therof be Atteynt, that euery suche persone so disorderyng hymself, shall, at the Fyrste tyme be Imprisoned by x dayes, and shall paye to the cominaltie for the contempte / x s / And at the ij^{de} tyme, he shall have Imprysonament by xx^{ti} dayes, And shall paye to the cominaltie xx s, And at the iij^{de} tyme, he shall have Imprysonement by xxx dayes, and shall paye to the Cominaltie xxx s / And at the iiijth tyme, he shalbe Imprysoned by xl dayes, And shall paye to the Cominaltie xl s /—//

¹ This is the last Clause before the Epilog or wind-up in South, p. 348-9.
² More omits 'or lyverye.'   ³ M. omits 'Courtes.'
⁴ in the Bedylls Skroll, M.   ⁵ frowardnes ne.   ⁶ M. omits 'and.'
⁷ aboue rehersed.

S 2

XIV. *Surgeons' and Barber-Surgeons' Wardens.*

The Wardens of the Surgeons and Barber-Surgeons, 1488—1491.

As we chanst to see some early entries of lists of Wardens of the Surgeons and Barber-Surgeons, we copied them, and here they are:

When Robert Tate was Mayor (Nov. 1488-9), the officers were (Journal 9, lf. 322 ink, 290 pencil):—

Rob*ertus* Palmer
Ric*ardus* Haymonde } Gard*iani* Art*is* de barbo*urs* Surgions, Jur*ati* 15 die
Jacob*us* Jugolby[1]   Sept*embris*
Andreu*us* Mayne   [1] or Ingolby

lf. 325 ink, 293 pencil:—

Th*omas* Ropesley
Th*omas* Thornton) } Gard*iani* Art*is* Cirurgic*orum*, Jurati 2 die Oct*obris* &c
Joh*annes* Hert

In 5 Henry VII (Aug. 1489-90) Wm. White, Maior (Nov. 1489-90), the officers sworn (Journal 9, lf. 312 ink, 280 pencil) were:—

Robertus Halyday, Magister
Ric*ardus* Snodnam) } Gard*iani* Art*is* de barbours-Surgeons, Jur*ati*
Joh*annes* Johnson), Jun*ior*   16 die Sept*embris*
Th*omas* Walton)

Will*elmus* Witwang
Robertus Taillour } Gard*iani* Art*is* de Surgeons, Jur*ati* iiij^to die Oct*obris*.
Joh*annes* Hert

[Journal 9, back of leaf 293 ink, 261 pencil, between an entry of 23 June, 6 Hen. VII (1491), and one of 6 Nov., 6 (*i.e.* 7) Hen. VII (1491), are lists of those Crafts who have paid their share of the cost of repairing the City Walls, and those who haven't. Among the latter are both the Barbers and the Surgeons.

Thise been the Craftes that haue doon) their Costes to the Reparacions of the walles.	Thise ben) the Craftes that must be desyrede to do theyr) Cost vppon) the Reparacion of the walles, And yit haue no thyng doon).
Mercers	Haberdasshers
Grocers	. . . . . .
Drapers	
Fisshmongers	Barbours ⎫
Goldsmythes	Surgeons ⎪
Taillours	Chesemongers ⎬]
Skynners	Stacyoners ⎪
	Vpholders ⎭

In 6 Henry VII (August 1490-1),—Jn. Mathewe, Mayor (Nov. 1490-1), —the following officers of the Barber-Surgeons, and Surgeons, were sworn (Journal 9, lf. 304 bk., 305 ink; 272 bk., 273 pencil):—

Joh*annes* Johnson, M*agister*
Jacob*us* Scot   } Gardiani Art*is* de barbo*urs*-Surgeons, Jur*ati*
Radu*lphus* Dowelle   12 die Sept*embris*
Nich*olas* Lyveryng

Will*elmus* Witwang
Rob*ertus* Taillour } Gard*iani* Art*is* Cirurgic*orum*, Jur*ati*, 4 die Oct*obris*
Th*omas* Ropesley
Nich*olas* Duraunt

App. XIV. *Barber-Surgeons the* 28*th Company.*

The Barber-Surgeons' right to the 17th Place in the Order of the City Companies.

Two years after the Barber-Surgeons had got their Ordinances revised and authorised by Sir Thomas More and his fellows, they claimd their old place of 17th Company in the City gatherings and processions, out of which they had been ousted; and it took them four years and a half to get the matter finally settled. The first document shows them 28th, in 1516; then they were 17th, then 18th; then they were stopt for a time; but at last they secured their old 17th place.

1516, Jan. 31. The Order of the City Companies in City Processions, &c.

(Letter-Book N, lf. 5, back.) Wille*l*mus Boteler, Maior.

Die Jouis, vlt*imo* Die Januar*ij* [7 Hen. VIII, A.D. 1516].

First, the disputes for precedence between the Salters and Ironmongers, and between the Shearmen and Dyers, are settled by declaring that the Salters shall go before the Ironmongers, and that the Ironmongers 'shall Charitably & louyngly Folowe next the' Salters; and that the Shearmen shall precede the Dyers, who 'shall Charitably & louyngly folowe next the' Shearmen. Then comes, on leaf 6, a cooler for the hot blood stirring in the Dyers:

Item, where the seyd Wardens of Dyers, this seyd Daye exp*ress*ely seid that they wold not goo in p*ro*cession, but absente theym Frome thens, Rather than they wold obey this Rule, Decree, & Jugment / Therfor nowe Iniunc*cio*n ys geuen to Jo*h*n Axe, & other his Felawes the Wardens, that they go to-morowe in the gene*r*all p*ro*cession accordyng⋅ to the order Abouetaken, vppon the payne of xx ti.

Then follows a General Order for all the Companies, putting the Barber-Surgeons 28th, instead of 17th, where they claimd of right to be (leaf 6).

Here Aft*er* ensuyth thorder & direc*cio*n taken at this Court by the Mayer & Aldremen aboueseid, of & for all the Craft*es* & Misteres ensuyng⋅, For their Goyng*es*, Aswell in all p*r*ocessions, as all other Goyng*es*, Standyng*es* [*leaf* 6, *back*] And Rydyng*es* for the busynessys & Causes of this Citie / The seyd order or direc*cio*n to be fromchensforth fermely obs*er*ued & kept / Eny other Rule, order, or direc*cio*n heretofore made to the Cont*r*ary, notwithstandyng⋅ / Prouided Alwayes, that the Felisshipe whereof the Mayer ys for the yere / Accordyng to the olde Custume, shall haue the p*r*eeminence in Goyng⋅ Afore All other Felisshippes, in all plac*es*, duryng⋅ the tyme of Mayralte, & c⋅

XIV. Barber-Surgeons' Company 28th or 17th.

Ordo processionum pro Misteris sequen*dis*.

Mercers		Bruers		Plummers	(32)
Grocers		Lethersellers		Inholders	
Drapers		Pewterers	(16)	Founders	
Fisshemongers	(4)	Cutlers		Pulters	
Goldsmythes		Fullers		Pastelers	(36)
Skynners and Tayl-		Bakers		Coupers	
ours Accordyng		Wexchaundelers	(20)	Tylers	
to thord*i*na*u*nce		Talughchaundelers		Bowyers	
therof made in the		Armorers		Flechers	(40)
tyme of M*aster*		Gurdelers		Blakesmythes	
Billesdon), in L,		Bochers	(24)	Joynours	
fol. 196		Sadelers		Weresellers	
Haberdasshers	(8)	Carpenters		Wevers	(44)
Salters		Cordeweners		Wolle pakkers	
Iremongers		Barbours	(28)	Sporiers	
Vynteners		Payntour Steynours		Felmongers	
Shermen)	(12)	Coriours		Fruterers	(48)
Dyers		Masons			

But early in 1532, the Barber-Surgeons have got their right old 17th place:

(Rep. 8, lf. 271, bk.) 4 Feb. 1532.

Also yt ys Agreed that, for diu*er*se Consideracions this Courte movyng, The Barbour Surgeons shall go in all processions.

On Feb. 9 (or 6[1]), lf. 272, bk., it is agreed and decreed that the Barbours Surgeons shall go 17th in all processions,

1 Mercers	10 Iremongers	
2 Grocers	11 Vynteners	
3 Drapers	12 Stokfysshmongers	
4 Fysshemongers	13 Clotheworkers	
5 Goldsmythes	14 Dyers	
6 Skynners	15 Brewers	
7 Merchauntayllo*u*rs	16 Lethersellers	Pewterers
8 Haberdasshers	17 Barbours Surgeons	Dyers
9 Salters		

and that at the next Assembly of the Livery, the Lord Mayor shall send one of his Serjeants to the Pewterers, to 'shewe theym that the seyd company of Barbours Surgeons be Restored ageyn) to their olde Rowme.'

[1] We think the clerk's *infrascripta* meant *suprascripta*; in which case, Feb. 6 is the date.

XIV. Barber-Surgeons' Petition for 17th Place.

But on May 13 'infrascripta' (or 8 'suprascripta'), 1532, lf. 287, bk. :

This day was made a Mocion to the Barbours Surgeons that they shuld be in the Rowme of the xviijth, Notwithstandyng the graunt made afore tyme therof to theym).

Next year the Barber-Surgeons petitiond the City Court to give them their old 17th place :—

1533, Feb. 4. The Barber-Surgeons' right to the 17th place in City Processions and Assemblies.

(Letter-Book O, lf. 213.) Pecok, Maior. [Nov. 1532-3.]

Memorandum, that the iiijth day of February, the xxiiij^{ti} yere of the Reigne of Kynge Henry the viijth, The Master, Wardens and Company of Barbours Surgeons of London), made humble sute and Request vnto the Right honourable sir Stephyn) Pecok, knyght, Mayre of the sayde Cytie, and hys worshipfull Bretherne Thaldremen) of the same, Shewyng & Alledging, that where they, the sayde Master, Wardeyns, and Company, yn thordre of goynges, standynges, Rydynges, syttynges, and other Assemblees of occupacions lawdablye vsed and contynued withyn this Cytye, for the worshippe of the same, haue vsed, and were wonte, tyme oute of mynde, to be taken) and accepted the xvijth Companye, tyll about xvjth yeres nowe passed / At whiche tyme, and alweys sythen) that tyme, they have been) putt farre back from theyr sayde Rowme and place accustumed, So that they be nowe the xxix or xxxth Companye yn thordre of suche goynges, Rydynges, standynges, syttynges, & other Assemblees, The cause whye, or by what occasyon), they been) nowe so vsed, they sayde they coulde not tell ; and prayed yn humble maner that yf no suche cause or occasyon) were/ That then) yt wolde please the sayde Mayre and Aldremen) to Restore and Admytt theym) vnto theyr sayde former place and Rowme of olde tyme Accustumed / Wherupon the sayde Mayr and Aldremen), consyderyng not onlye the sayd Request to be good and Reasonable, but also the good qualyties and humanytie whiche the sayde Companye have and shewe from) tyme to tyme yn

As the Barber-Surgeons have told the Lord Mayor and Aldermen

that their Company was always the 17th, till put back about 16 years ago,

and they have now askt for their old place;

the Court, considering the request reasonable,

and that the Barber-Surgeons have always paid their dues well,

264 App. XIV. Barber-Surgeons to be 17th Company.

Vide Jou*r*nal*em*
incipien*tem*
a *tem*po*re* Ed-
m*un*di Shaa
[A.D. 1482] in
fol*io* 18 ibidem,
et vlt*imo* fol*io*
eiusd*em*.

almaner Task*es*, contrybucyons, and other charges borne and leuyed of [and¹] among*es* the seuera# occupac*ions* of this Cytie, whery*n*) they be founde alweys Ryg#t tractable, redye and conformable / And also forasmoche as yt appereth by tholde Record*es* wit*h*yn this Cytie, that they have vsed to be yn the sayde xvij*th* Rowme, as on theyr behalf ys afore Alledged / Therfore, and for dyue*rs* other causes & consideracions theym) specyally mouyng / The vj*th* day of the sayde Moneth of February [1533], at and by A fu# Co*u*rte of Aldremen) then) beyng p*r*esent, the sayd lorde Mayr, M*as*ter John Baker, Recorder, s*i*r Wyll*iam* Butler, s*i*r Thomas Baldrye, s*i*r Nycolas Lamberd, knygh*tes* / M*as*ter John Hardye, M*as*ter John Champneys, M*as*ter Rafe Warre*n*), M*as*ter Wyll*iam* Forma*n*) / M*as*ter Wyll*iam* Roche, M*as*ter Wyll*iam* Denham, M*as*ter Migh*i*# Dorm*er*, M*as*ter Rychard Choppy*n*), M*as*ter Robert Paggett, And M*as*ter Water Champyo*n*), Aldremen), with good delyberacyo*n*) and aduyse-

the Court agrees

that the Barber-
Surgeons shall always
be the 17th Company,

after the Pewterers,
and before the Dyers
and Cutlers,

ment, fully Agreed and g*r*aunted, that fromhensforth, at a# tymes to co*m*me foreue*r*more / the Master, Wardeyns and Companye of the sayde Mysterye for the tyme beyng, shalbe accepted, taken) and Admytted the xvij*th* Companye, And so, at all tymes to co*m*me, sha# goo yn thordre of all suche goyng*es*, [*leaf* 213, *back*] Rydyng*es*, standyng*es*, syttyng*es* and other Co*m*en Assembl*es*, vsyng and contynewyng theyr sayde olde place and Rowme Accustumed, after thys maner and ordre ensuyng, that ys to wyte / Mercers / Grocers / Drapers / Fysshemongers / Goldsmythes / Skynners / merchaunttaylloúrs / Haberdasshers / Salters / Iremongers / Vynteners / Stokfysshemongers / Clothwerkers / ² Brewers / lethersellers / Pewte*rr*ers / Barboursurgeons / Dyers / Cutlers,² And so forth, by ordre, as more playnly yt appereth the last Daye of January, the vij yere of the Reigne of kyng Henry the viij*th*, yn the tyme of Mayraltye of s*i*r Wyll*iam* Butler, knyght, entred yn the booke of .N. folio Sexto.³ And to thentent that this p*r*esent g*r*aunte and Agreament sha# from) hensforth foreue*r*more stonde and be contynewed ferme and stable as concernyng the sayde Barbour-surgeons, the sayde Mayr & Aldremen) have com-

¹ MS of of. ²⁻² In a later hand, on an erasure.
³ This Order made the Barber-Surgeons 28t*h* instead of 17th.

App. XIV. Barber-Surgeons the 17th Company. 265

and that this Order shall be recorded in Letter-Book O, leaf 204. maunded yt here to be entred of Recorde yn the booke of O,[1] folio ij iiij, perpetuelly to be obserued and kept Accordynglye /—//

On Oct. 22, 1534 (Rep. 9, lf. 79), it was agreed that the above Act of 1533 should be 'vtterly Revoked, adnulled & repelyd,' and that an order made in the mayoralty of Sir Wm. Butler (Nov. 1515-16 : p. 261-2, above) as to the order of the Crafts in assemblies, &c. should be observd, so that the Barbers would be 28th again. And 'the Wardeyns of the mystery of the Barbour-surgeons of London' were 'orderd that theyre company shall no more goo yn processyons, standynges, Rydynges, goynges, & other assembles from hensfurth, tyll it be otherwyse orderd by thys courte.'

Nevertheless, on March 11, 1535 (Repertory 9, leaf ?) it is

'agreed that the sayd Company of barbours shalbe the xvij company, & immedyatly to goo afore ye companye of Cutlers, & after the Pewterers, as they be set yn order yn the tyme of ye mayoraltye of Master Butler, yn the Repertory N folio [6] vltimo die Januarii.'

But on March 16, 1535 (Letter-Book P, lf. 61), it was again agreed 'that the saide Company of Barbours shalbe the xvij Company, and ymmediatly to goo afore the Company of Cutlers / and after the pewterers.'

On July 29, 1535 (Rep. 9, lf. 118, and Letter-Book P, leaf 66, bk.), this last Act or Order is repeated, and the place of 'the barbour surgeons of london' settled as that of 'the xvij Company,' before the Cutlers and after the Pewterers. On Oct. 12, 1535 (Rep. 9, lf. 130, and Letter-Book P, lf. 71, bk.), 'the barbour surgeons' are again given the place of 'the sevyntenth company yn the order of the mysteryes of the companyes . . yn all theyre stondynges, goynges, Rydynges, & other comon assembles of thys Cytie.'

On Wednesday, Dec. 15, 1535 (Rep. 9, lf. 195, and Letter-Book P, lf. 78, bk.), a further Order again gives the barboursurgeons the 17th place : 1. Mercers, 2. Grocers, 3. Drapers, 4. Fysshemongers, 5. Goldsmythes, 6. Skynners, 7. Merchanttayllours, 8. Haberdasshers,

---

[1] On leaf 243 back, the Barbours have to provide 4 Bowmen to attend the Lord Mayor in the Watch of the Vigils of St. John and St. Peter. The 4 first Companies—Grocers, Mercers, Drapers, Goldsmiths—find 8 Bowmen each.

266 App. XIV. *Barber-Surgeons the 17th Company.*

9. Salters, 10. Iremongers, 11. Vynteners, 12. Clothworkers, 13. Brewers, 14. Lethersellers, 15. Pewterers, 16. Dyers, 17. Barbour-surgeons, 18. Cutlers, 'and so furthe, as apperyth yn the booke of O, fo. 204' (Rep. 9, lf. 195), p. 263-5, above. This Order was confirmd on March 30, 1536 (Rep. 9, lf. 166). Then on July 20, 1536 (Rep. 9, lf. 184), it was orderd that the Barber-Surgeons should have a new Boke or Charter made, under the Common Seal of the City,[1] granting them their 17th place for ever. This was duly made, and is enterd in full in Repertory 9, leaf 201 bk. to 203 bk. (headed 'xviij die Julij, anno 28 H. 8 [A.D. 1536], ante 184'), and in Letter-Book P, leaves 97 bk. to 98 front. It is dated Oct. 1, 28 Hen. VIII, A.D. 1536, the Order for sealing it with the Seal of Office having been made on Sept. 26 (Rep. 9, lf. 195). It is given under 'the seale of the offyce of Mayoraltye of the Cytie of London.'

1547. The Numbers of the Freemen of each Company who shall ride to meet K. Edward VI on his Coronation.

(Journal 9, lf. 18, bk.)

Commune Consilium tentum Die lune, xxj die Aprilis, Anno primo [Ed. VI, A.D. 1547].

[*Presentibus*] Maiore,[2] Recordatore, Tailour, Drope, Brome, Gardyner, Haryot, Stalbrow, W. Stokker, Hill, Billesden), Rawson), Colet, Warde, J. Stokker, Fisher, Tate, Hern), Pawson), Norlond, Nailer, Whit, Mathewe.
Concideratum est per Maiorem & Aldermannos, de qualibet Mistera subscripta, certe persone equitent erga Dominum Regem venientem ad Ciuitatem Londonie ad Coronacionem suam, indute Togis coloris Murrey [dark red].

---

[1] (Rep. 9, lf. 184) martis, xviij° die Julij, anno 28 H. 8 [1536] postea 201. 'Item that the Company of barbour-sugeons shall have a newe booke written, & the comen) seale of the Cytye to be setto the same. vide postea 201. scribe librum hic, vt intratur postea, fo. 201 /'
At the back of leaf 201 is the 'newe booke' written accordingly.
[2] Nov. 1546-7. Sir Henry Hobberthorne, Lord Mayor; Richard Jarveis and Thomas Curteis, Sheriffs.

XIV. City-Freemen at Edw. VI's Coronation. 267

numerus personarum eorundem subscribitur.

Goldsmythes ...	xxx	Wexchaundlers	iiij	Wolmen	vj
Mercers	xxx	Taloughchaundlers	vj	Plommers ...	ij
Drapers	xxx	Shermen)	viij	Stacioners ...	ij
Grocers	xxx	Fullers	viij	Founders... ...	iij
Fisshmongers ...	xxx	Gyrdellers ...	iiij	Paynters	ij
Skynners... ...	xx	Bochers	x	Staynours ...	ij
Salters	x	Bakers	vj	Wodemongers ...	ij
Vynters	viij	Bruers	x	Turnours	ij
Tailours	xxx	lepersellers ...	x	Curriours... ...	vj
Irmongers ...	x	Hurers¹	iij	Pulters	ij
Haberdasshers...	xx	Vphoiders ...	iiij	Pastelers	ij
Scryvaners ...	iiij	Cordewaners ...	iij	Coupers	ij
Diers	x	Joyners	ij	Wyremongers ...	ij
Peautrers... ...	vj	Masons	ij	Glasiers	ij
Cutlers	vj	Carpenters ...	ij	Tilers	iij
Sadlers	viij	Flecchers... ...	ij	lynyndrapers ...	ij
Barbours	viij	Bowiers	iij	Summa iiij x persones	
Armerers... ...	iiij	Inholders ...	iiij		

In the torn list of Companies or trades on the last page of Journal 9 (A.D. 1548), the names are not given in the order above, and the Surgeons are put before the Barbers. The complete names (after the torn ones) are Scryvaners, Diers, Peautrers, Cutlers, Surgions, Sadlers, Barbours, Armerers, Brasiers, Wexchaundlers. Glovers are put before the Hurers ; Coppersmiths follow the Founders ; Broiderers and Pouchemakers the Steynours. Between the Coupers and Wire-sellers (for Wyremongers) come the Greytawyers, Blaksmythes, Wevers, Sporiours, Lorymers, Horners; then the Lynyndrapers, Fuysters (saddle-tree makers), Fruterers, Chesemongers, Netters, Glasiers, Tapicers, Tylers, Felmongers, Whelwrightes, Shipwrightes, Pavyours, Corsers (horsedealers), P[astel]ers, Marblers.

On April 22, 1604 (Repertory 26, no. 2, leaf 327, 329 pencil, back), the Court orderd that—as the Stockfishmongers' Company had been 'wholly dissolved and abrogated, and noe Companye or corporacion [was] remayning within this Cittye, of that name,'—

yᵉ sayd Masters or governours of yᵉ sayd mesterye and Cominaltie of Barbors and Surgeons shall, from henceforth, be reputed, taken & placed, as yᵉ sixteenth Companye within this Cittye, in all their goinges, rydinges, sittinges, standinges and assemblies whatsoeuer.

---
¹ Makers of shabby caps, 'cappers & hurers.'

268 App. XIV. *Barber-Surgeons the 16th Company.*

This was to make amends for a snub to the dignity of the Barber-Surgeons five weeks before, when, though the Company was entitled to its old 17th place,—then practically the 16th,—

yet notwithstanding, at the royall passages of the king and quenes most excellent maiesties, and the Prince of Wales, attended by the nobilitye and gentrye of the land, through *this* Cittye on the xxth of march last past, through ignorance were misplaced by the Comittyes appointed by this Cittye for the managing of those affaires.

A FEW NOTES.

p. 64, *Zirbus.* See Lib. II, Cap. XIII, p. 78 of *Opera Chirurgica Ambrosii Paraei*, Frankfort, 1594. 'De Epiploo seu Omento, quod *Zirbum* etiam appellant. Post partes continentes, sequuntur contentæ, quarum prima est Epiploon, sic dictum, quòd intestinis omnibus innatet.' And on p. 79, in the references to the *icon* or woodcut, 'Omentum, seu *zirbum*, aut epiploon, in omnia intestina effusum, vnde & hoc epiploi nomen traxit.'

p. 80, *Perfection of the Fœtus.* For 18 and 46 days, Ambrose Paré allows 30 and 60 : *Op. Chirurg.* 1594, p. 667 :—'Cæterùm infans in vtero, vt ante trigesimum diem conformationem perfectam non adipiscitur, sic, non ante sexagesimum movetur : quod tempus sæpius etiam mulieres latet, propter motionis exilitatem.' He also insists that the soul comes to the fœtus, not from man, but from God, and quotes Augustine on the point. 'Itaque ab Adamo, aut parentibus, deriuari animam non est credendum : sed singulis momentis, & in ipso conformati fœtus articulo, à Deo creari, & in fœtum infundi.'

p. 153, *Gifts to Barts.* By his will of May 9, 1399, Thomas de Baumburgh, clerk, gives all his tenements in Holbourn to the Master and Brethren of the Hospital of S. Bartholomew de Smethefeld, for providing 2 Friars Regular of that order to celebrate divine service in the Hospital Church. See Dr. Reg-Sharpe's forthcoming *Calendar of the Wills in the Court of Husting*, Guildhall, London, Pt. I, p. 437. (A.D. 1888.)

p. 157, *Lazar Houses.* See the Order, Oct. 15 (3 Edw. VI), 1549, for yearly appointing Governors of them, in Letter-Book R, lf. 36, Guildhall Records.

p. 163, note 2. *The Plague of* 1563. Among those who must have died of it, and were buried at St. Giles's, Cripplegate, was the Rev. Richard Bullein, writer of a book on the Stone, brother of William Bullein, author of the *Bulwarke* 1563, *Dialogue* 1564, &c. *Dict. Nat. Biog.*, vii. 246/1. (William Bullein died in 1576.)

p. 177. *Archery*, 1633 Gerv. Markham. *Country Contentments*, p. 57.

The markes to shoote at are three, Buts, Prickes, or Roavers : the But is a levell Marke, and therefore would haue a strong Arrow with a very broad Feather ; The pricke is a marke of some compasse, yet most certaine in the Distance, therefore would haue nimble strong Arrowes with a middle Feather, all of one weight and flying ; and the Roaver is a marke incertaine, sometimes long, sometimes short, and therefore must haue arrowes lighter, or heavier, according unto the distance of place.

p. 188, *Vigo.* 'Other haue at hand, maister *Vygos* boke of *Chirurgj,* where ye shall finde, euen to the full, how to purge an humour. 1562-3, W. Bullein. *Bulwarke : Sorenes and Chyrurgi.* Fol. xxx.

XV.

THE

𝔄ncient 𝔒rdinary

OF THE

BARBERS AND SURGEONS OF YORK,

A.D. 1486,

AS REVISED AND AUGMENTED A.D. 1592;

TOGETHER WITH THE FRESH ORDINANCES OF 1614 AS TO THE MASTER OF ANATOMY, DISSECTIONS, READING OF LECTURES, ETC. ETC.

from the Egerton MS. 2572, in the British Museum.

[*inside the fly-leaf*]

[A.D. 1697.]

Civitatis Ebor- Ad Generalem Quarterialem Sessionem Pacis Domini
aci Sessio nostri Regis, tentam per Adjornamentum pro Civitate
Eboraci et Comite ejusdem Civitatis, apud Guildhall
in eadem Civitate, die Martis vltimo (?), xij die
Octobris, Anno Domini 1697, Coram Marco Gill,
Majore Civitatis Eboraci, Georgio Prickett, Serviente
ad Legem, Recordatore ejusdem Civitatis, Gilberto
Metcalfe, Militi, Ricardo Wynn, Armigero, de Consilio cum Civitate predicta, Johanne Foster, Samuel
Dawson, Georgio Stockton, Andrea Perrott, Roberto
Davy, et Rogero Shackleton, Aldermannijs, Custodi-
bus Pacis et Iusticiae dicti Domini Regis, ad pacem
conservandam assignatis, &c.

Ordered, that Mr Thomas Cundall and Mr John Gowland, Searchers for the Company of Barbers, doe give Notice to my Lord Major of the names of such persons of that Company as doe Shave or Trimm on the Sabbath days; And that they give Notice to the Company to forbeare to doe it, As they and the Company will Answer the Contrary :

per Curiam,
Tho. Mabe, deputatus Communis Clerici.

The contents of all y͡e Articles in this Ordinary. [leaf 3]

[A.D. 1592]

		PAG
1	Imprimis, The Election & Acoumpts of y͡e Searchers	27͡
2	Straungers to be Contributo*u*rs to the Company	274
3	Obstinate & Disobedient persons that will not come to y͡e Hall	274
4	Noe Master to take another Brothers Apprentice	274
5	Every Master new setting upp, to be Searched	274
6	Noe Brother of y͡e Company to Trimm on y͡e Sabbaoth Day	274
7	Searchers may take away Basings & Signes	275
8	None to practize Chyrurgery but vnder a Master	275
9	Noe Servant to worke vnsearched[2]	275
10	[1]Noe Apprentice to be taken for any lesse then Seaven yeares	275
11	Servants & Apprentices not to be Purloine*r*s	275
12	Noe Strange*r* to practize above five Dayes	276
13	Strangers founde faulty, to be fineable	276
14	Straungers to be Searched & to be Contributory	276
15	Noe Brother to take in hand to Deale with anothers Cure	276
16	Masters of y͡e Arte may Search all Cures	276
17	Misbehaviour one to another	276
18	Assembleing or meeting at y͡e Hall without their Gownes	277
19	Goeing to Tavernes or Alehouses on y͡e Sabbaoth Day	277
20	None to Intrude into anothers Cure; neither any Barbor to trymm any othe*r*s custome*r*	277
21	Every Brother to make a Dinner at his first being Searcher	277
22	Orders to be observed att y͡e Buriall of a Brother	277
23	Payments to be made quarterly, & Recording of Appre*n*tices	277
24	Every Master to pay att taking his Oath	277
25	Journeymen to pay Quarterly	277
26	Fee, for makeing Indentures	278
27	Inde*n*tures to be Enrolled	278
28	Noe man to be Admitted into y͡e Company before he be freed before my Lord Mayor	278
29	[1]Noe Master to haue any moe Appre*n*tices then one; & his first to be a freemans sonne	278

[1] 'Repealed' written in the margin.
[2] That is, all Assistants must be examind.

XV. Contents of all yᵉ Articles in this Ordinary.

[later]
(30) The antient Head Searcher to make a Dinner for the Company on the Election Day— 278

Additions and Alterations. [leaf 3, back]

[A.D. 1614]
(31) A Master in Anatomy to be chosen yearly 279
(32) Such Master to be a licensed Surgeon, & to read Lectures 279
(33) The whole Company to meet at every Dissection ... 280
(34) The Master to appoint who shall dissect 280
(35) And to describe to them the Part 280
(36) The Master and Searchers to examine strange Surgeons ... 280
(37) Every one of the Company professing Surgery, to read a Lecture in Surgery or Anatomy, if required 280
(38) Every new admitted Surgeon the like 281
(39) All Surgeons in the City to become free of the Company 281
(40) Penalty on unlicensed & unskilful Practicers ... 281
(41) The like on Persons employing them 281
(42) The Master in Anatomy Lecturer to have Precedency of the Searchers 282

[A.D. 1676]
(43) 10s. Penalty for shaving on the Lord's Day 282
(44) Searchers not to spend or waste the Company's Stock ... 283

[A.D. 1679]
(45) The Company to have all Fines and Forfeitures ... 283

[A.D. 1683]
(46a) The 10ᵗʰ & 29ᵗʰ Articles are vacated 284
(46b) None to take Apprentice for less than seven years ... 284
(47a) If any Master haue more than one Apprentice at a Time, one of them to be a Freemans Son 284
(47b) ¹The first Apprentice taken to be a Freeman's Son ... 284
(47c) The Company not to compound Fines above 6s. 8d. without Consent of the Corporation 284

[A.D. 1757]
(48a) The Order as to taking a freemans Son the first Apprentice repealed—And 284
(48b) 20s. Penalty for taking an Unfreeman's Son Apprentice 285

[A.D. 1768]
(49) Sixpence (instead of 3s. 4d.) Penalty for not attending Meetings 286

[Not dated]
(50) ²Agreement by each Member of the Company that, on Breach of any Article, the Forfeiture may be levied by Distress and Sale of his Goods 287

¹ '*Repealed*' written in margin.   ² leaf 4.

XV. York Barber-Surgeons' Ordinary, 1592.

[leaf 5]

This booke made in the yere of our lorde god A M CCCC lxxxvj, In the Seconde yeare of the Reigne of Kinge Henrye the Vij°; beinge Maior of this Cittie, William Chymney; Searchers that yeare, viz.

Adam Sigeswithe & George Kylede.

[*Oath of the Barber-Surgeons*]

Ye shall Sweare to bee trustie and trewe vnto the kinge our Sovereigne Lord, And to this Cittie of York, And also to the Science of Barbars & Chyreurgions within the same. And all good ordinances, statutes, vsages, and accustomes, heretofore made and vsed in the same arte or Science, ye shall kepe, supporte, and maynteine att all tymes to your power; and the secretes and counsell of the same arte, ye shall trewlie kepe and Layne,[1] So helpe yowe god, and by the contentes of this Booke.[2]

[leaf 14, back]

This Booke corrected and Augmented in ye yeare of our Lorde god 1592, in the xxxiij° yeare of the Reigne of our Soueraigne Lady Elizabethe, the Quenes maiestie that nowe is:
Thomas Harryson, Lorde Maiour the Seconde tyme;
Henrye Leache, and } Serchers[3]
George Dunnynge   } this yeare
This done att the costes and charges of the wholle companye.

The Auntiente Ordinarye of the Barbors and Surgions of this Cittie, att the requeste of the wholl companye, newlye perused, reformed, and Augmented, and this presente xxiij° daye of Iune, 1592, ratyfied, established & confirmed, to be from henceforthe obserued & kept, as hereafter is mencyoned.[4]

[leaf 14, back]
per ordines in
libro actium xiiij°
& xxiij° diebus
Iunij de Anno
predicto, scillicet
1592

---

[1] Conceal.

[2] On leaf 6 is a painting of the Barber-Surgeons Arms, the Barbers and Surgeons Quarterly, like those of London, with a lion or, on a red cross dividing the quarters. Underneath, the London Company's motto, 'De prescientia Dei.' On leaf 7 is a careful painting of Henry VII; on leaf 8 one of Henry VIII; on leaf 9, one of Edward VI; on leaf 10, one of Queen Mary; on leaf 11, one of Elizabeth; on leaf 12, one of James I; on leaf 13, a less careful one of Charles I; on leaf 14, one of Charles II; on leaf 14 back, the text begins again. Most of these Portraits are extremely well done.

[3] Examiners.

[4] Portraits of James II. (leaf 15) and of William and Mary (leaf 16) take up the next two leaves.

VICARY.

274   XV.   *York Barber-Surgeons' Ordinary*, 1592.

[leaf 17]
The election and accomptes of the Searchers.

1 Inpr*i*mis, that y*e* S*er*chers and Maisters of the saide arte or science be chosen euerye yeare vpon the Mondaye nexte after the feaste of the Natyvitie of S*ainc*t Iohn Baptiste : and the same Mondaye the Searchers of the yeare before, there to render vp theire accompt*es* vnto the Maisters of the saide arte, of all thinges belonginge to them, vpon payne of vj s viij d to the chamber and companye.

Straungers to be contributors.

2 Item, that all Aliaunt*es* and Straungers that vses the arte or Science of Phisicke or Chireurgerie within this Cittye, and takes moneye for the same, to be contributorie to the companie of the same arte, yearelie vj s viij d, to be paide to the Searchers of the same companye for the tyme beinge, in manner and forme aforesaide.

[leaf 17, back]
Obstinate and disobediente parsons.
Altered by Order in folio 24.

3 Item, If anye man of the saide arte be founde obstynate, and will not come to the hall of theire assemblie, beinge lawfullye warned by the Searchers or theire deputie, or els aske leeve of the searchers, or the one of them, upon lawfull busynes, shall forfett to y*e* companye iij s iiij d, to be deuided in manner and forme aforesaide.

No maister to take an other brothers aprentice.

4 Item, if any M*aiste*r of the saide arte, receyue or take into his service, anye aprentice or servante of any other M*aiste*r, vnto[1] suche tyme and tearme betwixte them agreed, be fullye ended, the offender so convicted herein shall forfett, as is aforesaide, to the chamber and companye vj s viij d.

Euerie M*aiste*r setting vpp new, to be searched.

[² leaf 19]

5 Item, that everie man of the saide arte, when he firste settes vp, to kepe shoppe as a m*aiste*r, shall be first a fre man of this Cittie, and then searched by the Searchers of the saide arte, whether he be able to[2] [³]occupie as a M*aiste*r or no ; And if the Searchers approue him able, then att the firste settinge vp as a M*aiste*r in the arte, he shall paye xiij s iiij d in manner and forme abouesaide (excepte 'the sonnes of franchesed men). And if he be founde vnable, then he shall serue suche a conveniente tyme withe some brother of the said Science, as shalbe appointed and sett downe by the Searchers of y*e* companye for the tyme beinge.

6 Item, it is ordered and set downe that none of the saide Barbors shall worke or kepe open theire shoppe

---

[1] until.   [2] On leaf 18 is a Portrait of Queen Anne.

XV. York Barber-Surgeons' Ordinary, 1592.

vpon the Saboathe daye (exceptinge tuo sondayes nexte before the assize weekes, nor affter, in this Cittye; and if any Barbour Presume to do the contrarye, for euerye tyme so founde [he] shall forfet x s to the vses, as is aforesaide. *No brother of the companie to work upon the Saboathe daye.*

Item, if anie man, after his yeares of aprentishippe be expired, do presume to sett up as A Maister, not beinge admitted of the Searchers of that companie, it shall be lawfull for the saide Searchers to take awaye his Basinges, or other signes whiche he hathe towards the strete to shewe his arte, and to carrye them to the chamber on owsebridge[1] to the then Lorde Maior, and to paye suche fyne as the saide Lorde Maior shall set downe, to the vses aforesaide. **7** [leaf 19, back] *Searchers to take awaye Basinges & Signes.*

Item, that no person or parsons within this Cittie, or Suburbes of the same, practizinge Chierurgerye, or drawinge of teethe, or anye other thinge belonginge to the saide arte, vnles theye be vnder the gouernance of A maister, ahd approued able to vse and occupie the saide arte; and if anye of them do the contrarie to this ordinarye, and be convicted vpon the same, [he] shall forfett and paye vj s viij d to be equallye devided as is aforesaide.[2] **8** *None to practize Surgerie but under A Maister.*

Item, that no Maister of the saide arte, hier,[3] or sett to worke in his howse, any seruauntes to occupie in ye saide arte aboue the space of vj° daies, vnles the Serchers for the tyme beinge have Serched the saide servante, and so licensed by the saide Serchers, vpon payne or forfeyture of vj s viij d, to be paide as is aforesaide. **9** [leaf 21] *No servante[4] to worke vnsearched.*

Item, that none of ye saide arte shall take anye aprentice for lesse tearme than vij° yeares; and that to be done by Indentures, and recorded by the clarke of our companye, vpon payne or forfeyture of vj s viij d; and the saide Indentures to be made (within viij° dayes after the takinge of the saide aprentice) by our Clarke, vpon payne and forfeyture of the some aforesaide, and devided as is aforesaide. **10** *No aprentice to be takne for anie Less tearme then vij° yeares; and ye Indentures to be made by our clarcke. Vacated by Order in Folio 23.*

Item, if anie servante or aprentice do purloyne or stealle from his Maister, anye of his goodes, to the value of vj d, the offender so convicted, shall be clearlie dis- **11** *Seruaunts & Apprentices not to be purloiners.*

[1] Bridge over the river Ouse at York.
[2] Leaf 20, Portrait of George I.   [3] hire.
[4] Assistant.   See p. 190, 208, and p. 271, note 2, above.

T 2

XV. *York Barber-Surgeons' Ordinary*, 1592.

chargded forthe of the saide companye for euer, at the discretion of the then Lorde maior.

[leaf 21, back] 12 **Item,** that no alianntes nor stranngers come into the saide Cittie to exercise the arte of Chireurgerie, or other thinges belonginge to the Barbors, Presume to occupie the same (not admitted by the saide Searchers) over the space of v° daies: whiche fyve daies beinge expired, for euerye daye after, the offender so convicted shall forfett and paye ij s euerie daye, as is in forme aforesaide.

No strannger to exercise aboue v° dayes.

13 **Item,** that all suche aliantes and straungers beinge founde withe a faulte by the saide searchers in the saide arte, shall be fyneable accordinge to the ordinannces and Statutes made in the saide arte.

Stranngers founde faltie.

14 **Item,** that the Searchers of the same arte of Barbors and Chirerugions [so] for the tyme beinge, shall haue full power att all tymes to searche all[1] [2]manner of cures which the saide Aliauntes and Strangers shall haue in hande, remayninge and abidinge within this Cittye, or the libertyes thereof. And also that all suche Aliantes shall be contributors to all manner of charges belonginge to the saide arte.

Straungers to be searched & to be contributarie.
[2 leaf 23]

15 **Item,** that no Maister of the Arte, or his Seruauntes,[3] shall dresse the patient of any other Maister, vntill suche tyme as he whiche haithe the patiente alreadye in hande to cure, be fullye satisfied, contented, and agreed, with-all; vpon forfeyture and payemente of xiij s iiij d, as aforesaide.

No brother to take in hand to deale with an-others cure.

16 **Item,** that the Barbors and Chireurgions of this Cittie, shall haue power att all tymes, & especiallye ye Searchers, to searche all cures whatsoeuer. And if anye Maister of the saide arte be requested or commaunded by anye aucthoritie to searche, then shall he [4]make it knowne to the searchers, and to haue their assistance; and if anye of the arte do contrarye to this ordinarie, [he] shall forfett to ye chamber and companye, vj s viij d.

To searche all Cures.
[4 leaf 23, back]

17 **Item,** if anye brother of this companie, att the tyme or place of our assemblie, or anye other place elsewhere, do vtter or giue anye vndecente wordes, to the searchers, or to anye brother of the saide companie,— but orderlye vse them, accordinge as they oughte to do, —whosoeuer shall offende herein, shall forfett and paye iij s iiij d to the vses aforesaide.

Misbehavioure one to another.

---

[1] **Leaf 22, Portrait of George the Second.**   [3] **Assistants.**

XV. York Barber-Surgeons' Ordinary, 1592. 277

Item, if any brother of the saide companye do come to the hall att anye tyme, that is, or hathe bene searchers of the companye havinge gownes, and comethe without them, [he] shall forfett and paye for euerie offence, vj d, to the vse of the saide companye onelye.[1] — 18 Assemblinge or metinge att hall without theire gownes.

Item, that none of the saide companie shall resorte to anie Inne, Tauerne, or ailehowse, vpon the Saboathe daye or other holidaye, in tyme of devyne service or sermon, vpon payne of euerye one offendinge, xij d; thone halfe to the comon chamber, and thother halfe to the presentor.[2] — 19 [leaf 25] Tavernes or Aile-house[s.]

Item, that none of the saide companie, intrude hym selfe into y*e* companye of anye other brother beinge dressinge of anye patient, either wounded or hurte, excepte he be speciallie requested by the patiente or by some frende of his, vpon payne of vj s viij d to the vses as aforesaide. And also that no Barbor shall powle, tryme, or shave, anie of his brothers customers, vntill suche tyme as the saide brother be fullie contented and paide; vpon payne and forfeyture of the some aboue saide, conteyned in this article. — 20 None to intrude into an others cure, neither anie Barbor to receiue another brothers customer.

Item, that euerie M*aiste*r, at his firste beinge searcher shall make the companye a dynner, and shall paye att the same tyme towardes the encrease of the Stocke, v s, as, accordinge to auntiente custome, hearetofore hathe bene vsed. — 21 [leaf 25, back] Euerie brother to make a dynner at his firste being Searcher.

Item, it is agreed that, att the buriall of anie brother, the whole companye to be there. And if anye be absente, beinge lawfullye warned, and haue not A lawfull excuse, [he] shall forfett and paye iij s iiij d in forme as is aforesaide. — 22 Orders to be obserued at the buriall of A Brother.

Item, it is agreed by the Barbors and Chireurgions, that euerie one of them shall paye quarterly iij d towardes the encrease of the Stocke; And also att the recordinge of anie aprentice into our ordinarie, xij d. — 23 Payment*es* to be made quarterlie; and recordinge of aprentices.

Item, that euerie one of the saide arte beinge allowed M*aiste*r by the Searchers and company, shall paye, att the receyvinge of his oathe, xij d. — 24 [leaf 26] paye at taking oathe.

Item, if anye M*aiste*r of the saide companye sett anye seruante on worke, beinge not prentice within this Cittie, that saide servante or Iourneye-man, shall paye quarterlye to the saide companye, vj d. — 25 Iourneymen to paye.

[1] Leaf 24, Portrait of George III. [2] informer, complainant

XV. York Barber-Surgeons' Ordinary, 1592.

Fee for Indentures making.

26 Item, it is agreed amongste our whole companye, that our clarck, Iohn Rawden, shall haue the makinge of all Indentures for aprentices within our companye; and to haue for euerie paire, xxxv [s], and for his yearelye waiges, x s.

[leaf 26, back] *Indentures to be enrolled.*

27 Item, that euerie Maister shall enrolle the Indentures of his aprentice in the comon clarkes office, within one moncthe nexte after the takinge of the same aprentice; and shall paye for the same, viij d, to thuse of the comon chamber and the saide comon clarcke, to be equallye deuided; vpon payne of euerie one makinge defaulte, to forfett for euerie offence vj s viij d; thone halfe to the comon chamber, and thother halfe to thuse of the saide companye.

No man to be Admitted into the companie, before he be freed before the Lorde Maior.

28 Item, it is agreed that y⁵ Serchers of the saide companie shall not admytt, nor receyue, anye person to be a fre brother of the saide companye, before the same person be made a freman of this Cittie, and do showe the coppie of his fraunchessed oathe under the Clarckes hande vnto the same Searchers; vpon payne of the saide Searchers admyttinge or allowinge any suche, contrarie to thintente and meaninge of this order, to forfett for euerie person so admitted or allowed, iij li vj s viij d to the comon chamber.

[leaf 27] *No maister to haue or take any mo apprentices then one at once at his first settinge up as maister; and that same one to be the sonne of A freman.*

[later] Vacated by Order in folio 23.

29 Item, it is Agreed that euerie Maister of the Companye nowe beinge, or which heareafter shalbe, havinge more Apprentices then one at once, at anye tyme or tymes heareafter, shall alwayes haue A fremans sonne one of the same apprentices; and that euerie Maister of the companye which shall heareafter newlye sett vp, shall take to his firste Apprentice, A fremans sonne; vpon payne that euerye Maister doinge contrarye, shall forfett for euerye tyme so doinge, iij li vj s viij d, to be paide, thone halfe to the comon chamber, and thother halfe to the saide companye. Prouided that euerye Maister whiche att this presente hathe two or more apprentices, maye kepe the same vntill theire tearmes be expired, So as he take no other apprentice in the meane tyme, contrarye to this order.

[*In a somewhat later hand and ink.*]

[30] *On Election-Day,*

Item, it is agreed by a generall consente of the companye of Barbor-Surgions, that from henceforthe the Auntiente heade Searcher, vpon the Election daye,

XV. York Barber-Surgeons' Ordinances, 1614. 279

shall make the whole companye A dynner ; and euerie person payinge vj d a pece of there owne chardge ; and the Surplussage (yf anye suche be) to be paide out of the Stocke. _{the eldest Searcher shall give the Company a Dinner, every one paying 6d. for it.}

The Fresh Ordinances of 1614, as to the Master of Anatomy, Dissections, Reading of Lectures, &c.

[leaf 27, back] In Camera Consilii Super pontem vse,[1] Civitatis Eboraci, octavo die Iunii, 1614.
tempore Maioratus Leonardi Besson, Maior Ciuitatis predicte.

Inprimis, that the companye of Chirurgions, euerye yeare shall choise one of the saide companye to be the Maister in Anatomie ; which saide Maister shall haue the disposinge of all thinges belonginge to the saide Anatomie, as also the kepinge of all thinges perteyninge to the dissection of the same ; and to make accompte of those things at the endinge of his yeare, and to delyuer them up to the companye, and theye to the nexte Maister elected.   1   [31] A Master of Anatomy shall be elected yearly, who shall take charge of the Dissecting Instruments, &c.

Item, the Maister so chosen, shall be A licenced Chirurgion ; and twyce in the tearme of the saide yeare, the saide Maister shall reade a lecture, either in Anatomie or Chirurgerie ; and if he so refuse to do, he shall paye for euerye suche refusall, x s to the use of the Lorde Maior and cominaltye of the saide Cittye, to be levyed by distresse, or to be recouered by accion of debte by the towne Clarcke of the saide Cittie for the time beinge, in the Kinges Maiesties courte to be *Holden before the Sheriffes of the saide Cittie, wherein no Essoigne or wager of lawe[2] shalbe allowed for the defendant.   2   [32] This Master shall be a Licenst Surgeon, and give 2 Lectures a year on Anatomy or Surgery.

[* leaf 28]

---

[1] The River Ouse.
[2] *Essoin* (*Essonium*, Fr. *Essoine*), Signifies an Excuse for him that is summoned to appear and answer to an Action, or to perform Suit to a Court-Baron, &c., by Reason of Sickness and Infirmity, or other just Cause of Absence. It is a kind of Imparlance, or craving of a longer Time, that lies in Real, Personal and Mix'd Actions.—1744. Jacob, *Law Dict.*, ed. 5.

Wager of Law : by this, a Debtor who swore that he owed his Creditor nothing, and also got 6 friends to swear that they believd him, got clear of any debt not witnest by deed or record. Says Jacob, "The Manner of *Waging Law* is thus : He that is to do it [the Debtor], must bring six Compurgators with him into Court, and stand at the End of the Bar towards the Right-hand

[33] 3 Item, that att euerye dissection, ye whole companye
The whole Company shall attend every Dissection.
shall meete; and those that shall either willinglye or wilfullye at anye tyme, (if in anye sorte he professe Chirurgerye) absent them selues, not havinge a reasonable excuse, shalbe fyned for euerye defaulte iij s iiij d to thuse afore saide, and to be levyed and recouered in manner and forme aforesaide.

[34] 4 Item, the saide M*aiste*r att euerye dissection, shall appointe such of the licenced Chirurgions as he shall like best of, to dissecte the saide Anatomy; and if theye so refuse to do, to paye for everye tyme theye so denye, v s. to thuse aforesaide, and to be levyed and recouered in manner and forme aforesaide.
The Master of Anatomy shall appoint Dissectors.

[35] [leaf 28, back] 5 Item, the saide M*aiste*r shall describe to such as he shall appointe to dissect (if they be vnskillfull in ye dissection of that part) the risinge, circumference, site, and insertion of the saide parte; wh*i*ch if he do not, they requestinge him therevnto, he shall paye iij s iiij d to thuse aforesaide, and to be levyed and recovered in manner and forme aforesaide.
The Master shall describe the 'Part,' to unskild Surgeons.

[36] 6 Item, that the saide M*aiste*r, and twoe Searchers for the tyme beinge, shall call before them (havinge suche other companye as they thincke fitt to assiste them) all suche as be Straungers, and others vnlicenced, practizinge Chirurgerie within the Cittie of Yorke, to examyno them; and findinge them insufficient, or refusinge te be examyned, to forfett and paye for euerye tyme offendinge, contrarie to the effecte of this ordynance, xx s to thuse aforesaide, and to be levyed and recouered in manner and forme aforesaide.
He and 2 Searchers (or Examiners) shall examine unlicenst practitioners, and fine the incapable ones.

[37] [leaf 29] 7 Item, euerye one of the saide companye professinge Chirurgerie, shall reade a Lecture, either in Chirurgerye or Anatomie, to the whole companye, out of
Every Surgeon shall read a Lecture on

of the Chief Justice; and the Secondary asks him, whether he will *wage his Law?* If he answers that he will, the Judges admonish him to be well advised, and tell him the Danger of taking a false Oath; and if he still persists, the Secondary says, and he that *Wageth his Law* repeats after him: *Hear this, ye Justices, that I* A. B. *do not owe to* C. D. *the Sum of,* &c., nor *any Penny thereof in Manner and Form as the said* C. D. *hath declared against me: So help me God.* Though before he takes the Oath, the Plaintiff is called by the Crier thrice; and if he do not appear, he becomes nonsuited, and then the Defendant goes quit without taking his Oath; and if he appear, and the Defendant swears that he owes the Plaintiff nothing, and the Compurgators do give it upon Oath that they believe he swears true, the Plaintiff is barred for ever; for when a Person has *waged his Law*, it is as much as if a Verdict has passed against the Plaintiff.—1744. *Law Dict.* This *Wager of Law* was 'abused by the Iniquity of the Times,' and was therefore done away with.

XV. York Barber-Surgeons' Ordinances, 1614. 281

Some Aucthor in Chirurgerye or Anatomye, as shalbe appointed by the Maister of Anatomie, and of one of the Searchers, beinge a licenced Chirurgion, whiche if he refuse to do (havinge had reasonable warninge to prouide for the readinge of the saide lecture), from suche tyme not to practize the arte of Chirurgerie, till he performe the readinge of the same lecture, vpon payne to forfett and paye for euerie tyme not readinge a lecture as aforesaide, xx s to thuse aforesaide, and to be levyed and recouered in manner and forme aforesaide. *Surgery or Anatomy to the whole Company. If he refuses, he shall be suspended till he lectures.*

[38]

Item, euerye Chirurgion, within A monethe after he is made free, shall likewise reade a lecture vnto the whole companye, out of some Aucthor, either in Chirurgerie or Anatomie, as shalbe appointed vnto him by the Maister and one of the Searchers, beinge a licenced Chirurgion, vpon payne to forfett and paye for not readinge thereof, xx s to thuse aforesaide, and to be levied and recovered in manner and forme aforesaide. 8 [leaf 29, back] *Every Surgeon made a freeman, shall, within a month, read a Lecture on Surgery or Anatomy.*

Item, that euerie one professinge Chirurgerie, and livinge within this Cittie, or others cominge to this Cittie, beinge licenced or otherwise, shall either become fremen of the saide Cittie and companye, within thre monethes after there saide cominge, or els to avoide the Cittie; and to paye for euerye monethe after remayninge in this Cittie, and practizinge Chirurgerie, xl s to thuse aforesaide, and to be levyed and recouered in manner and forme aforesaide. 9 [39] *All Surgeons in York shall join the Company or leave the City. [MS.] Professors of Chirurgions, to avoyd, or to become free within three monthes!*

[40]

Item, that none vnlicenced, or suche as can giue no reason for the cure theye vndertake, as to haue knowledge of the causes and signes thereof, or none that vnderstande not the vertues of suche medicines as they applie, whether theye be simple or compounde, takinge moneye for theire medicines, shall practize Chirurgerie, vpon payne to forfett for euerye tyme they shall practize Chirurgerie within this Cittie, xx s to thuse aforesaide, and to be levyed and recouered in manner and forme aforesaide. 10 [leaf 30] *No unlicenst or ignorant man who takes money for medicines, shall practise Surgery.*

Item, that euerie freman or woman of this Cittie, either takinge, or vsinge or sufferinge theire children or servantes to take or vse the counsell or helpe of anye straunger, or anye other vnworthie professor, or vnlicenced Chirurgion, havinge not firste had and vsed the counsell and helpe of the fre licenced Chirurgions of this Cittie (Bone-Setters excepted) shall forfett for euerye tyme so doinge, xl s to the vse aforesaid, and to be levied as aforesaide, &c. 11 [41] *Every person going to an unlicenst Surgeon before consulting a free licenst one, shall forfeit 40s.*

[*The* **Master** in Anatomy is to take precedence of the Searchers.]

[ƒ. 30, bk]

xiij° daye of Septembre
Anno Domini 1614.

[42]

In Camera Consilij super pontem Vse, Ciuitatis Eboraci,
Coram Leonardo Besson, Maiore Ciuitatis Eboraci,
Aldermannis & alijs.

As the Master in Anatomy,

and the Searchers, dispute who shall have precedence,

This Court orders

that the Master of Anatomy shall have it.

And whereas there is at this presente, controuersie arisen betwene the Master in Anatomye, Lecturer, on thone partye, and the Searchers of the Companye of Barbour-Surgions on thother partye, wheather the saide Master, or the Searchers of the same companye for the tyme beinge, sholde, in all the assemblies of the saide companyes, have the place or precedencye; vpon consideration had by this courte, it is thoughte mete, and so ordered by the saide Lord Maior, Aldermen, Sheriffes, and pryvye Counsell of the saide Cittie, that, for the endinge of the same Varyaunce, the saide Master in Anatomye, Lecturer, shall, as it is verye fittinge, have the place or precedencye of the Searchers of the same companye for the tyme being, in all there assemblies.

per me, Willelmum Scott, Communem
Clericum Ciuitatis predicte.

[leaf 31]

Att the councell Chamber
on Ouze bridge, ye xxth of June, Anno Domini 1676.

In the maioralty of the Right honourable Yorke Horner, lord Maior of the Citty of Yorke.

[43]
Whereas Barber-Surgeons have been shaving and cutting hair on the Lord's day,

We order, that if any of them do it hereafter, in any place, public or private,

This Court, takinge notice of seuerall irregular and vnreasonable practices committed by the Company of Barbor-Chirurgions within this Citty, in Shavinge, trimminge and cuttinge of Seuerall Straingers, as well as Cittizens, haire and faces vpon the Lords day, which ought to bee kept sacred, Itt is ordered by the whole consent of this Court; That if any Brother of the said Company shall att any time hereafter, either by himselfe, Servant, or Substitute, tonse, barbe, or trim any person on the Lords day, in any Inn, or other publique or private house or place; or shall goe in or out of any such house or place on ye said Day, with instruments vsed for that purpose, albeit the same cannott bee

App. XV. *Ordinance against Sunday-Shaving.* 283

positively proued, or made appeare; butt in case yᵉ
Lord Maior for yᵉ time beinge shall, vppon good cir- (of which the Lord Mayor shall
cumstaunces, conceive and adiudge any such Brother to judge)
haue trimmed or barbed (as is aforesaid); that then
euery *such offender shall forfeite, and pay for euery [* leaf 31, back]
such offence, the summe of Ten shillings; yᵉ moyty he shall be fined
thereof to yᵉ Lord Maior, and the other to th'use of 10s.
the said Company; vnlesse such Brother shall voluntarily purge himselfe by oath to the contrary: and the
Searchers of the said Company for the Time beinge, And the Com-
are to make diligent search in all such publique & pany's Searchers are to look up
private houses as aforesaid, for discovery of such offenders.
offenders.

per me, Willelmum
Kitchingman, Clericum Communem
Civitatis Eboraci.

29ᵗʰ September 1676 / [leaf 32] [44]

Item, that noe searcher of the Companie shall here- Searchers not to
after spend or waist the moneye or stock belong- pany's money in
ing to the said Company, Comitted to his keeping, feasting.
either in ffeasting or any other way, but onely as it
shall be Judged ffitt by the Generall consent of the
whole or Major parte of the Company; & that every
Searcher soe offending, shall be lyable to pay all debts
Contracted over and aboue what the said stocke will
discharge: / not Exceeding the summe of Three pounds.

[Ordinances of 1679 as to the Company
keeping Fees.]

xxvjᵒ Junij, Anno Domini, 1679.
Richard Shaw, lord Mayor. [45]

Ordered, with Consent of the said Company, that Searchers (on paying 10s. a year)
the Searchers thereof for the time beinge (vpon con-
sideration of payinge the yearly Composition of Ten
Shillings of lawfull English mony to the Mayor and
Comonalty of this City) doe from henceforth Take and may take all small Fines, for
receive to the vse of the said Company, all such fines the Company's use.
and forfeitures as shall hereafter become due by breach
of any Artickle of this Ordnary; Fines, dues, or
forfeitur's taken of Doctours or Montebankes only excepted; of whitch the said Mayor & Comonalty are to
have the moyety, or one halfe.

Kitchingman.

App. XV. York Barber-Surgeons' Ordinances.

[Ordinances of 1683 as to Apprentices.]

[46] Att The Counsell Chamber vpon Owse bridge, the 24th of September 1683,

In the Maioralty of the Right Honorable Edward Thompson, Lord Mayor of the Citty of Yorke

No Apprentice to be taken for less than 7 years,

under a fine of £5.

Ordered, that the Tenth Article and the nyne and Twentith Article mentioned in this booke,[1] be Vacated and made Void; and that for the future, none of the said Arte shall take any Apprentice for lesse tearme then seauen years, and that to be done by Indentures, and recorded by the Clarke of the company, vpon fforfeiture of ffiue pounds; and that the said Indentures be made within eight dayes, vpon the penalty of vj s viij d.

[47]
No Master shall have 2 Apprentices, unless 1 is a Freeman's son.

[[2] leaf 33, back]

The 1st Apprentice of every Master henceforth setting up, must be a Freeman's son. Penalty £10.

Fines above 6s. 8d. not to be taken without the Lord Mayor's consent.

Ordered also, that euery Master of the company now beinge, or which hereaf[t]er shall be, haueinge more apprentices then one at once, at any tyme or tymes hereafter, shall alwayes haue a freemans sonne one of the same apprentices; and that every Master [2] of the company which shall hereafter newly sett vp, shall take to his first apprentice a ffree mans sonne, vpon paine that euery Master doeinge contrary, shall forfeit for euery tyme soe doeinge, the sume of tenn pounds, to be paid, thone halfe to ye comon Chamber, and thother halfe to the said Company. Prouided that euery Master which at this present hath two or more apprentices, may keepe the same vntill their tearmes be expyred: and it is further Orderd, that none of the said company of Barbers and Chirurgions presume to take or compound for the future, any fines aboue six shillings eight pence, without the consent of the Lord Mayor for the tyme beinge.

Kitchingman,
Communis Clericus Ciuitatis.

---

[1] Pages 275 and 278 above.

App. XV. Order for the York Barber-Surgeons.

[*Alteration of last-named Penalty of £10 to 20s.*]

28th January 1757.

City of York Assembled at the Council Chamber upon Ouzebridge in the said City, the Twenty Eighth day of January, One Thousand Seven Hundred and ffifty Seven, when and where (amongst others) the following Order was made.

present
Richd ffarrer Esqr Lord Mayor.

James Barnard Esqr
Richd Lawson Esqr
John Mayor Esqr
Willm Coates Esqr
Thos: Matthews Esqr
In{o} Allanson Esqr
In{o} Telford Esqr
In{o} Greggs
R{d} Garland } Gen[1] Sher'

Joseph Buckler.
Edward Wilson.
Henry Richmond.
Auby Taylor.
Chris{t} Rawden.
Richd Dawson.
Chas Wightman
Thos Spooner
Geo: Thompson
ffrancis Ingram
Edward Thwing
ffrancis Stephenson
Iohn Skilbock
William Baker
Thomas Hungate
Henry Iubb
Thos Marfitt
Iohn Bradley
Willm Dunn

Ald{n}. Upon the Petition of the Searchers of the Company of Barber Chirurgeons, It is Ordered, that the By-Law made by this House the Twenty fourth day of September, One Thousand Six Hundred and Eighty three, whereby every freeman of this City who should newly set up and take for his first Apprentice an Unfreemans Son, should forfeit the sum of Ten pounds, shall be, and the same is by these presents, repealed. And it is further Ordered that, for the future, Every freeman of the said Company who

Gen[1] of 24 shall take an Unfreemans Son Apprentice, shall forfeit the sum of Twenty Shillings to the Mayor and Commonalty of the said City; One half thereof for the use of the common Chamber, and the other half thereof for the use of the said Company of Barber Chirurgeons.

Examined by me, John Raper, Comon Clerk.

---

[1] 'Gen' means 'gentlemen.' The Aldermen being 'esquires,' the Sheriffs and Common-Council are of the next class, 'gentlemen.'

286 App. XV. *Order for the York Barber-Surgeons.*

[*Fines of* 3s. 4d. *for not attending Meetings, reduced to* 6d.]

[leaf 34, back]   9th May 1768.   [49]

City of York.
Present :
Iames Rowe Esq.ʳ
second time
Lord Mayor.

George Eskricke
Iohn Allanson
Fraˢ. Stephenson
Francis Bacon
Iohn Wakefield
Esqʳˢ Aldermen.
Iohn Hardisty
Samˡ. Wormald
Gent. Sheriffs.

Thomas Norfolk ⎫
Iohn Bradley
Edward Wallis
Hale Wyvill
Iohn Stow   ⎬ of the 24
Christopher Oldfield
William Siddall
Thomas Wilson
Thomas Varley. ⎭

Assembled at the Council Chamber upon Ousebridge, the ninth day of May, one thousand seven hundred & sixty eight, when and where (amongst others) the following Order was made.

Upon the Petition of the Searchers of the Company of Barber Surgeons of this City, It is ordered that the penalty of Three Shillings and four pence inflicted on Members of that Company for Nonattendance at their Meetings, by an Order of this House of the twenty third day of Iune, One thousand five hundred and Ninety two, shall be henceforth reduced to the Sum of Sixpence.
Examined by Iohn Raper,
Comon Clerk.

This Book came
into the Posession of
Mʳ. F. N. Alexander
by Purchase
A.D. 1817.
[*Stampt in gilt letters.*]

[The Brit. Mus. bought the MS. of W. H. Richardson, 9 April, 1881.]

App. XV. York Barber-Surgeons' Pledge. 287

(*The York Barber-Surgeons' Pledge to the City Corporation to keep the Company's Rules.*)

[leaf 35, back]

Memorandum, that wee whose names are subscribed, Freemen of the Citty of Yorke, and of the Company of Barber-Chirurgions, doe hereby promise and ingage our selues to the Maior and Commonality of the saide Citty, to performe and obserue all and singuler the Orders and Ordinances made for the good Gouerment of the said Company, contained in the book of Ordinances; And if wee, or any of vs Respectiuely doe Faile in any one of them, Then we are content and doe promise, Euery one of vs, for himselfe seuerally and respectiuely, to pay to the said Maior and Commonalty the seuerally summes and Forfeitures mencioned in the respectiue Orders, to bee Levyed by the Searchers of the said Company, or such as the Lord Maior for the tyme being shall appoint, vpon our respectiue goods, by distresse and sale thereof, rendring the ouerplus to the owner.

[50] We York Freemen of the Barber Surgeons' Company,

promise the Corporation to keep all our Ordinances;

and in default,

to pay the Fines named therein.

George Matthews   John Anderson
Tho. Hall   Nathaniell Nelson

and about 7 other columns of signatures.

Then follow names of Members of the Company, with entries of their Apprentices, the last seemingly in 1666. Then comes a Calendar of the 12 months; a sketch of a man with his bleeding-points shown, and the labels printed at p. 229, above; 3 astrological and other figures, with tables, prose treatises of the Elements, &c., the influence of the Planets on Man, John of Burdus's (Bordeaux) medicine against the Pestilence, the Poem on Blood-letting printed above, p. 228-9. Follow, names of the York Barbers and Surgeons, and their Apprentices to 1784 (or past); then a stampt Agreement of Feb. 2, 1777, that the Barber Surgeons won't shave or dress wigs, &c. on the Lord's Day, save for strangers at the Assizes and Races (?) under a Penalty of £5. Then another Order of 6 May, 1701, that Searchers shall enter into a Bond not to spend more than 2s. 6d. without authority; and then more Members' names.

XV. Sunday-Shaving, 1413. Prices of Meat, 1545.

Sunday shaving in 1413.

On July 24, 1413 (1 Hen. V), in consequence of a letter from the Archbp. of Canterbury of July 23, the Lord Mayor & Aldermen issued an Order enterd in Latin in Letter-Book I, leaf Cxxv, enjoining that the London Barbers should no longer, against the Law of God, the Canon law, & public decency (*honestatem*) keep open their houses & shops on Sunday, the 7th day which God made holy, & on which He rested after His six days' work; that neither they, their wives, sons, daughters, apprentices or assistants (*seruientes*) should, in or out of their houses & shops, ply their shaving or barbing trade on Sundays, under a penalty of 6s. 8d. for every default, of which 5s. was to go to the work (*ad opus*) of the Guildhall [building the present one], & the other 20d. to the Masters or Wardens of the London Barbers' chest, for their use. (The Archbishop's Letter is englisht in Riley's *London Memorials*, 1276-1419, p. 593. London, 1868.)

Prices of meat in London in 1545.

At a Court of Common Council held on May 15, 1545 (36 Hen. VIII), present the Mayor (Waren), Recorder, Forman, Dormer, Cotes, Laxton, Hoberthorne, Amcottes, Sadler, Wylforde, Lewen, Judde, Hyll, Barne, and Tolos and Dobbys, sheriffs, it was stated that 'as the Bochers of this Cytye, blynded in Averyce & syngler geyne & lucre, haue nowe of late dayes, so furre inhaunsed the prices of all kyndes of vytayles that they medle withall & putt to sale / that nott onely the Comons of the sayd Cytye & others repayryng' to the same, haue beyen) gretly greved therby, but Also that Complaynt therof hath & is comyn) vnto the kynges most honourable Counsayll, to the no lyttyll dyspleasure of the lorde Mayer & Aldermen) of the sayd Citye //' & as the Butchers would not sell at the reasonable prices fixt by the Lord Mayor, 8 Mercers were appointed to visit the flesh-markets[1] from 5 to 11 a.m., & 1 to 5 p.m., & see that only the proper prices were charged :

'That ys to sey / the pounde of Beoffe, from Crystmas to Mydsomer, for ob. q^a (3 farthings); the pounde of Mutton) j d / The pounde of veale ob q^a & dimidium quadrantæ (3½ farthings) / And from) Midsomer to Crystmas. the pounde of Beoffe for ob & dimidium quadrantæ / Mutton for j d / the pounde of Veale for j d the Pounde / The best lambe[2] for ij s / The seconde lambe for xx d, & the meanest lambe for xvj d, & the halfe of euery suche lambe, & also the quarters, after the same rate Att all tymes of the yere / And . \`ke att all tymes of the yere for ob dimidium quadrantæ the pounde /' (Repertory 11, leaf 155).

---

[1] seynt Nycholas Shambles / The Stokes / Leaden Hall / & Est chepe
[2] The whole lamb.

XVI.

The Ordre of

the Hospital of .S. Bartholomewes in Westsmythfielde in London.

¶ i. Epist. Iohn .ij Chap.

He that sayeth he walketh in the lyght, and hateth his brother, came neuer as yeat in the lyght. But he that loueth his brother, he dwelleth in the lyght.

LONDINI
ANNO
1 5 5 2.

The Contentes of this Booke.

[I]

	[Page]
A preface	[291]
A deuision[1] of the .xii. Gouernours into their particuler gouernaunce	[295]
A Charge geuen to the Gouernours at the tyme of their admission	[296]
The President, and his authoritie	[297]
The Threasaurour and his charge	[297]
iiii. Surueiours and their charge	[298]
iiii. Almoners and their charge	[299]
ii. Scrutyners and their charge	[300]
An admonition to the Auditours	[301]
An Order for the saufe kepyng of the Euidences and Wrytynges of the Hospitall	[302]

[II]
Officers of houshold, with their charges particularly [303]

The Renter clerke and his charge	[303]
The Hospiteler and his office	[307]
The Butler and his dutie	[308]
The Matrone and her office	[309]
The Sisters and their dutie	[310]
The Chirurgiens and their duty[2]	[311]
The Porter and his dutie	[312]
The Biddilles and their dutie	[313]
The visitour of Newgate and his duty	[314]

[III]

The Estimate of the Charges and Expences of the Hospital	[316]

[IV]

A dayly Seruice for the Poore	[319]
A Prayer to be sayd at their delyuerie out of this House	[335]
A Passe-porte for the Poore at their deliuerie	[336]

[The Writer has a few peculiarities of spelling: toguether, yearth, officiers, theim; the Northern *awne* for *owne*, &c.: *the* for *they* was customary.]

---

[1] The *Deuision* is put after the *Charge* in the original.
[2] The *Chirurgiens* are put after the *Visitor of Newgate* in the Original.

A Preface to the Reader. [A. ij.]

He wickednes of reporte at thys Daie, good reader, is growen to such ranckenes, that nothing almost is able to defend it selfe against the venyme thereof, but that, either with open slaunder or priuie whisperyng, it shalbe so vndermyned, that it shall neither haue the good successe, whiche otherwise it myght, ne the thankes whiche for the worthines it ought.

It is better knowen by reaporte vnto the nombre, then weyghed in effect almoste to any, that for the relief of the sore and sicke of the citie of London, *It pleased the Kinges Maiestie, of famous memorie, Henry the eight (father to this our moste drad souereigne lorde nowe reignyng) to erecte an hospitall in West Smithfield, for the continual relief & help of an .C. sore and diseased. And the same endowed with the yerely reuenues of v. c. Markes, to geue vnto y* sayd Citie and Citezeins condicionally, that they also, for their part, should adde other .v. hundred Markes by the yere. Whiche thyng, with al due thanckefulnesse, thei receiued at his maiesties handes: And (for that thei sawe it prociede from his highnesse, aswell of moste charitable zeale toward the afflicted membres and his brethren in Christ, as of a singuler fauour toward *the Citie) very gladly embraced the condicion. Thinkyng it for their partes rather to litle then enough.

But when they had taken suche suruey therof as was conuenient for them in this case to do: Although the Kynges maiesties endowment was after the rate of his hyghnes moste gracious gifte, yeat founde thet the nature of the same, and the state of the whole, farre vnder that that they at the first had hoped. The raysing of this .v. hundred marke rent, to lie only in a certeine of houses, some in great decaye, and some rotten ruynous; And some other to whom better tenauntes had happened, alreadie leased out at terme and rent, skant reasonable for the behofe of y* poore. So that first to ma*ke them againe worth the wonted

Slander is so rank, that nothing is safe from its venom.

It hinders good deeds, and stops gratitude for them.

[* A. ij. back.] To relieve the London poor and sick, Henry VIII (in 1544-6) founded Barts for 100 patients,

with 500 marks a year, the City finding another 500.

[* sign. A. iij.]

But the City found that

Henry's 500 marks were to come only from houses in ruin, or let at very low rents.

[* A. iij. back.]

reuenue, and then to continue them in the same, was no smal charge; & the helpe therunto, whiche oute of the better repaired might have growen, was by the former leases and rentinges preuented. In thospitall it selfe (besyde the pencions yssuyng out of the sayd .v hundred markes, and graunted by the letters patentes of his said highnes to the Hospiteler there, and to other the ministers of the same[1]) was founde so much of housholde ymplementes and stuffe towarde the succouryng of this hundred poore, as suffised thre or foure harlottes, then lieng in chyldbedde, and no more, yea, barely so muche, if but necessary clenlinesse ware regarded, so far *had the godly meanyng of the gracious Kyng bene abused at those daies, & yet was litle then smelled, and lesse talked of. The good citizeins neuerthelesse, not so muche discouraged with others euill doynges, & the great falle of their hope, as moued with ye duetie of their entreprise & godly regard, not to their own poore and afflicted only, but to al other pore and diseased, which daily out of all quarters of the Realme resort to the Citie (as in to a commune receipt and refuge of their miserie), proceded with suche spied as they could, to the redresse of al these decayes, disordres and defaultes, and bestowed thereabout, aboue their couenaunt of .v. hundred markes yerely, for their welcomyng and *beginnyng, not muche lesse then a thousand poundes;[2] wherby (toguether[3] with other their good endeuours) when ther had wonne it to such poynct that it was fitt to receiue the nombre, and to succour the same with all necessaries requisite and in suche case nedeful, and had in deade receiued and daily mainteyned it at the full, certeyne busie bodies, more ready to espie occasion how to blame other, then skilful how to redresse thynges blame-worthie in diede, yea, I feare me, hauing al their zeale in their tongue only, not contented priuately, one and another, emong their neighbours, to hynder the profette of the poore, and to slaunder the good Citizeins occupied thereabout, rounded into the ea*res of the preachers also, their tender consideracion. Who being lesse circumspect in crediting their matter-ministrers, then to men of suche calling apperteineth, and thynkyng peraduenture if the

---

[1] See Forewords, the Section on the Hospital.

[2] Sir Hy. Hubbathorne, merchant-tailor, was Lord Mayor in 1546, and Sir John Gresham (sheriff in 1537) in 1547, when the first Surgeons at Barts were appointed.

[3] A *u* is generally in this word in the Orders. See also p. 221.

App. XVI. Barts Order, 1552. Hospital Work. 293

citie had done their dutie herein, this Hospital should haue made a generall swiepe of all poore and afflicted,— As though this priuie backebityng could not so sufficiently and weyghtely set forth this enormitie of the Citezeins, as somed behouefull for the querele of charitie,—toke vpon them to geue spiede and aucthoritie to the thyng, eche after his maner. So that the good Citizeins, which nowe for these .v. yeares space haue shonned for no lothesomenes, to administer the relief without other gayne *then that Iesus Christe, God & man, promiseth, & will vndoubtedly paye, haue here receyued nothyng elles, but for a commune benefight, an open detraction, and the pore (as shal afterward appiere) a larger hynderaunce. Where in the meane season notwithstandyng, there haue bene healed of the pocques, fystules, filthie blaynes and sores, to the nombre of .viij. hundred, and thence saufe deliuered, that other hauyng nede myghte entre in their roume; Beside eyght skore and .xii. that haue there forsaken this life, in their intollerable miseries and griefes, whiche elles might haue died, & stoncke in the iyes & noses of the Citie, for all these charitie-tenderers, if thys place had not vouchedsaufe to be*come a poompe alone, to ease a commune abhorryng. Wherein, although they haue at all handes so well deserued, that harde it ware with the moste fauourable reporte to requite it, yet for that they loke for their rewarde another where, contented to passe that in silence : It may iustely be aunswered to all suche charitie-proctours, that if they well weighed these thynges already alleaged, and the wages of the Cyrurgiens, and such officiers and seruauntes as nedefully are attendaunt about the poore, the charges of beddyng and shifte for so many sore and diseased, & the excessyue prices of all thynges at this day, thei might both merueile how so many are there relieued and daily mainteyned, *and with repentaunce of that they haue myssayde, endeuoure them selues, with asmuch good reporte and prayse, to aduaunce both the died and the doers, to wipe away the slaunder, as they haue to hinder them both by the contrary.

But, forasmuch as it is doubtful whether thef wil do as they maie, and of conscience are bounden, and the slaundre is so wide spred, that a narowe remedy cannot amend it : It is thought good to the Lord Mayour of thys Citie of London,[1] as chief patrone and

[1] Sir George Barnes, haberdasher (sheriff in 1545), was

[marginalia:] These Preachers wrongly made public the back-biters' slanders; and the good Citizens, for their 5 years' nauseous work done for Christ's sake, [* A. v. back.] receivd only detraction.

During these 5 years (1547-1552), 800 sick folk were heald in the Hospital,

and 92 died,

who else would have stunk in the noses of the City.

[* sign. A. vj.] if the Hospital had not acted as a pump to this nuisance. Yet, instead of praise, slander has come. But the Citizens have been silent, looking for their reward in Heaven.

The Hospital Surgeons and servants have been paid, and bedding, &c. found, tho' prices have been excessive.

[* A. vj. back.] The slanderers ought to repent, and praise and help the good-doers.

But as they may not,

the Lord Mayor (Sir G. Barnes), as Patron and Governor of Barts,

294 App. XVI. *Barts Order*, 1552. 1000 *Patients*.

[marginal notes:]
now publishes the Officers and Orders appointed by him and 12 of the oldest Citizens, [* sign. A. vij.] both to stop the slander,

and to let all men know how the Hospital is administerd.

If further reform is found needful,

the Hospital men will gladly adopt it. [* A. vij. back.]

And let all folk know that, though

at first the number of poor was kept to 100,

the City wish to enlarge it to 1000.

The City wish too that all other Hospitals and the Savoy [* sign. A. viij.] may be stird up by their example to help the poor, specially now, when their misery is so great. May Christ kindle in us all the Faith that works by Love!

[main text:]

gouernour of this Hospitall, in the name of the Citie, to publishe at this present the officiers and ordres by hym appoincted, and from time to tyme practysed and vsed by twelue of the Citizeins moste *auncient, in their courses, as at large in the processe shal appier, partly for the staye and redresse of such slaundre, and partly for that it myght be an open wytnesse and knowledge vnto all men, howe thynges are administred there, & by whom. Wherein, if any man iudge more to be set forth in woorde, than in diede is folowed, there be meanes to resolue him.

But if there be not so muche set furth as is expedient (as what thyng at the first can atteyne to the toppe of perfectnesse), or that any manne spieth ought in this ordre worthie to be refourmed, he shall not nede to crie it at the Crosse,[1] but shall fynde those at the Hospital, that both gladly will & may refourme it. And where yet by suche *meanes, occasion is founde, as tofore was sygnified, to withdrawe mennes charities, by reason that it is thought but folly to bestowe more relief where there is enough for the nombre already: The Citie, of their endlesse good wil toward this most necessarie succour of their pore brethren in Christ, although at y⁵ first they semed bounde to the precyse nombre of an hundred, and no more, wyshe al men to be most assuredly perswaded, that if by any meanes possible thei might, they desire to enlarge the benefyght to a thousand, as ordinarie as at this daie the hundred are.

Finally, they wyshe that all Almoisners and houses of Almoise, knowen either by the name of hospital, or Sauoy,[2] might, *by these their doynges, be prouoked to lyke endeuour & benefyght to the poore, that what one is not able alone to succour, the other myght in felowshippe supplie, at this tyme namely, when the mysery of the poore moste busily semeth to awake.

The Lorde Iesus, kyndle in vs all, that faith that worketh by loue, that we may in diede put on Christe,

---

Lord Mayor in 1552, and Sir Thomas White, merchant-tailor, in 1553.

[1] Paul's Cross, in the Cathedral Yard.

[2] The Savoy Hospital was suppresst by Edw. VI on June 10, 1553, just before his death, and its furniture and part of its income used for Bridewell and St. Thomas's. Mary refounded it in Nov. 1556; the court-ladies and maids of honour gave it beds, &c.; and it was confirmd by patent on 9 May, 1558.—Stow's *London*, p. 166, col. 1, ed. 1842.

our ryghteousnesse before God, and not suffre him to lye vp in presse, that sieketh to be worne, to the glory of his father, and ours, and to the testimony of our hope layd vp in hym. Amen.

*The diuision of the Gouernours, and officers: the names, and nature of them both.

[* sign. B. j.]

T behoueth first to vnderstande for the more euidentnesse of that that foloweth, that there are in this administracion, two sortes or kyndes of menne. The one called Gouernours (by a name proper to their aucthoritie) placed there by the lorde Maiour, as patrone of this Hospitall: And the other called officiers, that for wages are hyred, for to haue ye necessarie doynges *in the seruice of the house and the poore.

1. Governors.

2. Paid Officers.
[* B. j. back.]

The gouernours so chaunge, that thone haulfe remayneth .ii. yeares in their gouernaunce to helpe and enstructe the later elected, whiche also become enstructours to their folowers. And these are in nombre twelue, whereof foure are Aldremen, & the residew Communers; and accordyng to their gouernaunce, thus are they named:

Governors serve 2 years.

They are 12 in number;
4 Aldermen,
8 Commoners.

The President, alway the Seniour Alderman.

President, the Senior Alderman.

Surueyours foure, two Aldremen and two Communers.

4 Surveyors.

Almoisners foure, one Aldreman, and thre Communers.

4 Almoners.

The Threasaurour, a Commoner.

1 Treasurer.

Scrutyners, two, both Communers. *The officiers are .vii. in nomber, continuable or remouable, as the gouernours shall fynde cause, and be thus called:

[* sign. B. ij.]
2 Scrutineers.
7 Paid Officers,

  The Hospiteler.
  The Renter Clerck.
  The Butler.
  The Porter.
  The Matrone.
  The Sisters .xii.
  The Byddles .viii.

the Chaplein first,

Porter fourth,

Beadles last.

296 XVI. *Barts Order*, 1552. *Charge to new Governors.*

3 Surgeons, who get wages, and attend daily.

The Visitor of Newgate.

The Governors, the City yearly [* B. ij. back.] elect six: 2 Aldermen and 4 Commoners.

The 12 old Governors make their Clerk read to the 6 new Governors, this Charge:

There are also as in a kynde by them selues .iii. Chirurgiens in the wages of the Hospitall, geuyng daily attendaunce vpon the cures of the poore,

And a minister named the visitour of Newgate, accordyng to his office and charge.

The Gouernours are alwayes elected by the lorde Maior and his brethren, who 3erely *electeth vj,[1] that is to saye, two Aldermen, and .iiii. Commoners, which are admitted into the hospitall, after this maner.

The whole companie of the xii. olde Gouernours, sittyng in assembly toguether, cause their clerck to reade vnto the .vj. newly elected, the charge hereafter folowyng:

The Charge.

'You are elected Governors for 2 years;

and, under the Lord Mayor's [* sign. B. iij.] Orders,

you shall (setting your own business aside) attend to the Hospital with loving diligence.

Having set hand to the plough,

you must not turn back, [* B. iij. back.] for work for the poor is work for Christ.

On God's behalf, then, do your utmost to comfort the poor of this Hospital,

as faithful Stewards

IT may please you to vnderstand, that ye are here elected and chosen, as fellowe gouernours of this hospitall, to continue by the space of two yeares: By all whiche tyme, accordyng to such · laudable decrees and ordinaunces, as haue bene & shalbe made by the aucthoritie of the lorde Maiour, *chief patrone hereof, in the name of the Citie, and the consent of the gouernours for the tyme beyng, (all your other businesse set aparte, asmuche as you possibly may,) ye shall endeuoure your selues to attende onely vpon the nedeful doynges of this house, with suche a louyng and careful diligence, as shal becomme the faithfull ministers of God, whom ye chieflie in this vocation are appointed to serue, and to whome, for your negligences or defaultes herein, ye shall render an accompt. For truly ye cannot be blamelesse before God, if after you haue sette hande to this good ploughe, and promysed your diligence to the poore, ye shall contrarywyse tourne your head backwarde, & not perfour*me the succour that Christ loketh for at your handes, & hath witnessed to be done to hymself, with these wordes : "Whatsoeuer ye do to one of these nedy persones for my names sake, the same ye do vnto me. And contrary wyse, if ye neglecte and despyse them, ye despise me." We therfore require and desire euery of you, on Goddes behalfe, and in his moste holy name, that ye endeuour your selues, to the best of your wittes and powers, so to comfort, ordre and gouerne this house and the poore therof, that at the last daie, ye maie appere before the face of God, as true and faithfull Stewardes and dis-

---

[1] *Orig.* 'vp,' with the body of the p scratcht out.

XVI. Barts Order, 1552. President & Treasurer.

posers of all suche thynges as shal, for the comfort and succour of them, (duryng the tyme *of your office) be committed to your credite and charge. And this to do, we require you faithfully to promes, in the syght of God, and hearyng of your brethren. And so doing, we here admitte you into our fellowshyp. [* sign. B. iiij.] *in the sight of God! Thus we admit you into our Fellowship.*

THAT done, & the new elected consentyng and yelding them selues to the charge, the haulf of the gouernours that haue already fulfilled their two yeares gouernaunce, to stand apart: and the other haulf that shall remayne with the newe elected, to take them by the handes, after their degrees, and so admitt them, and not to depart felowshyppe before thei haue dyned togueather all wholy, aswell those that come newe, as those that haue gouerned their tyme, and those *that remayne, euery man at hys awne cost and charge.[1] *Then the 6 Governors who've servd 2 years shall stand aside; and the 6 one-year men shall take the new ones by the hand, and all 18 shall dine together,* [* B. iiij. back.] *each at his own cost.*

The President.

THe President of this Hospitall, is chief ruler and gouernor of the same,[2] vnder the lord Maiour, who hath aucthoritie from tyme to tyme, to conuocate and cal together al the gouernours for matters concernyng the maintenaunce and good orderyng of the poore, and to demaunde of euerie of theim, the accompt of their doynges in their seuerall offices, & with the assent and consent of the sayd gouernours, to graunte leases and fees, & make necessarie decrees and ordinaunces. *The President is chief ruler, calls the Govenors together, asks for an account of their doings, grants leases, and makes Ordinances.*

*The Thresaurour and his charge.

ALl the Treasure of thys house, is committed to your charge, that is to saye, all suche money as shall ryse and growe, either by rentes or by giftes to the vse of thys house, of the whiche ye shall kepe a true and a iust accompte. And it shal not be lauful for you to pay any maner of persone, any some or sommes of money, (excepte it be to the Stewarde of this house, for the victuallyng of the same, and the ordinary fees and wages that goeth out thereof): but ye shall first haue the names of those persones subscribed to the said some of money, vnder whose office and charge suche [* sign. B. v.] *The Treasurer takes charge of all money, keeps account of it, and pays none away (save to the Steward for food and wages) unless the officer responsible signs his name to the bill.*

---

[1] N.B. No guzzling out of poor folk's funds.
[2] The first specially-chosen President of the Hospital was Sir John Ayliff, appointed in 1553. Till then, the Senior Alderman, under the Rules above, acted as President.

XVI. Barts Order, 1552. Treasurer & Surveyors.

[* B. v. back.]

The Treasurer is to keep a separate Rent Account,

to check the Renter, and show the rise or fall of rents.

To hand-in a yearly Cash Account on Oct. 20, which is

to be audited by 4 Auditors, [* sign. B. vj.] and verified by the Treasurer at 8 a.m. every Nov. 2 at the Hospital.

He shall then and there tell the new Treasurer the whole state of the Hospital affairs,

and hand him the balance of cash, and all documents.

[* B. vj. back.]

The Treasurer's reward is Christ's promises.

The Surveyors

shall see to the Hospital lands and leases,

and register all Leases in the Repertory Book.

[* sign. B. vij.]

pay*ment shall happen to ryse and growe, or the names of the most part of them.

Ye shal also kepe one seueral accompte betweene the Renter & you, by whiche maie appere, not onely the charge of the said Renter and his arrerages, but also whether the rentes of the landes perteinyng to the said house, encrease or decaye.

Ye shal also yerely the .xx. day of October (within this Hospitall) yelde and geue vp in wrytynge vnto the President and gouernours of the same, a true & a perfect accompte of your whole charge, duryng the yere of your treasorourship, and then the said President and gouernours shall name and appoint emong theim selues .iiii. to be auditours for *the same. And the second daie of Nouember nexte folowyng, ye shall likewyse resorte to the said Hospitall, at the houre of eight of the clock in the forenone, that ye may then aunswere and clere your accompte, if any doubtes or faultes shall happen to arise or be found by the auditors of the same. And the same daie, then and there ye shall declare vnto the newe treasaurer that shalbe appointed, the whole course & state of the affaires, profites & commodities of this house, in as large sorte as ye possibly canne, and deliuer vnto hym all suche somes of money due to ye house, as shal then rest in your handes, and al suche acquitaunces, rentalles, and other writynges, as necessarily shall apperteyne, to *the affaires of the sayde house. And thesame daie to dyne within the said Hospitall, with the gouernours therof. And in recompence of your paines, ye shalbe assured of the mercies laied vp for you in the promises & bloud of Iesu Christ our Sauiour.

Surueiours.

VNto you is committed the viewe of all the landes & leases perteinyng vnto this house, aswell suche as heretofore haue bene graunted, as also hereafter shalbe graunted; and ye shall cause thesame to be regested[1] in the repertory booke by the Clerke, from time to tyme, when and as often as you shall assygne hym, to thentent that ye gouernours of this house *may alwaies be assured, what grauntes haue passed them; and both whereunto thei haue bound them selues, and also wherunto their tenauntes are bounde, that the landes and tenauntes maie be loked vnto accordingly. And ye

---

[1] See the verb 'regeste,' in the 'Scrutyners,' p. 301, below.

XVI. Barts Order, 1552. Surveyors & Almoners. 299

shal adioyne vnto you yᵉ treasorer of this house for the tyme beyng, as a necessarie ayde in all youre doynges, for that he moste chefely hathe experience of all the affaires and doynges of this house. And for the better accomplisshyng hereof, you or the greatest parte of you, shall mete euery .xiiii. daies in thys house, on the Wedensdaie, at whiche tyme ye maie warne the Tenauntes that haue made defaulte in none doyng of reparacions, or none paiment of their *rentes or other to be before you, to take order with them, accordyng to the couenauntes expressed in their leases. And youre graunte, with the particulers of suche reparacions as by you shalbe allowed, to be entered into a boke with the name of the tenaunte and tenement, wherunto you or the moste parte of you shall subscribe your names, and then committe the ouersight therof to the Renter, so that it be agreed that one or mo of you may visite & peruse¹ the same in suche wise as the greatenes or quantitie of the thynge wyll require.

Also euery yere at the feast of Saint Michell tharchaungell .ii. newe Surueiours to be chosen, and the old with yᵉ new to make the .xii day of October folowing *or with-in two daies before or after, a generall view and suruey of al the landes apperteinyng to this house, and truly to kepe a boke of the defaultes therof; and for youre paines takyng here, God hath promised to geue you rest and pleasure in heauen perpetually.

They shall join the Treasurer with them,

meet fortnightly on Wednesday, and summon defaulting tenants before them.
[* B. vij. back.]

They shall enter in a book all repairs authorised by them,

after examination on the spot.

Every Michaelmas 2 new Surveyors shall be chosen, who, with the old, shall about Oct. 12,
[* sign. B. viij.] view the Hospital property, and enter defects in a book. God will reward Surveyors.

Almoners.

YOw shal euery Mondaie come vnto this house, or oftener if you shall think good, but at the least ones in the weeke: Alwaies prouided, not on the Saterdaie, for that daie specially shalbe reserued & kept for the session of the President and Gouernours of thys house, for the generall affayres of the same.

*And at euery tyme of youre being here, if there be cause why, ye shall call before you euery particuler officer of this house, and enquier if euery man do his dutie therein accordyng to hys charge, & whether there be peace and quietnes mainteyned in the same. And if ye shall at any time fynde any disordred persone or persones, then to take suche order with hym or theim for their better reformacion, as to you shal seme most mete. And if any refuse to be ordered by you, then

Almoners

shall be at the Hospital every Monday, or once a week, (but not on Saturday,
which is Governors' day,)

[* B. viij. back.] shall call up every Officer, and ask if all is right and quiet;

if not, correct the offender,

and if he disobeys,

¹ examine.

report him to the Governors.	to make suche persone knowen to the President and the rest of the gouernours, that further order may be taken by the whole house.
Also see that the Surgeons do their [* sign. C. j.] duty, call them up to report the weekly cures, give the cured (?) some money,	Ye shal also diligently enquire if the Chirurgiens of this house *do their duetie toward the pore, without corrupcion or parcialitie, and callyng them before you, ye shall enquire what nombre there were healed that weke, and the same deliuer, and reward, accordyng to your discrecions; and of the same rewardes to haue your allowance of the Threasaurour, so that ye deliuer vnto hym the particulers therof, sygned with the handes
and admit other poor in their stead.	of two of you at the least. And in the places of the poore so departed, to admitte other, in suche sorte and maner, as in the charge of the Hospiteler is mencioned and declared.
Also keep an Inventory of the [* C. j. back.] utensils, &c. of the Hospital, and provide wood, coal, &c., and report to the Governors any needed enlarging of rooms, fresh beds, &c.	Ye shal view from time to time this house, keping one entier and perfecte Inuentarie of the vtensiles and necessarie imple*mentes therof, in a boke, aswell that prouision may be made in due tyme, for supplieng that whiche shalbe founde to lacke, as also in due tyme to prouide for wood, cole, and other necessarie furniture. And whatsoeuer elles shall seme nedefull vnto you for the benefitte of the poore, as ye enlargyng of roumes, or encreasyng the nomber of beddes, the same ye shall sygnifie to the president and gouernours, that by one assent it maie be decreed, & by you finished & performed.
Also keep the poor sweet, visit them weekly, and see that their food is duly supplied. [* sign. C. ij.] God will reward Almoners.	Ye shall also se vnto the kepyng swete of the poore; and in your proper persons visite them once euery Wieke at the least, and to see that their seruice of bread, meate and drinke, be truly and faithfullie deliuered vn*to them. And for your laboures and paines, ye shalbe sure of the rewarde that God hath promised to all them that succour hys members.

Scrutiners.

The Scrutineers are to search for the gifts to the Hospital,	YE shalbe ready and diligent to make searche and enquiry from time to time for al suche giftes, legacies, and bequestes, as haue bene or shall be geuen or bequethed to the succour and comforte of the poore
get them from the givers, with a bill of the amount,	of this house; And the same receiue at the handes of the gyuers or executors, toguether with a bille of the somme, subscribed with their names that make payment or deliueraunce therof; the whiche bille and
[* C. ij. back.] to hand to the Treasurer,	money, ye shal furthwith deliuer vnto the *Threasaurour of this house, receiuyng his acquyetaunce for the

XVI. Barts Order, 1552. Scrutineers & Auditors. 301

same; kepyng neuerthelesse a boke your selues, wherin ye shal entre & regeste al suche charitie, the giuers, the time, & the somme. And for al suche somme or sommes of money, as by you, or any of you, shalbe procured, had, or receiued, ye shall (if it be required) make vnto the geuers, or deliuerers therof, an acquitaunce in your owne names, as the gouernours and scrutiners of this house. *[but entering the same in their own Book, and giving the Donors a receipt.]*

And yerely at the Election of the newe gouernours into thys house, shalbe elected one newe Scrutiner; and the olde Scrutiner that shalbe remoued, shall make deliuery vnto the newe Scrutiners, of al such recordes, *billes and writinges, as concerne the affayres of this house. And also at the audite of the Treasorers accompt, the Scrutiners booke of giftes and bequestes shall in like maner be examined and allowed. *[Every year 1 new Scrutineer shall be elected, and the old one shall hand him all his documents. [* sign. C. iij.]]*

Finally, ye shall in euery place where you shall haue occasion to come in the company of good, vertuous, and welthy men, to the vttermoste of youre power, commend and set furth the good order of this house, and how rightelie the goodes geuen to the poore, are here bestowed, to the encouragement of other to extende their charitie therunto. Ye shall also, as occasion and oportunite serueth, moue those that haue the Office of Preachyng committed to them, that they *may the rather prouoke the deuocions of the people, to the help and comforte of this house. And thus doyng, you shall not lose the reward that God hath promised to all them that seke to glorifie and reuerence hys name in hys poore members. *[Scrutineers shall praise the Hospital to all folk, to encourage gifts to it; and shall specially ask Preachers to [* C. iij. back.] stir up people to give donations.]*

An Admonition to the Auditours.

INto youre audite muste be brought these sortes of Bookes: first, the Hospitall boke, beyng in the custody of the Hospiteler, to whiche also ye shall loke, that euery page or totall somme therof be subscri*bed with two of the handes of the Almoners: And this booke shal ye conferre with the Stewardes boke, who first maketh the prouisions. Ye must also haue the Scrutiners booke, to examine the accompte of the Treasourer for money deliuered vnto hym by giftes & bequestes. Also the booke of Surueye, to conferre the Bylles brought in by the Treasourer with the alowaunces of reparacions, expressed in the sayd booke. Also ye shall demaunde of the Renter, his rental for that yere, not forgettyng alwaies to charge hym with *[The Auditors must audit the Hospital Book, [* sign. C. iiij.] the Steward's Book, the Scrutineers' Book, the Treasurer's, the Surveyors', and the Renter-Clerk's Book,]*

302 App. XVI. *Barts Order*, 1552. *Hospital Deeds, &c.*

[* C. iiij. back.]

and the Journal or Order Book.

the arrerages that remaine the yere before (if any be), and to conferre the sommes of money receiued by the Treasourer, with the charge and accompte of the sayd *Renter. And lastly, to haue speciall regarde, if any somme of money haue bene paied by the Treasourer, by any decre or general order of this house, to loke in the Iournal for the same. And thus in the whole affayres of this house, shall ye perfectly be instructed.

The Hospital Deeds and Documents

An order for the saufe kepyng of the euidences and writinges apperteining to the Hospitall.

shall be kept in a chest

with 3 locks and 3 keys;
 [* sign. C. v.] the President having one key, the Treasurer one, and a Commoner the third.
No Deed, &c. shall be taken out of the House,

but only a copy of it.

THere shall one fayre and substanciall chest be prouided, and the same be set in the moste conuenient and surest place of the house, the which shal haue .iii. seueral lockes, and iii. keyes, whereof the President *alwaies to haue one, & the Treasurer one, and a Commoner appointed by the whole house, to haue the thirde. And it shal not be laufull to any of the Gouernours to haue any specialtie, euidence or writyng, out of the said chest, neither any other persone, to cary any of them out of the house (no, though it be for the affaires of the said house), but only a copie therof, which shalbe taken in the presence of the .iii. persones aboue named, that haue the keyes
 & the original forthwith
 to be locked up
 agayne.

Officers of Housholde
with their particuler charge.

[C. v. back.]

The Renter Clerck and his charge.

YOure office is, with all care and diligence to collecte and gather the rentes dew of the landes and tenementes apperteinyng to this house, and of all sommes of money so by you collected and gathered, to make deliueraunce and payment to the Treasourer of this house for the tyme beyng, receiuyng his acquitaunce for your discharge.

The Renter-Clerk is to collect the Hospital rents and pay them to the Treasurer;

You shal also, once euery weke at the least, resorte vnto the President of this house, or to the *Treasourer therof, for the knowledge of the affaires of the same; and at euery of the ordinary sittynges of the Gouernours in this house, for the affaires therof, aswel at the daies appointed for the assembly of the Surueiours and Almoners, as also when the President and all the masters shall assemble, ye shall geue your attendaunce, that from tyme to tyme ye maie enter and regeste all suche decrees, order and determinacions as by them, and euery of them, in their seuerall charges shalbe decreed, ordeined, and determined.

to attend the President or Treasurer [* sign. C. vj.] surer weekly, and be present at all Meeting of the Governors, to register their Orders.

And for that the good order and gouernaunce of this house may the better appere, aswell to the gouernours nowe beyng, as to all other worthy personages *that hereafter shall gouerne, or shall desire the certeintie therof, it shalbe requisite that ye kepe diligently .iiij. seuerall bookes, the names wherof, and the vse, are here described:

[* C. vj. back.] The Renter-Clerk shall keep 4 Books,

A Reportory.
A Booke of Suruey.
A Booke of Accomptes.
A Iournall.

And first you shall note, that before euery of these Bookes ye must haue a Calendre, into the whiche ye may entre, by order of letters of the .A. B. C. all proper

and start each with an Alphabetical Index of its Contents,

names & matters, that shall be conteyned in every of them. And for the better accomplishyng hereof, ye shall, with your penne, in the heade of the lefe, nombre the pages of euery lefe, in euery of these bookes, and then ad*dyng in your Calendre the nombre of the page, where the name or matter is entred in your boke, the reader without any difficultie may tourne to the same.

Side-notes: and shall number the pages, [* sign. C. vij.] and add page-numbers to the Index.

The Vse of the first boke
called a Repertory.

INto this booke shall ye first entre the foundacion of this Hospital, and also al dedes, leases, obligacions, acquitaunces, and other specialties: vsyng alwaies in the margent of the sayde boke, to note in a fewe Englyshe wordes, the somme and content of euerye article of those wrytynges that shall appiere noteworthie; and the same notes particularly to enter into their seuerall and propre places of your calendre, ac-*cordyng to the order of the .A. B. C.

Side-notes: In book I, the *Repertory,* enter the Founding of the Hospital and all Deeds, with side-notes stating their purport, which shall be entered also in the Index. [* C. vij. back.]

The vse of the second
booke, called a booke
of Suruey.

FIrst, in a seuerall lefe, yerely before ye enter any other thyng into this boke, ye shall make an abstracte of the names and surnames of euery of those tenauntes, to whome this house is bounde to doe reparations, and also of them that are bounde to fynde their owne reparacions, notyng in the margent, the leafe of your repertorie, where euery of their leases is entred. Also euery yere, when the Surueiours shall Suruey the landes of this house, ye shal be attendaunt uppon them, and *aptly & playnelie enter into this booke all suche defautes as by them shalbe founde, in the tyme of their view, makyng a distinct difference betwene tenauntes at wil and tenauntes by lease; and also betwene those to whom this house is bounde to finde reparacions, and such as haue bound them selues to reparacions.

Also ye shall diligently enter into this booke all suche orders and grauntes of reparacions or other, as the Surueiours from tyme to tyme shall make or take with the tenauntes.

Side-notes: II. The *Book of Survey.* Make an Index of the Names of Tenants for whom Repairs are to be done, and of those who do their own repairs. Attend the Surveyors in their yearly Survey; [* sign. C. viij.] and note defaults; distinguishing the different classes of tenants. Enter the Surveyors' orders for repairs.

And euery yeare when the Treasourer shall bryng in his accompte, and before the Auditours, shewe suche billes of reparacions, sygned with two of the handes of the Surueiours, as *he hath paied, ye shall, after the admission of the sayde billes by the Auditours, entre euery of theim into this booke, particularlie, vnder this title. <small>At the yearly Audit of the Treasurer's accounts, bring in your Bills of Repairs, [* C. viij. back.]</small>

"Reparacions doone in the yere that .A.B. was treasaurour of this Hospitall (that is to saie), from the feast of saint Michaell in the fyfth yeare, &c." <small>and enter them under a special title,</small>

And then shall ye write first the name and surname of the tenaunt, the tenement, and the daie of the moneth; and then the reparacions. And thus shall ye do with all other. And it is to be noted, that in your Calendre must be entred the name of euery treasourer, & the lefe wher the reparacions brought in his accompt are entred. And next after the reparacions, ye shall entre yerely *your whole rentall, beyng first examined by the Surueiours, and hauyng two of their names at the least, subscribed therunto. And in a particuler and playne maner ye shall expresse and declare the encreace of rentes that yere; and that shall ye entre into your Calender vnder this title, *Augmentacion of Rentes*, titlyng from leafe to leafe, where the said encreacinges be noted. And in lyke maner shal ye do with rentes decayed, entring them into your Calender by this worde, *Decayed Rentes*. Lykewyse with tenementes or rentes altered or chaunged, by this name, *Alteracion of Rentes*. <small>with names of tenant, tenement, date, and repairs.</small>

<small>[* D. j.] Then enter your whole Rental, with its year's increase, under *Augmentation of Rents*; and the lessened rents, under *Decayed Rents*; the changed ones, under *Alteration of Rents*;</small>

*The vse of the third booke, called a Booke of Accomptes.

<small>[* D. j, back.] III. The *Book of Accounts*.</small>

IN this booke ye shal first entre all the Accomptes (being allowed by the auditours) of al the treasaurours that hath bene sence this Hospitall was first committed to the Citie of London. And from hencefurth, at the fote of euery accompte made by the Treasaurour, ye shal expresselie & playnly adde and entre the arrerages of the renter for that yere, which also first by the Auditours shal be examined, and subscribed as aforesayd. <small>In it, enter all the audited Treasurers' Accounts since 1546. Hereafter, put the Arrears at the foot of every Account.</small>

And forasmuche as in all accomptes, diuers and many thinges at sondry times are requisi*te to be knowen, ye shall therefore in your Calender first note the name of the Treasaurour, with the leafe where his <small>In your Index [* D. ij.] enter the Treasurer's name and the leaf of his account,</small>

VICARY. X

306 Barts Order, 1552. The Rent-Clerk's Journal.

and that of the Survey-Book where the Repairs are put.

accompte is entered, and also in the margent at the enteraunce of the saied accompte, ye shall note the leafe of your booke of Suruey, wher the reparacions mencioned in the same accompte, are particularly entered.

And for the ready fyndyng of euery matter conteined in euery accompt, ye shal, in the margent of this boke, vse as is aforesaid, to note dyuers generall wordes,

Keep accounts under separate headings,

Accomptes, prouisions, liueries, giftes, legacies, revvardes, agreementes, Surrenders, Bargaynes, Sutes, recoueries, pencions, Fees, &c., Addyng to euery of these, beyng

[* D. ij, back.] *with references to the leaf of each.*

placed in your calender, the lefe *wher euery of them is mencioned in any of the accomptes conteined in this booke, that at a woorde may be sene what hath bene done in all these thynges, from the first Treasaurour to the last.

And for a perfect declaracion of the whole affaires

From the Hospital Book kept by the Almoners,

of thys house, ye shal also, out of another booke (which shall conteyne the doynges of the Almoners, and shalbe called the Hospital boke) entre into this

enter all Implements in the Hospital,

booke of accomptes, aswell a perfect Inuentarie of all suche Implementes as then shalbe founde within thys

and what is left of the Provisions and Victuals.

hospital; as also a ful remainder of all the prouisions and victualles, fyrst subscribed by twoo of the said Almoners. And in the ende ye shall manifestly declare

Also the names of all sick folk
[* D. iij.]
cured and discharged every year;
the names of all who've died,

the names and sirenames of so *many diseased persones, as that yeare haue bene cured and deliuered out of this house, and also the names and sirenames of so many as that yeare haue died in the house. The names and

and of those still in the Hospital,

sirenames also of as many as then shall remaine sycke and diseased in thys house, toguether with the name of

with their birth-counties and occupations.

the shier where-in eche was borne, & their faculties,[1] exercise, or occupacions.

IV. *The Journal.*

The Use of the .iiii. boke
called a Iournall.

It must have an Index too.

THis Booke must also haue a Calender; & it shal alwaies be brought furthe at suche tyme as the President and moste parte of the Gouernours shall

[* D. iij, back.]
In it, enter all the Orders of the Governors,

sit within this Ho*spitall, for the generall affaires of the same. And into this booke shall ye entre all suche orders & decrees, as from tyme to tyme shall by the sayde Gouernours, or greatest parte of theim, be decreed

with side-notes

and ordeined. And in the margent thereof ye shall do

---
[1] Professions, trades.

as before is assigned in the Booke of Repertory: in fewe wordes set furth the somme of euery decree, order, &c. conteyned therein. And chiefely ye shall vse the generall woordes before described in the booke of accomptes, that by the enteraunce of them into your calender, euery matter may easelie and readylie be founde. And ye shall not fayle, but in fyue daies next after the enteraunce of any thyng into this booke, to enter the same by a generall worde in*to the Calendre, that as wel when you are absent, as present, the gouernours may without difficultie be satisfied of that they seke for therein.

stating their effect.

Make your entries in the Index within 5 days of [* D. iiij.] the Orders passing, so that the Governors may easily find what they look for.

The office of the Hospiteler.

YOur office is chiefely and moste principally, to visite the pore in their extremes and sickenesses, and to minister vnto them the moste wholsome and necessary doctrine of Gods comfortable worde, aswel by readyng & preaching, as also by ministring the sacrament of the holy Communion at tymes conuenient.

To receiue also into this house, of the Stewarde, to the vse of the same poore, suche victualles and other prouision as by hym *shalbe prouided, entryng the same into your boke, and saufelie to kepe them to their vse.

Also to deliuer vnto the cooke of this house, from time to time, so muche of the same victualles as shalbe nedefull for the present tyme, to be dressed for the poore. And the same beyng dressed, to see seasonably and trulie deliuered, and distributed, vnto them.

Also, whensoeuer any poore persone shalbe here presented or sued for, to be admitted into this house, you shall receiue the same presentacion, callyng vnto you, two of the Chirurgiens of this house, to view and examyne the disease of the said persone, whether it be curable or not curable : if they Iudge it curable, then *you, by a bill of your hande, to certifie the name and sirename of the sayd diseased persone, vnto the Almoners, or two of them at the least, desiring them to subscribe their names thereunto; & that beyng done, you to kepe vpon a file the same byll for your warrant. And then ye shall committ the same pore to the matrone of this house, to be placed accordingly as ye case shal require.

Also at the admission of euery poore person into this Hospitall, ye shall enquire what money, or other

The Hospitler or Chaplain.

Visit the Poor in their sickness,

and comfort them with God's word.

Take from the Steward all food needed for the Poor [* D. iiij, back.] (entering it in your Book,)

and hand the Cook what he is to prepare for the Poor:

then see it given them.

When any poor person comes with an Order,

see him, get 2 Surgeons to examine him;

and if he is curable, [* D. v.] certify his name

to 2 Almoners;

get their signatures; file this;

and hand the Patient to the Matron.

Find out what valuables he has,

[marginal: and keep 'em for him till he's discharged. [* D. v, back.]]

thynges of valewe, he or she hath; and the same, to-gether with his or her name, to enter into your booke; and you to receiue & saufelie kepe the same, to the vse of the same poore, to be deliuered againe vnto hym, her, *or them, when they shalbe cured out of this house   And monethly to deliuer to the sayd Almoners, a copie of your boke of enteraunces, that they maie regest the same in the booke of their ordinary doynges.   And if any suche pore fortune do decease and die in this house, then you to deliuer all suche money and other thynges as shalbe in your custody, to the Treasaurour of this house for the tyme beyng, enteryng the same into your booke, to be committed and disposed to the vse of the poore.

[marginal: Give the Almoners a monthly list of Entrances, for them to register.]

[marginal: If any Poor die,]

[marginal: hand their money, &c. to the Treasurer for the use of the other Poor.]

And as often as any of the poore shalbe cured and made whole, you, with the Chirurgiens, to present them to the Almoners of this house, at their next assemblie here, & to regeste into your *booke the names and sirenames of them, and euery of them, with the daie and yere of their deliuerie and departure out of this house.   And at their departure, to geue vnto them a passeporte, to be made accordyng to the President and fourme that is expressed in the end of this booke.

[marginal: When Patients are cured, do you and the Surgeons present them to the Almoners, [* D. vj.] and, after registering their names, &c.,]

[marginal: give 'em a Passport, after the precedent at the end of this Book.]

This is your charge; and ye haue not to doe with any other thyng in this house. Howbeit, if ye shal perceyue at any time any thyng doone by any Officer of this house, or other persone, that shal maynteyne disorder, or procure slaunder, to this house, that ye then declare the same to some one or two of the gouernours of this house, & to none other persone, and no furder to meddle therein.

[marginal: If you see any wrong going on,]

[marginal: report it to 1 or 2 of the Governors.]

*The office of the Steward and Butler.

[marginal: [* D. vj, back.] The Steward and Butler.]

YOure charge is, faithfully and trulie to make prouision of such nedeful victualles, as from time to time ye shalbe appoynted by the Almoners to prouide for the poore of this house, remembryng alwaies that, wherein so euer you shall hynder, or negligentlie burden this house, either with excesse prices, or not makyng your prouision in due tyme, the same dammage and hurte you do vnto GOD, whose members the poore are;[1] & therfore ye ought the rather to study to serue

[marginal: Buy all food that the Almoners tell you to.]

[marginal: If you neglect your duty,]

[marginal: you hurt God, whose the Poor are.]

---

[1] Compare Chaucer, *Parson's Tale*, Works, ed. Bell, iii. 72 :—"Thilke that they clepe thralles, ben Goddes people; for *humble folk ben Cristes frendes;* they ben contubernially with the Lord."

in this house with feare of God and conscience, as one that manifestly and plainly walketh before the face of God, *who perfectlie seeth and beholdeth the very thoughtes of your harte.

Your charge is also to kepe a true and perfect accompt of al suche victualles as by you shal be bought, and to make deliuerance of the said victualles vnto the Hospiteler of this house, declaring vnto him the iust weight, nomber, and prices, of the same, that he may make due & true enteraunce and accompte therof.

Also, at al such tymes as shal be nedefull for the poore to be serued of their ordinary meales or otherwyse, eyther of Bread or drynke, ye shal not be absent, but with all diligence & redines ye shall geue your attendaunce.

Ye shall haue to do in none other mannes office in this house, *but only with your owne, in maner as is aboue described. But if ye shall perceiue at any tyme, any thyng doone by any officer or other persone of this house, that shalbe vnprofitable therunto, or that may be occasion of any disorder, or shal engender slaunder to the same, That then ye declare the thyng to some one or two of the Gouernours of thys house, and to none other persone, nor farther to meddle therin.

Margin notes: Serve here with feare of God and Conscience. [D. vij.] Keep a true account of all food you buy, hand it to the Hospitler, and tell him its real price. Attend at all the poor folk's Meals. [* D. vij, back.] If you see any wrong done in the Hospital, tell 1 or 2 Governors of it.*

The office of the Matrone.

The Matron.

YOure office is to receyue of the Hospiteler of this house, all suche sicke and diseased persones as he, by hys warraunt sygned from the Almoners of this house, shall pre*sent vnto you; and the same persones to bestowe in suche conuenient places within this house, as you shall thynke mete.

You haue also the charge, gouernaunce, and order of all the Sisters of this house, to see from tyme to tyme, that euery of them in the wardes committed to their charge, do their dutie vnto ye pore, as wel in makyng of their beddes, & keping their wardes, as also in wasshyng & purgyng their vnclene clothes & other thinges. And that the same Sisters euery nyght, after the houre of .vij. of the clocke in the wynter, and ix. of the clock in the Somer, come not out of the womans ward, excepte some greate and speciall cause (as the present daunger of death or nedefull succoure of *some poore persone.) And yet at suche a speciall tyme, it shall not be laufull for euery Sister to go furth to any person or persones (no though it be in her

Margin notes: You're to receive the sick from the Hospitler, [D. viij.] and put them in fit places. You govern the Sisters, see that they mak. the Patients' beds, wash their clothes, &c.; and that the Sisters don't leave the Women's Ward after 7 in winter, and 9 in summer, save in case of [* D. viij, back.] death, &c.,*

Barts Order, 1552. The Matron and Sisters.

[margin: and then only to godly patients, for a short time.]
warde,) but onely for suche as you shall thinke verteous, Godly & discrete. And the same Sister to remayne no longer with the same sicke persone, then nedefull cause shall require.

[margin: In spare time, make the Sisters spin;]
Also at suche tymes as the Sisters shall not be occupied about the poore, ye shall set them to spinning, or doyng of some other maner of worke that maie auoyde ydlenes, and be profitable to the poore of this house.

[margin: get flax from the Governors, [E. j.] and return it, when spun, for the Weaver.]*
Also ye shall receiue the flaxe prouided by the gouernours of this house, and the same beyng sponne by the Sisters, ye shall *committe to the sayde Gouernours, that they may bothe put ordre for the weyghyng of the same to the Weauer, and for the measuryng of it at the returnyng thereof.

[margin: Take special care of Sheets, Blankets, Beds, &c.]
You shal also, as the chief gouerneresse and worthy Matrone of this house, haue speciall regarde to the good orderyng and kepyng of all the Shetes, Couerlettes, Blankettes, Beddes, and other implementes, committed to your charge, that now do, or hereafter shal, apperteine vnto the poore.

[margin: Let no poor Patient sit and drink in your house. And never send drink into the [E. j, back.] wards.]*
Also ye shall suffre no poore persone of this house to sitt and drynke within your house at no tyme; neyther shall ye so sende them drynke into their wardes, that thereby dronkenesse myght *be vsed and continued among them; but as much as in you shal lie, ye shall exhorte them to vertue and temperaunce,

[margin: This Hospital is for members of Christ, not for drunkards.]
declaring this house to be appointed for the herboure and succour of the dere members of Christes body, and not of dronkardes and vnthankefull persones.

[margin: If you see any wrong doing,]
Herewith ye are charged; and not with any other thing. But if there shalbe any thyng done by any officer or other persone of this house, that shalbe vnprofitable thereunto, or that may be occasion of any disorder, or shal engendre slaundre to the same, that ye

[margin: tell it to 1 or 2 of the Governors.]
then declare it to some one or two of the Gouernours of this house, & to none other persone, nor no further to meddle therein.

[margin: [E. ij.] The Sisters.]*

*The Sisters.

Your charge is, in al thinges to declare and shewe your selues gentle, diligent, and obedient to the Matrone of this house, who is appointed & aucthorised to be your chief gouerneresse and ruler.

[margin: are to obey the Matron,]

[margin: and tend the Poor,]
Ye shall also faithfully and charitably serue and helpe the poore in al their grieues and diseases, aswell

Barts Order, 1552. The Sisters & the Surgeons. 311

by kepyng them swete and cleane, as in geuyng them their meates and drinkes after the moste honest & comfortable maner. Also ye shall vse vnto them good and honest talke, suche as may comforte & amend them; and vtterly to aduoyde all lyght, wanton, and foolishe wordes, gestures and maners *vsyng youre selues vnto theim with all sobrietie and discretion. And aboue all thynges, se that ye auoyde, abhorre and detest, skoldyng and dronkenesse, as moste pestilent and filthie vices. *[keep them sweet, give them their food, talk improvingly to them, avoiding all wanton words [* E. ij, back.] and gestures, and specially detesting scolding and drunkenness.]*

Ye shall not haunte or resorte to any maner of persone oute of this house, except ye be licenced by the Matrone; neither shal ye suffre any lyght persone to haunt or vse vnto you; neither any dishonest persone, eyther man or woman; and so muche as in you shall lie, ye shall auoyde & shonne the conuersacion and company of all men. *[You're only to visit folk authorised by the Matron, and you're to shun the company of men.]*

Ye shall not be out of the womans warde, after the houre of vii. of the clocke in the nyght, in the wynter tyme, nor after .ix. of the clocke at nyght, in the So*mer, except ye shalbe appointed and commaunded by the matrone so to be, for some greate and speciall cause that shall concerne the poore (as the present daunger of death or extreme sicknes); and yet so beyng commaunded, ye shall remaine no longer with such diseased persone, then iust cause shall require. *[You're not to leave the Women's Ward after 7 in winter, [* E. iij.] or 9 in summer, unless orderd by the Matron, and then you're not to stop long.]*

Also if any iust cause of grief shal fortune vnto any of you, or that ye shall see lewdenes in any officer or other person of this house, whiche maie sounde or growe to the hurte or slaunder therof, ye shall declare thesame to the Matrone, or vnto one or two of the Gouernours of this house, that spedy remedy therin may be had, & to none other persone; neither shall you talke or *meddle therin any furder. This is your charge; and with any other thyng you are not charged. *[If you see lewdness in any Officer, tell the Matron and 1 or 2 Governors, [* E. iij, back.] but no one else.]*

The Chirurgiens. *[The Surgeons.]*

YOure charge is, faythfully and truelie, to the vttermoste of your knowlege & connyng, to helpe to cure the greues and diseases of the poore of this Hospitall, settyng aside all fauoure, affection, gayne or lucre; and that as well to the poorest, destitute of all frendes and succours, as to such as shal peraduenture be better frended, ye shall, with al fauour and frendship, procure the spedie recouery of their health. *[Do your very best to cure the diseases of the Poor, without favouring those with good friends.]*

Also for your stipend and fee, geuen & payd out of

312 Barts Order, 1552. The Surgeons & the Porter.

[* E. iiij.]
Be always ready, when bidden by the Almoners and Hospiteler, to examine Patients. Then give your honest judgment on them,

this house, ye shalbe redy at the commaun*dement of the Almoners of this house, & Hospiteler of the same, to view and loke vpon such diseased persones as here from tyme to tyme shalbe presented. And after your view, to signifie to the sayde Almoners or Hospiteler, your Iudgement of the said diseased persone, without all affection, whether he or she be curable or not, to the entent there may be none admitted into this house

and if they're incurable, don't admit them, so as to keep out the curable ones.

that shalbe incurable, to y⁰ great lette and hinderaunce of the curing & helping of many other; ne none reiected and put back that are curable, to the greate slaunder of this house, and displeasure of God.

When you dress a Patient,
[* E. iiij, back.]

Also, at all suche tymes as ye shall go to the dressyng of any diseased persone in this house *as muche as in you is, ye shall geue vnto hym or her, faithfull and

advise him to sin no more, but to thank God.

good cou*n*saill, willing theim to mynde to sinne no more, and to be thankefull vnto almighty GOD, for whose sake they are here comforted of men. And

Take no gift from any poor men or their friends.

aboue all thyng, ye shal take nor receyue of no persone, any gifte or rewarde for the curyn*g* or helping of them, either of them or their frendes; but ye shall first make the same offer or reward know*en* vnto y⁰ Almoners of this house.

And never burden this House with a Patient for whom you've been paid.

Also we vtterly forbidde and commaunde you, that ye by no coloure,¹ pester or burde*n* this house with any sicke or diseased persone, for the curyng of whiche persone, ye before haue receiued a somme or sommes of money, vpon paine to be dimissed thys house.

[* E. v.]

*This is your charge and office, with the whiche ye haue to do, and not with any other thing, neither with any other office, in this house. But if you shal per-

Report any wrong-doing you see

ceiue at any tyme, any thynge done by any officer or other persone of this house, that shalbe vnprofitable therunto, or that maie be occasion of any disorder, or shal engender slaunder to the same, that ye then declare

to the Almoners.

it to the Almoners, or one of them, & no farther to medle therein.

The Porter.

The Office of the Porter.

Keep the doors,

YOur charge is, to kepe the dores, openyng and shuttyng them in due time, and to geue good

[* E. v, back.]
and look to all folk going in and out. Don't let the food of the Poor be stolen,

hede to all suche *persones as shall at any tyme passe to & fro out of this house, as wel for the conueighing or embesillyng of any thyng that apperteyneth to the poore of thys house, as Wood, Cole, Bread, meate or drynke,

¹ pretence.

as also for all suspicious persones, as men to resorte to the womens warde, or women to the mens wardes, or such suspicious men to resorte vnto the men, or women to the women, as shalbe thought to bee petie pickers, or persones otherwise of naughtie disposition. *[margin: or suspicious men go to the Women's ward.]*

And also euery nyght, at the houre of .vii. of the clocke in the Somer, ye shall goo into euery warde where the poore men be, and see them in good order, and suffer no Sister nor other woman to remayne among them (ex*cepte iust cause be declared by the Matrone) and cause them to saie the appointed praiers. *[margin: Every night at 7 in summer, go into every Men's Ward, and turn out the Women. [* E. vj.]]*

And whatsoeuer poore persone shalbe founde a swearer, or an vnreuerent vser of his mouth, toward God or his holy name, or a contempner of the Matrone or other officer of this house, or that shall refuse to go to bedd at the lauful houres before appointed, hym shall ye punyshe (after ones warning geuen) in the stockes, and further declare his folie vnto the Almoners of thys house, that they maie take suche order with him or theim, as shal seme mete by their discretions. *[margin: And if any Patient swears, or abuses the Matron, or won't go to bed, put him in the Stocks, and report him to the Almoners.]*

Ye shall also be diligent and redy from tyme to tyme, to doe such other thinges as the gouernours of this house shal assigne *and appointe you. This is your charge, and more you haue not to do; but if ye perceyue at any time, any thyng done by any officer of this house, or other persone that shall mainteine disorder, or procure slaunder to this house, that ye then declare the same to some one or two of the Gouernours of this house, and to none other persone, and no furder to medle therin. *[margin: Do whatever the Governors bid you, [* E. vj, back.] and report any disorder to them.]*

The Biddelles.

[margin: The Beadles.]

YOure office and charge is to geue attendaunce from tyme to tyme, vpon the gouernours of this house, and to do suche busines as they shall assygne you. *[margin: Attend the Governors when they're at the Hospital.]*

And also all suche daies as the Gouernours of this house *shall not sitte in thys Hospitall for the affaires of the same, ye shall separate and deuide youre selues into sondrie partes of the citie & liberties therof, euery man takyng his seuerall walke. And if in any of your walkes ye shall happen to espie any persone infected with any lothelie grief or disease, whiche shall fortune to lie in any notable place of thys Citie, to the noiaunce and infection of the passers by, and slaunder of this house, ye shall then geue knowlege therof vnto the Almoners of this Hospital, that they maie *[margin: [* E. vij.] When they're not, let each Beadle patrol his district, and if he sees any diseased man, report him to the Almoners.]*

314 Barts Order, 1552. Beadles & Newgate Visitor.

take suche order therein as to them shalbe thought mete.

[margin: Also watch that no cured [E. vij. back.] Patient shum disease, and beg.*

[margin: If he does, put him in the Cage,]

[margin: and report to the Governors.]

Ye shal also haue a speciall eye and regarde vnto all suche persones, as haue bene cured, & *healed in this house, that none of them counterfeicte any griefe or disease, neither begge within the Citie and liberties thereof. And if ye shall fortune to fynde any so doyng, ye shal immediatly committe hym, or them, to some Cage, and geue knowledge thereof to the Gouernours of this house, that they maie take furder order, as they shal thinke best.

[margin: Beadles must not drink with beggarly folk in pothouses,]

[margin: or take bribes from them to let them beg.]

[margin: [E. viij.]]*

Ye shall not haunte nor frequente the company of any poore and beggarlie persones (that is to saie), to drinke or eate with them in any victuallyng house or other place, neither shall ye receiue any bribe or reward of any of theim, least by occasion thereof ye should wyncke at them, and so lewedly licence them to begge, * vpon paine to be dimissed this house.

[margin: Beadles mustn't let any idle vagabonds beg,]

[margin: but must put them in jail, and report them to the Alderman or the Lord Mayor.]

Also ye shall not suffer any sturdy or ydle begger or vagabounde, to begge or aske almoise within this Citie of London, or suburbes of the same; but ye shal forthwith committe all suche to warde, and immediatly signifie the name and sirename of hym or theim, to the Alderman of that warde where ye shal apprehend any suche begger, or els to the Lorde Maiour, that execucion may be done, as the lawe in that case hath prouided. This is your charge.

The Visitour of Newgate.

[margin: The Visitor of Newgate.]

[margin: [E. viij. back.] is to visit the poor Prisoners,]*

YOur charge is, faithfully and diligently to visite all *the poore and miserable captiues within the pryson of Newgate, and minister vnto them suche ordinary seruice at times conuenient, as is appointed by the kynges maiesties booke for ordinary praier.

[margin: and learn texts to comfort them with.]

Also that ye learne, without booke, the most wholsome sentences of holie Scripture, that may comforte a desperate man, that redilie ye may minister them to suche persones as ye shal perceyue them moste nedefull to be ministred vnto.

[margin: He is to act justly, take no bribes,]

[margin: [F. j.] but exhort the Prisoners to restore their thefts,]*

Also ye shall faithfully and truelie vse and beare youre selfe betwene partie and partie, excludyng brybes and all other corrupcion, that is to saie, betwene the prisoners and the parties to whome they haue offended, ex*hortyng them to the vttermoste of your connyng, to make restitucion of their thynges falsely gotten, shewyng them the burden of conscience depending thervpon.

And that also thei disclose all suche other persones as they knowe liuyng, whiche by robberie or murther maie hurte a common weale. And in al their extremes and sickenesses, ye shal be diligent and redy to comforte them with the most pitthie and frutefull sentences of Goddes moste holy worde. *and tell of other thieves.*

And whatsoeuer persone you shall perceiue to haue substaunce, and to be mynded to bestowe somwhat thereof in dedes of charitie, ye shal exhorte him or them to bestowe some parte to the relief of the nedy and diseased *persones of this house. And of al suche giftes from tyme to tyme, to geue knowledge to the Almoners or scrutiners of this house. *When he sees a charitable man, he is to ask him to give to the Hospital poor. [* F. j, back.]*

And forasmuche as you are nombred among the ministers of Christes churche,¹ ye shal therfore, foure tymes in the yeare at the least, (that is to saie) euery quarter ones, do suche seruice in the said churche as is requisite for suche a Minister to do. *He is to officiate at Christ Church once a quarter.*

> This is your charge, which
> see that ye do ; and with
> any other thynge ye
> are not char-
> ged.

¹ Christ Church, Newgate St., founded by Hen. VIII. on the dissolution of the Grey Friars Monastery. He put together the parishes of St. Nicholas and St. Ewin, and so much of St. Sepulchre's as was within Newgate, for his new parish and its Grey Friars Church which he cald Christchurch. The present church is from Wren's designs, and was finisht in 1705.—Cunningham. See page 131, above.

*The estimate of the yearely charges of this Hospitall.

[* F. ij.]
The Hospital yearly expenses.

No account is taken here of the foundation expenses of the Hospital,

IT is first here to be considered, that although the charges ware very great, to bryng the endowment of the Hospitall, into suche poynte as behoued, and to furnysshe the house with necessary Implementes and beddyng for suche nombre (as hath bene afore touched in the beginnyng[1]) yet is there of all these charges, no parcel here vnder mencioned, but the yearely expences onely, susteined for the maintenaunce and continuaunce of the same. And albeit these charges folowyng, be all and euery of them ordenary, and of necessitie, yeat, for that there *is a difference in the certeintie of the one and the other, they are deuyded into twoo kyndes, with these titles: Charges certeine, & Charges vncerteine.

but only of the Maintenance charges,

[* F. ij, back.]
1, certain,

2, uncertain.

Charges certeine.

1. *Fixt charges.*

i. Wages and Fees.

Are firste, the yearely wages and fees of those Officiers and Seruauntes, that necessarilie serue and attende for the poore, as ensueth; and after them the charges of housholde, Reparacions, and suche lyke.

Hospitler, £10;

To the Hospiteler	x. l.
To the Renterclerck	x. l.
To the butler	vi. l. xiii. s. iiii. d.
To the Cooke, for his meate, drincke, and wages	vi. l.
To the Porter	vi. l.
To .iii. Chirurgiens	lx. l.
To .viii. Biddles	xxvi. l. xiii. s. iiii. d.
*To these and to the other, for their liueries	x. l.
To the Matrone & .xii. Sisters, for their wages	xxvii. l. vi. s. viii. d.
To the Matrone, for her boord wages, at .xviii. pence the wieke	iii. l. xviii. s.
To the .xii. Sisters, for their boord wages at .xvi. d. the wieke for euery of them,	xl. l. xii. s.
To the Matrone for her liuerie ...	xiii. s. iiii. d

Cook, £6;

3 Surgeons, £60.

[* F. iij.]
Liveries, £10.
Matron and Sisters' wages,

board,

liveries.

[1] p. 293.

Barts Order, 1552. Yearly Hospital Expenses.

To the Sisters for their liueries ... ... vi. l.	
To the ministers of Christes churche, by the kinges maiesties assignement, that is to saie, a vicare, a visitour of Newgate, v. priestes, two clerckes, and a sextein, yearely[1] ... ... C. and .vi. l.	ii. Ministers of Christ Church, £106.
To the ministers of the chur*che within the Hospitall,[2] by the same assignement, that is to saie, to a Vicare, a clercke, & a sextein ... ... xxiii. l. vi. s. viii. d.	[* F. iij, back.] Ministers of Lit. St. Bartholomew's, £23 6s. 8d.
To certeine men of Law and other persones, geuen in fees by the kynges sayd maiestie, yerely by patente ... ... xxviii. l. iiii. s.	iii. Lawyers, &c., £28 4s.

Charges of houshold.

iv. House Charges:

For the dietes of an .C. persones, at twoo pence the persone for euery daie, iii. C. l. vi. s. viii. d.	Food,
For .lxviii. lode of Coles, at xvi. s'. the lode ... ... liiii. l. viii. s.	Coals,
For woodd yerely ... ... xxiiii. l.	Wood,
For candles yerely ... ... v. l.	
For yerely reparacions of the Hospital, and tenementes apperteinyng to thesame ... xl. l.	Repairs.
*Somme of the charges certein vii. C. lxxx. viii. l. ii. s.	[* F. iiij.] Total, £798 2s.

The charges vncerteine (forasmuche as it cannot certeinly be knowen to what they may amounte) are here sette forth without Sommes, onely to sygnifie vnto you, that there are many charges more to be considered, then certeine accompte can be made of.

Charges vncertein.

2. Varying Charges.

For Shirtes, Smockes, and other apparell for the poore, niedefull, either at their commyng in or departure. For Sugre & Spices for Cawdelles for the sicke, Flaxe for shetes, and Weuyng of the same ; Soltwhiche[3] cloth for winding shetes, bolles, bromes, baskettes, encence, Iu*niper, asshes to boocke[4] their clothes. And

Clothes, Sugar, Caudles, Flax, Weaving, Winding-sheets, [* F. iiij, back.]

---

[1] See Forewords, § on Vicary at St. Bartholomew's.
[2] Little St. Bartholomew's.
[3] The only Saltwick in Bartholomew's *Gazetteer* 1887, is 'Saltwick, hamlet, Stannington parish, Northumberland, 4 miles S. of Morpeth.' This can hardly be the place meant.
[4] buck, wash.

318 Barts Order, 1552. Yearly Hospital Expenses.

<small>leaving and journey-money. Last year, 1551, £60.</small>

also money geuen to the poore at their departure, whiche is measured accordyng to their Iourney and nede. The whiche vncertein charges amounted one yeare to the some of .lx. l.

<small>Total, fixt charges, £798 2s., varying [? £100 : say £900 the two].</small>

So cometh the certeyn charges of this house yearely to the somme of vii. C. lxxx. xviii. l. ii. s. besyde the vncertein expences, and other extraordinary charges, whiche can not be rated ne accompted.

<small>To meet this, are Hen. VIII's £333 6s. 8d., and the City's £333 6s. 8d., total, £666 13s. 4d.,</small>

Toward the whiche, is yerely receiued by the endowment of the kynges maiestie .iii. C. xxxiii. l. vi. s. viii. d. And by the like endowment of the Citie of London, iii. C. xxxiii. l. vi. s. viii. d. The whiche, in the whole, is .vi. C. lxvi. l. xiii. s. iiii. d.

<small>[* F. v.] leaving £131 8s. 8d.</small>

*So is the Hospitall charged yerelie of certeine (besyde the vncerteine expences) ouer & aboue the somme of their reuenues .C. xxxi. l. viii. s'. viii. d.

<small>and all unfixt charges, to the charity of merciful Citizens. For the increase of which, we pray to Christ.</small>

Whiche onely ryseth of the charitie of certeine mercyfull citizeins ; for whose continuaunce, with the encrease of moe, we earnestly praie vnto the founteine of mercie, Iesus Christe, the lord of all, to whome for euer apper-

teigne, the kyngdome, the
power, and the glory,
worlde without
ende.

Amen.

¹A daily seruice
for the poore.

T the Houre of eyght of the Clocke in the mornyng, and .iiij. of the clock at the afternoone, throughout the whole yeare, there shal a bel be rong the space of halfe a quarter of an houre, and immediatly vpon the seassyng of the bell, (the poore liyng in their beddes that cannot aryse; & kneling on their knees, that can aryse in euery ²warde, as their beddes stande,) they shal by course, as many as can rede, begyn these praiers folowyng. And after that the partie whose course it shalbe, hath begon, all the rest in that warde shal folow and. aunswere, vpon paine to be dimissed out of the house. And thryse in the weke, that is to saie, Sondaie, Wedensdaie, and Fridaie, they shal saie the letany in maner and forme as it
is thende of this
booke.

The minister shal begyn
and the rest shal folowe.

OUre Father whiche arte in heauen, hallowed be thy name; thy kyngdom come; thy wil be done in earth as ³it is in heauen. Geue vs this day our dailie bread, and forgeue vs oure trespasses, as we forgeue theim that trespasse against vs.

And leade vs not into temptacion.
The poore.
But deliuer vs from euel. Amen.
The minister.
O Lord fauourablie here vs!
The poore.
And mercifullie graunte oure peticions!
The minister.
We confesse thy goodnes.
The poore.
For we haue tasted of thy mercy.

---

¹ F .vj.   ² F .vj. back.   ³ F .vij.

The minister.
Blesse thine own people O God!
The poore.
Whiche succour vs for thy names sake.
The minister.
Remember not our wickednesse, O Lorde!
[1]*The poore.*
And pardon all our synfulnes!
The minister.
Let vs geue prayses vnto the Lorde!
The poore.
We will praise hym in his holy woorde.
The minister.
Glorie be to the father, and to the sonne, & to the holie ghost!
The poore.
As it was in the beginnyng, is now, and euer shalbe worlde without ende. Amen!

The v. Psal. *Verba mea auribus.*

POnder my wordes, O Lorde, consider my meditacion! O herken thou vnto the voice of my callynge, my kyng and [2]my God, for vnto the wil I make my praier.

My voice shalt thou here be-tymes O lord; early in the mornyng, wil I directe my praier vnto the, and will looke vp.

For thou art the GOD that hath no pleasure in wickednes: neither shal any euill dwel with the.

Suche as be folishe, shal not stande in thy syght: for thou hatest al them that worke vanitie.

Thou shalt destroie them that speake leasyng: the lord will abhorre both the bloud-thirstie and deceiptfull man.

But as for me, I will come vnto thy house, euen vpon the multitude of thy mercy: and in thy feare wil I worship toward thy holy temple.

[3]Leade me, O Lorde, in thy righteousnes, because of myne enemies: make thy waie playne before my face!

For there is no faithfulnes in his mouthe: their inward partes are very wickednes.

Their throte is an open Sepulchre: they flatter with their tongue.

Destroie thou them, O God; let them peryshe through their owne ymaginacions: cast them out in the multitude of their vngodlinesse, for they haue rebelled against the.

And let all them that put their trust in thee reioyse: they shall euer be geuyng of thankes, because thou defendest them; they that loue thy name shalbe ioyful in the.

[1] F .vij. back. [2] F .viij. [3] F .viij. back.

[1] For thou, Lord, wilt geue thy blessyng vnto the ryghteous: & with thy fauourable kyndnesse wilt thou defende hym, as with a shylde.

The Psal. *Domine dominus.*

O Lorde our Gouernoure, how excellent is thy name in all the worlde: thou that haste sett thy glorie aboue the heauens.

Out of the mouthe of verie babes and sucklinges hast thou ordeyned strengthe, because of thine enemies: that thou mightest stil the enemie and the auenger.

For I will consider the heauens, euen the workes of thy fingers the Mone and the starres whiche thou hast ordeyned.

[2] What is man, that thou arte myndefull of him? and the sonne of man, that thou visitest hym?

Thou madest hym lower then the aungels: to crowne him with glorie and worshippe.

Thou madest him to haue dominien of the workes of thy handes: and thou hast put all thynges in subiection vnder his fete:

All shepe and oxen: yea, & the beastes of the fielde;

The foules of the ayre, & the fishe of the sea: and whatsoeuer walketh through the pathes of the Seas.

O lorde our gouernour: how excellent is thy name in all the worlde!

Glory be to the father, &c.

As it was in the beginnyng, is now, and euer, &c. Amen.

[3] Then this antheme.

BEyng made the seruauntes of God by faith in the merites & bloudsheddyng of his moste deare sonne, our sauiour Iesu Christe, we are certayne and sure to be saued, and that no dampnacion can happen vnto vs, so that we walke not in the wicked desires of the fleshe, but in the heauenlie & verteous life praysed and commended of God.

Then this Psalm. *Miserere.*

HAue mercy vpon me (o god) after thy great goodnes: & accordyng vnto the multitude of thy mercies, doe awaie myne offences!

Washe me throwlie from my wickednesse: and clense me from my synne!

[4] For I knowledge my faultes; and my synne is euer before me.

Against the, onely, haue I sinned, and doone this euill in thy syght: that thou myghtest be iustified in thy saiyng, and cleare when thou arte iudged.

[1] G .j. [2] G .j, back. [3] G .ij. [4] G..ij, back.
VICARY.

Behold, I was shapen in wickednesse: and in synne hath my mother conceiued me.

But lo, thou requirest truthe in the inwarde partes: and shalt make me to vnderstande wisedome secretly.

Thou shalt purge me with Isope, and I shalbe cleane: thou shalt washe me, & I shalbe whiter then Snowe.

Thou shalt make me heare ioye & gladnes: that the bones which thou hast broken may reioyce.

Turne thy face from my sin[1]nes: and put out al my misdedes!

Make me a cleane harte (O God): and renue a ryghte spirite within me!

Cast me not awaie from thy presence: and take not thy holie spirite from me!

O geue me the comfort of thy helpe agayne: and stablishe me with thy fre spirite!

Then shall I teache thy waies vnto the wicked: and sinners shalbe conuerted vnto the.

Deliuer me from bloud-giltines, O god, for that thou art the God of my helth: & my tongue shal syng of thy ryghteousness.

Thou shalt open my lyppes, O lorde: my mouthe shal shewe thy prayse.

For thou desirest no sacrifice; els wold I geue it the: but thou [2]delitest not in burnt offeryng.

The sacrifice of god is a troubled spirite: a broken and a contrite harte, O GOD, shalt thou not despice.

O be fauourable & gracious vnto Sion: builde thou the walles of Ierusalem!

Then shalt thou be pleased with the sacrifice of ryghteousnes, with the burnt offerynges and oblacions: then shall they offer younge bullockes vpon thyne Aultar.

The Lesson.

LEt vs walke in the holy spirite of God, & abhorre the lustes and desires of oure filthy fleshe; for our fleshe is contrary to our spirite, and the spirite contrary to the fleshe: these [3]are so contrary, one to another, that we cannot do what we wold. But if we be led by the spirite of God, then are we not vnder dampnacion. The dedes of the fleshe are these, aduoutrie, fornication, vnclennes, wantones, worshyppyng of ymages, witchecrafte, hatred, variaunce, zeale, wrathe, strief, sedicious sectes, enuieng, murther, dronkenesse, glotony, & suche lyke. And whatseuer he be that committeth these thinges, shall not enherit the kyngdome of God. But the frutes of the holy spirite of God, are contrarie, whiche are these: Loue, peace, long suffering, gentlenes, goodnes, faythfulnes, mekenes, temperauncie,

---

[1] G .iij.   [2] G .iij, back.   [3] G .iii[i].

and such like; against the whiche there is no lawe. And if we be the chyldren of God, we [1]must crucifie our fleshe, with all the lustes and affections therof.

The Psalme of *Benedicite*.

O Al ye workes of the lord, speake good of the lorde! prayse hym and set hym vp for euer!

O ye aungelles of the Lorde, speake good of the lorde! prayse him and set hym vp for euer!

O ye heauens, speake good of the Lorde! prayse hym and sett hym vp for euer!

O ye waters that be aboue the firmament, speake good of the Lorde! prayse hym and set hym vp for euer!

O all ye powers of the Lorde, speake good of the Lorde! praise hym, and set hym vp for euer!

O ye sunne and mone, speake [2]good of the Lorde! prayse hym and set hym vp for euer!

O ye starres of heauen, speake good of the Lorde! praise him & set hym vp for euer!

O ye showers & dewe, speake good of the Lord! prayse hym and set hym vp for euer!

O ye windes of God, speake good of the Lorde! prayse hym and set hym vp for euer!

O ye fyre and heate, speake goode of the Lorde! praise hym & set hym vp for euer!

O ye Winter and Sommer, speake good of the Lorde! praise hym and set hym vp for euer!

O ye dewes & frostes, speake good of the Lorde! prayse hym and set hym vp for euer!

O ye froste and colde, speake good of the Lorde! prayse hym [3]and set hym vp for euer!

O ye yse and Snowe, speake good of the Lorde! prayse hym and set hym vp for euer!

O ye lyght and darkenesse, speake good of the Lorde! praise hym and set him vp for euer!

O ye lightenynges and cloudes, speake good of the Lorde! praise him & set him vp for euer!

O let the yearth speake good of the Lorde! yea, lette it prayse hym and set hym vp for euer!

O ye mountaines and hilles, speake good of the Lord! praise him and set him vp for euer!

O all ye grene thinges vpon the yearth, speake good of the Lorde! prayse hym and set hym vp for euer!

O ye welles, speake good of the Lorde! prayse hym and sette [4]hym vp for euer!

[1] G .ii[i], back.   [2] G .v.   [3] G .v, back.   [4] G .vj.

O ye seas & flouddes, speake good of the Lorde! prayse hym and set hym vp for euer!
O ye whales, and al that moue in the waters, speake good of the Lorde! prayse hym & sette hym vp for euer!
O all ye foules of the ayre, speake good of the Lord! praise hym and set hym vp for euer!
O all ye beastes and cattell, speake good of the Lorde! praise hym and set hym vp for euer!
O ye chyldren of men, speake good of the Lorde! prayse hym and set hym vp for ever!
O let Israell speake good of the Lorde! prayse hym and sette hym vp for euer!
O ye priestes of the Lorde, speake good of the Lorde! praise [1]hym and set hym vp for euer!
O ye seruauntes of the Lord, speake good of the Lorde! praise hym and set hym vp for euer!
O ye spirites & soules of the ryghteous, speake good of the Lorde! prayse hym and set hym vp for euer!
O ye holy and humble men of harte, speake ye good of the Lorde! praise ye him, and set him vp for euer!
Glory be to the father, and to the sonne, and to the holy ghost! As it was in the beginnyng, is now, and euer, &c. Amen.

  The minister.
Lorde, haue mercy vpon vs!
  The poore.
Christ, haue mercy vpon vs!
  The minister.
Lord, haue mercy vpon vs!

[2]Then shall all saie together.

I Beleue in God the father Almyghtie, maker of heauen and yearth; and in Iesus Christ, his onely sonne oure Lorde, which was conceiued by the holy ghost, borne of the virgin Mary, suffered vnder ponce Pilate, was crucified, dead, and buried; he descended into hel; the third daie he rose agayne from the dead; he ascended into heauen, and sytteth on the ryght hande of God the father Almightie; from thence shall he come to Iudge the quicke and the dead. I beleue in the holy Ghost, the holy Catholike churche, the communion of sainctes, The forgeuenes of synnes, The resurection of the body; And the lyfe euer[3]lastyng. Amen.

Our father, whiche art, &c.
  The minister.
Deale fauourably with vs, O Lorde!

[1] G.vj, back.   [2] G.vij.   [3] G.vij, back.

The poore.
For we be very miserable.
The minister.
Heare vs, O Lord, when we cal vpon the!
The poore.
For in the, onely, is all our trust.
The minister.
O Lorde, saue the kyng!
The poore.
And blesse oure gouernours!
The minister.
Power fourth thy great mercy, O Lorde,
The poore.
Vpon all thy poore membres in this house!
The minister.
Let vs praie!

[1] For the Kyng.

ALmighty and euerliuyng God, we moste humblie & hartelie beseche the, for the precious bloud sake of our Sauiour Iesu Christe, thy onely sonne, to gouerne, protecte, and defende, our moste innocent and dreade souereigne lorde, Kyng Edwarde the sixte, thy seruaunt, and our gouernour and defendour, that he maie so rule & gouerne al thy people of England committed to his charge, as shall be to the honoure of thy holy name, and proffit of all his louyng subiectes & commons of the same. Indue hym also, O Lorde, with encrease of grace, and nombers of yeares, that he may long reigne ouer vs in thy feare; and graunt hym victory ouer all his aduer[2]saries & enemies! This we beseche the to graunt, O Lorde, for Iesu Christes sake, our mediatour and aduocate. Amen.

We beseche the, O Lorde, to prospere and kepe the Gouernours of this house, and, accordyng to thy moste holie promes, to blesse and encrease all suche as helpe to fede and heale oure hungry and sicke bodies, not only with the encrease of goodes in this worlde, but also with the life euerlastyng, whiche, of thy great mercie, thou hast promised them, through Iesu Christe our Lorde. Amen.

GRaunte, moste mercifull Lorde, vnto euery one of vs, beyng diseased persones, to haue in remembraunce the bitter peynes that thy sonne [3]suffered for vs in his moste holy passion, and to arme oure selfes with pacience, knowyng that for sinne this hath happened vnto vs. Graunt vs also that we may be plentifully indued with thy spirite, that in all our troubles and paines we may extolle & praise thy holy name, with a stedfast mynde and purpose, neuer

---

[1] G .viij. [2] G .viij, back. [3] H .j.

more to offende thy dyuine maiestie. And this we desire of the, for thy moste blessed sonnes sake, our sauiour Iesu Christe: To whom, with the and the holy ghost, be all prayse and glorie for euer and euer. Amen!

[1] After-noone praier.

Our father whiche art, &c.

The minister.
Heare vs, O Lorde, that call vpon the.
The poore.
And encline thine eares vnto our praiers.
The minister.
For we are very miserable.
The poore.
Be mercifull vnto vs, O Lord.
The minister.
For daie and night we will praise thee.
The poore.
Thy name is to be praised for euer and euer.
The minister.
Glory be to the father, &c.
The poore.
As it was in the beginnyng, is now and euer, &c. Amen.

The .lxxxxvi. Psal. *Inclina domine.*

Bowe downe thine eare, O Lord, and heare me: for I am poore and in miserie.

Preserue thou my soule, for I am holie: my God, saue thy seruaunt that putteth his trust in *thee*!

Be merciful vnto me, O Lord: for I will call daily vpon the.

Comfort the soule of thy seruaunt: for vnto the, O Lorde, do I lifte vp my soule.

For thou, lorde, arte good and gracious: and of great mercy vnto all them that cal vpon the.

Geue eare, Lorde, vnto my praier: and ponder the voice of my humble desires.

In the tyme of my trouble I will call vpon the: for thou hearest me.

[2] Among the Goddes there is none like vnto the, O Lord: there is not one that can doe as thou doest.

---

[1] H .j, back. [2] H. ij.

Barts Order, 1552. *Daily Service for the Poor.*

All nacions whom thou hast made, shall come and worshyp the, O Lorde: and shall glorifie thy name.

For thou art great, and doest wonderous thynges: thou arte God alone.

Teache me thy waie, O Lorde, and I will walke in thy truthe: O knitte my harte vnto the, that I may feare thy name.

I will thanke *th*ee, O lord my god, with all my harte: and will praise thy name for euer.

For great is thy mercy toward me: and thou hast deliuered my soule fro the nethermoste hell.

O God, the proude are rysen [1] against me: and the congregacion of naughty me*n* haue sought after my soule, and haue not sett the before their eies.

But thou (O lorde God) arte full of compassion and mercie: long suffering, ple*n*teous in goodnesse and truthe.

O turne the, then, vnto me, & haue mercy vpon me: geue thy strength vnto thy seruaunt, and helpe the sonne of thyne hande mayde.

Shewe some good token vpo*n* me, that they which hate me, may se it, & be ashamed: because thou, lorde, hast helped me and comforted me.

Glory be to the father, and to the sonne, and to the holy Ghost. As it was in the beginning, is nowe, and euer shalbe, &c. Amen.

[2]The .xcvi. Psal. *Cantate domino.*

O Syng vnto the Lorde a newe song! sing vnto the Lord, al the whole earth!

Syng vnto the Lorde, and prayse his name! be tellyng of his saluacion from daie to daie!

Declare his honour vnto the Heathen: and his wonders vnto all his people!

For the Lorde is greate, and cannot worthely be praysed: he is more to be feared than all the Goddes.

As for all the Goddes of the Heathe*n*, thei be but Idolles: but it is the lord *tha*t made the heaue*n*s.

Glory and worship are before hym: power and honour are in his sanctuary.

Ascribe vnto the Lorde (O ye [3] kinredes of the people:) ascribe vnto the Lorde, worshippe and power!

Ascribe vnto the Lord, the honour due vnto his name! bryng prese*n*tes, & come into his courtes!

O worshippe the lorde in the bewtie of holines! let the whole earth stande in awe of hym!

Tell it out among the heathe*n* that the lorde is kyng: and that

[1] H .iij.   [2] H .iij, back.   [3] H .iiij.

328 *Barts Order*, 1552. *Daily Service for the Poor.*

it is he whiche hath made the round worlde so faste that it can not be moued : and howe that he shall iudge the people ryghteously.

Let the heauens reioyce, and let the earth be glad ! let the Sea make a noyse, & al that therin is !

Let the fielde be ioyfull, and all that is in it ! then shall all the trees of the wood reioyce before [1]the Lorde.

For he commeth, for he commeth to Iudge the yearth : & in righteousnes to iudge the world, and the people with his truthe.

Glorie be to the father, and to the sonne, & to the holie ghost !

As it was in the beginnyng, is now, and euer shalbe worlde without ende. Amen.

The lesson. *Roman* .vi.

KNow ye not *tha*t al we whiche are baptised in Christ, are baptised to die with hym ? We are buried with hym by baptisme for to die, that likewyse as Christe was raysed fro*m* death by the glory of the father, euen so we also shold walke in a newe life ; for if we be graffed in death like vnto hym, euen so shal [2]we be partakers of the resurrection ; knowyng this, that our old man is crucified with hym also, that the body of synne myght vtterly be destroied, that he*n*cefurth we should not be seruau*n*tes vnto sinne. Wherfore, if we be dead with C*h*riste, we beleue that we shal also liue with him : knowing that Christe, beyng raised from death, dieth no more, death hathe no more power ouer him ; for as touching that he died, he died co*n*cerning once. And as touchyng that he liueth, he liueth vnto God. Likewise consider ye also, that ye are dead as touchynge sinne, but are aliue vnto God, through Iesus Christ our lord. Let not sinne therefore reigne in youre mortall body, that you should thereunto obey by the lu[3]stes of it. Neither geue you your members as instrume*n*tes of vnrighteousnes v*n*to sinne, but geue ouer your selues vnto God, as they that, of deathe, are lyue. And geue ouer your me*m*bers, as instrume*n*tes of ryghteousnes vnto God, for sinne shall no haue power ouer you, because ye are not vnder the lawe, but vnder grace : what then ? Shal we sinne because we are not vnder the lawe, but vnder grace ? (God forbid !) Knowe ye not how that to whom soeuer ye committ youre selues as seruau*n*tes to obey, his seruauntes ye are, to whome ye obey, whether it be of sinne vnto death, or of obedience vnto righteousnes ? God be thanked that, though ye were the seruauntes of sinne, ye haue yet obeyed with [4]harte vnto the rule of the doctrine that ye be brought vnto ; ye are then made fre from synne, and are become the seruauntes of righteousnes. I speake groselie, because of the infirmitie of your fleshe : as ye haue geue*n* your members seruauntes to vncleanes and to iniquities (from one iniquitie to another), eue*n* so now geue ouer youre members seruauntes vnto righteousnes, that ye maie be sanctified.

[1] H .iiij, back. [2] H .v. [3] H .v, back. [4] H .vj.

The .lvii. Psal. *Miserere mei.*

BE merciful vnto me (o god) be mercifull vnto me, for my soule trusteth in the : & vnder the shadowe of thy wynges shalbe my refuge, vntil this tiranny be ouer past.

I wil cal vnto the moste high [1]God : euen to the God that shall performe the cause which I haue in hande.

He shall send from heauen : & saue me from the reprofe of him that would eate me vp.

God shall sende furth his mercy and truthe : my soule is among lions.

And I lie euen among the children of men (that are set on fier) : whose tethe are speres and arrowes, and their tongue a sharpe swerde.

Set vp thy selfe (O God) aboue the heauens : and thy glory aboue all the yearth!

They haue laied a net for my feete, and pressed down my soule : and haue digged a pitte before me, and are fallen into the middes of it them selfes.

[2]My harte is fixed (O God) my harte is fixed : I will synge and geue prayse.

Awake vp, my glorie ; awake, lute and harpe : I my selfe wyll awake right early.

I will geue thankes vnto the (O Lorde) among the people : & I will syng vnto the among the nacions.

For the greatnes of thy mercy reacheth vnto the heauens : & thy truthe vnto the cloudes.

Set vp thy selfe (O God) aboue the heauens : & thy glorie aboue all the yearth!

 Glorie be to the father, &c.
 As it was in the, &c. Amen.

 Lord haue mercy vpon vs!
Christ haue mercy vpon vs!
 Lorde haue mercy vpon vs!
[3]I beleue in God the, &c.

& so furth, with all the suffrages and Collectes, vsed in the mornyng praier.

The Euensong praier
at .vii. of the clock at nyght.

Our father whiche art, &c.
The minister.
Praise we the Lorde!
The poore.
Let vs geue hym thankes for euer and euer!

[1] H .vj, back.   [2] H .vij.   [3] H .vij, back.

Barts Order, 1552. Evening Service for the Poor.

The .cxxi. Psal. *Leuaui oculos.*

I Will lifte vp myne iyes vnto the hilles : from whence my helpe commeth.

My helpe cometh euen from the Lorde : which hath made heauen and yearth.

[1] He will not suffre thy foote to be moued : and he that kepeth the, will not slepe.

Behold, he that kepeth Israel : shall neither slomber nor slepe.

The lorde hymself is thy keper : the lord is thy defence vpon thy right hande.

So that the Sunne shall not burne the by daie : neither the moone by nyght.

The lorde shall preserue the from all euell : yea, it is euen he that shall kepe thy soule.

The lorde shall preserue thy goyng out and thy comyng in : from this tyme forthe for euermore.

Glory be to the father, and to the sonne, and to the holy Ghost.

As it was in the beginnyng, is nowe and euer, &c.   Amen.

[2] Let vs praie.

O Almighty God, kyng of kynges, and lorde of lordes, that onely gouernest and kepest all them that put their trust in the, kepe vs, thy poore members, this present nyght, that we maie rest and slepe in the remembraunce of thy moste holy name : To whom, with the sonne and the holy ghost, be al honour, glorie and praise, worlde without ende.   Amen !

GOd saue our souereigne lorde the kyng, al the Gouernours of this house, & the holie churche vniuersal, and graunt vs peace in Christ, and grace for euer.   Amen !

The letany and Suffrages.

God the father of heauen, haue mercie vpon vs miserable sinners !

O God the Father of, &c.

O God the Sonne, redemer of the worlde, haue mercy vpon vs miserable sinners !

O God the sonne, &c.

O God the holy ghost, procedyng from the father and the sonne, haue mercie vpon vs miserable sinners !

O God the holy ghost, proceding, &c.

O holy, blessed, and glorious Trinite, thre persones and one God, haue mercie vpon vs miserable sinners !

O holy, blessed, and glorious, &c.

[1] H .viij.   [2] H .viij, back.   [3] I .j.

Barts Order, 1552. Evening Service for the Poor.

Remember not, lorde, our offences, nor the offences of our forefathers, neither take thou vengeaunce of oure sinnes; spare us, good lorde; spare thy people, whom thou hast redemed with thy moste precious bloud, and be not angrie with vs for euer!
Spare vs, good lorde!
From all euil and mischief, from sinne, [1] from the craftes and assaultes of the deuill, from thy wrath, and from euerlastyng dampnacion,
Good lorde, deliuer vs!
From blyndnesse of harte, from pride, vainglory and hipocrisie; from enuie, hatred, and malice, and all vncharitablenesse,
Good lorde, deliuer vs!
From fornicacion, and all other dedly sinne; and from all the deceiptes of the worlde, the fleshe, and the Deuill,
Good lorde, deliuer vs!
From lightenyng, and tempest, from plague, pestilence and famine; from battaill and murder, & from sodein death,
Good lorde, deliuer vs!
From all Sedicion and priuey conspiracie, from the tyrannie of the Byshop of Rome, and all his detestable Enormities, from all false doctrine and heresie, from all hardnesse of harte, and contempte of thy worde and commaundement,
Good lorde, deliuer vs!
By the misterie of thy holie incarnacion, by thy holie natiuitie and Circumcision, by thy Baptisme, fastyng, & temptacion,
[2]*Good lorde, deliuer vs!*
By thyne Agonie and bloudie sweate, by thy Crosse and passion, by thy precious death & buriall, by thy glorious resurrection and Ascension, by the commyng of the holy ghost,
Good lorde, deliuer vs!
In all tyme of our tribulacion, in all tyme of our wealthe, in the houre of death, in the daie of Iudgement,
Good lorde, deliuer vs!
We sinners do beseche the to heare vs, O lord God; and that it may please the to rule and gouerne thy holy churche vniuersall in the right waie.
We beseche the to heare vs, &c.
That it may please the to kepe Edward the sixt, thy seruant, our Kyng and gouernour.
We beseche the to heare vs, &c.
That it may please the to rule his hart in thy faith, feare and loue, that he may alwaies haue affiaunce in the, and euer seke thy honour and glory.
We beseche the to heare vs, &c.
That it may please the to be defender and keper, geuing him the victorie ouer all his enemies.
[3]*We beseche the to heare vs, &c.*
That it maie please the to illuminate all Byshoppes, Pastours,

[1] I .j, back. [2] I .ij. [3] I .ij, back.

332 Barts Order, 1552. Evening Service for the Poor.

and Ministers of the churche, with true knowledge and vnderstandyng of thy worde, and that both by their preaching and liuyng, they may set it furth, and shewe it accordyngly.

We beseche the to heare vs, &c.

That it maie please the to endue the Lordes of the counsail, and all the nobilitie, with grace, wisedome and vnderstandyng.

We beseche the to heare vs, &c.

That it maie please the to blesse & kepe the magistrates, geuyng them grace to execute Iustice, and to maintein truth.

We beseche the to heare vs, &c.

That it maie please the to blesse and kepe all thy people.

We beseche the to heare, &c.

That it may please the to geue to all nacions, vnite, peace, and concorde.

We beseche the to heare vs, &c.

That it maie please the to geue vs an harte to loue and dreade the, and diligently to lyue after thy commaundementes.

[1] We beseche the to heare vs, &c.

That it maie please the, to geue al thy people encrease of grace, to heare mekely thy worde, and to receiue it with pure affection, and to bryng furthe the fruites of the spirite.

We beseche the to heare vs, &c.

That it may please the to bryng into the waie of truthe, al suche as haue erred, and are deceiued.

We beseche the to heare vs, &c.

That it maie please the to strengthen suche as do stande, and comforte and helpe the weake harted, and to raise vp them that fall, & finally to beate doune Sathan vnder our fete.

We beseche the to heare, &c.

That it may please the to succour, helpe and comfort all that be in daunger, necessitie and tribulacion.

We beseche the to heare, &c.

That it may please the to preserue, all that trauayle by lande or by water, all women labouryng of chylde, all sycke persones and younge chyldren, and to shewe thy pitie vpon all prysoners and captiues.

We beseche the to heare, &c.

[2] That it maie please the to defende and prouide for the fatherlesse children and widowes, and all that be desolate and oppressed.

We beseche the to heare vs, &c.

That it maie please the to haue mercy vpon all men.

We beseche the to heare, vs, &c.

That it may please the to forgeue oure enemies, persecutours & slaunderers, and to turne their hartes.

We beseche the to heare, &c.

That it maie please the, to geue & preserue to our vse, the kyndly fruites of the yearth, so as in due tyme we maie enioye theim.

We beseche the to heare vs, &c.

[1] I .iij.   [2] I .iij, back.

That it maie please the to geue to vs true repentaunce, to forgeue vs all our sinnes, negligences and ignoraunces, &[1] to endue vs with the grace of thy holy spirite, to amend our liues accordyng to thy holy worde.
We beseche the to heare vs, &c.
Sonne of God, we beseche the to here vs!
Sonne of god we beseche the to heare vs!
O lambe of God, that takest away the sinnes of the worlde,
[2] Graunt vs thy peace!
O lambe of God, that takest away the sinnes of the worlde,
Haue mercy vpon vs!
O Christ, heare vs!
O Christ, heare vs!
Lorde, haue mercy vpon vs!
Lorde, haue mercy vpon vs!
Christ, haue mercy vpon vs!
Christ, haue mercy vpon vs!
Lorde, haue mercy vpon vs!
Lorde, haue mercy vpon vs!
Our father, whiche art in heauen, &c. And leade vs not into temptacion. But deliuer vs from euill.
The versicle.
O Lorde, deale not with vs after our sinnes!
Aunswere.
Neither rewarde vs after our iniquities!

Let vs pray.

O God, merciful father, that despisest not the sighinges of a contrite harte, nor the desires of suche as be sorowfull, mercifully assist our praiers, that we make before the, in all our troubles and aduersities, whensoeuer thei oppresse vs : [3] And graciously heare vs, that those euilles whiche the craft and subtiltie of the Deuill or manne worketh agaynst vs, be brought to naught, and by the prouidence of thy goodnesse, may be dispersed, that we thy seruauntes, beyng hurt by no persecutions, may euermore geue thankes vnto the, in thy holy Churche : thorowe Iesus Christ our Lorde.
O Lorde, arise, helpe vs, and deliuer vs for thy names sake!
O God, we haue heard with oure eares, and our fathers haue declared vnto vs, the noble workes that thou diddest in their daies, and in the old tyme before theim.
O Lorde, aryse, helpe vs, and deliuer vs, for thine honour!
Glory be to the father, & to the sonne, and to the holy ghost!
As it was in the beginning, is nowe, and euer shalbe, worlde without ende.
Amen.
From our enemies, defende vs, O Christe!
Graciously loke vpon our afflictions!

[1] *orig.* & and   [2] I .iiij.   [3] I .iiij, back.

Pitifully beholde the sorrowes of oure [1] hartes!
Mercifully forgeue the sinnes of thy people!
Fauourably with mercy heare oure praiers!
O sonne of Dauid, haue mercy vpon vs!
Both nowe & euer, vouchesafe to heare vs, O Christ!
Graciously heare vs, O Christ! graciously heare vs, O Lorde Christ!
The Versicle.
O Lorde, lette thy mercy be shewed vpon us!
The aunswere.
As we do put our trust in the.

Let vs praie.

WE humbly beseche the, O Father, mercifully to loke vpon oure infirmities; and for the glory of thi names sake, turne from vs those euils, that we moste ryghteously haue deserued: graunt this, O lorde God, for our mediatour and aduocate Iesus Christes sake!

ALmightie God, whiche hast geuen vs grace at this tyme, with one accorde to make our common supplicacions vnto the, & doest [2] promise, that when two or thre be gathered together in thy name, thou wilt graunt their requestes; fulfill now, O Lorde, the desires and peticions of thy seruauntes, as may be moste expedient for theim, grauntyng vs in this worlde, knowlege of
thy truthe, and in the worlde to come,
life euerlastyng.
Amen.

[1] I .v. [2] sig. I .v, back.

[1]A thankesgeuyng vnto

Almyghtie God to be said by the poore that are cured in the hospital, at ye time of their deliuery from thence, vpon their knies in the hall before the Hospiteler, and twoo masters of this house, at the least.

And this the Hospiteler

shal charge them to learne without the booke, before they be deliuered.

WE magnifie and prayse thee, O Lorde, that so mercifully and fauourably haste loked vpon vs miserable & wretched synners, whiche so hyghely haue offended thy diuine maiestie, that we are not worthy to be nombred among thy elect & chosen people : our synnes beyng [2]great and greuous, is daily before our eyes ; we lament and be sorie for them ; and with sorowful harte, and lamentable teares, we call and crie vnto the for mercie ; haue mercy vpon vs, O Lorde, haue mercy vpon vs ; and accordyng to thy great mercie, wype awaie the multitude of our synnes ; and graunt vs now, O lorde, thy moste holie and workyng spirite, that settyng a-syde all vice and ydlenes, we maie, in thy feare, walke and go foreward in all vertue and godlines. And for that thou hast moued, O Lorde, the hartes of godly men, and the Gouernours of this house, to shewe their exceding charite towardes vs, in curing of our maladies & diseases, we yelde moste humble and hartie thankes to [3]thy maiestie, and shall incessauntlie laude and praise thy moste holy and glorious name ; Besechyng the, moste gracious and mercifull Lord, according to thy most holy woorde and promes, so to blesse this thyne awne dwellyng house, and the faithful ministers thereof, that there be here founde no lacke, but that their riches & substaunce may encrease ; that thy holy name maie thereby be the more praysed and glorified ; to whom be al laude honour, and glory, worlde without ende.
Amen !
(∴)

[1] sig. I .vi.   [2] sig. I .vi, back.   [3] sig. I .vii.

A passeport

to be deliuered to the
Poore.

To all Maiours, Bailiefes,
Constables, &c.

Know ye, that .A. B., tailour, borne in the towne of .S. T. in the countie of Northampton, beyng cured of his disease in the Hospital of. St. Bartholomews in West smithfielde in London, and from thence deliured the .xiii. daie of August, in the syxt yeare of the Reigne, &c. hath charge by vs, A. B. C. the gouernours of the [2]same, to repaire within   days next ensuyng the date hereof, to his sayd place of natiuitie, or to Westhandfield, the place of his last abode, and there to exhibite this present passe porte to. the head officer, or officers, in either
of the places appointed, that
they maie take further
order for his de-
meanour.

(∴)

Imprinted at London by Rycharde Grafton, Printer to the Kynges maiestie.

Cum priuilegio ad imprimen-
dum solum.

---

[1] sig. I .vii, back. The Passport was needed to preveut the arrest of the Patient as an idle and masterless Vagabond roving about the country.
[2] sig. I .viii. (The back of this leaf is blank.)

The manufacturer's authorised representative in the EU for product safety is Oxford University Press España S.A. of El Parque Empresarial San Fernando de Henares, Avenida de Castilla, 2 - 28830 Madrid (www.oup.es/en or product.safety@oup.com). OUP España S.A. also acts as importer into Spain of products made by the manufacturer.
Printed and bound by CPI Group (UK) Ltd, Croydon, CR0 4YY

20/03/2026

02075336-0005